Business Intelligence Tools for Small Companies

A Guide to Free and Low-Cost Solutions

Albert Nogués
Juan Valladares

Apress®

Business Intelligence Tools for Small Companies: A Guide to Free and Low-Cost Solutions

Albert Nogués
Barcelona, Spain

Juan Valladares
Barcelona, Spain

ISBN-13 (pbk): 978-1-4842-2567-7
DOI 10.1007/978-1-4842-2568-4

ISBN-13 (electronic): 978-1-4842-2568-4

Library of Congress Control Number: 2017943521

Managing Director: Welmoed Spahr
Editorial Director: Todd Green
Acquisitions Editor: Susan McDermott
Development Editor: Laura Berendson
Technical Reviewer: Marcos Muñoz
Coordinating Editor: Rita Fernando
Copy Editor: Karen Jameson
Compositor: SPi Global
Indexer: SPi Global
Cover image designed by FreePik

Distributed to the book trade worldwide by Springer Science+Business Media New York, 233 Spring Street, 6th Floor, New York, NY 10013. Phone 1-800-SPRINGER, fax (201) 348-4505, e-mail orders-ny@springer-sbm.com, or visit www.springeronline.com. Apress Media, LLC is a California LLC and the sole member (owner) is Springer Science + Business Media Finance Inc (SSBM Finance Inc). SSBM Finance Inc is a **Delaware** corporation.

For information on translations, please e-mail rights@apress.com, or visit http://www.apress.com/rights-permissions.

Apress titles may be purchased in bulk for academic, corporate, or promotional use. eBook versions and licenses are also available for most titles. For more information, reference our Print and eBook Bulk Sales web page at http://www.apress.com/bulk-sales.

Any source code or other supplementary material referenced by the author in this book is available to readers on GitHub via the book's product page, located at www.apress.com/9781484225677. For more detailed information, please visit http://www.apress.com/source-code.

I want to thank all my family and friends for the support they always shown me, including when writing this book. I want to especially dedicate it to my two grandmas, which recently passed away, as they have always been an inspiration to me.

—Albert

To my wife for all those times she has been waiting for me to go to sleep. To my kid for all those times he has been waiting for me to help him with his homework. To my parents that always have supported me in all my initiatives. To my whole family, friends, and colleagues just to be there.

—Juan

Contents at a Glance

Contents

About the Authors

Albert Nogués is a BI Project Manager/BI Architect/BI Developer for Technology2Client, a consultancy to the BI and DWH teams at Danone. He also manages BI projects for Betvictor, a sports gaming company; and designs APIs and interfaces for market pricing services and traders. He has deep knowledge of the full BI stack and holds Oracle certifications in OCA, OCP, and Performance Tuning. Albert received multiple MS degrees in computer science and ICT from the Universitat Politècnica de Catalunya and Universitat Oberta de Catalunya.

Juan Valladares is the founder and CEO of Best in BI Solutions, a Business Intelligence consultancy whose clients include T2C, everis, and other consultancy companies. Juan has been collaborating with end customers such as Zurich, Danone, and Mondelez. A telecommunications engineer by training, Juan has 15 years of experience in business intelligence, data modeling, and BI system administration. He is specialized and certified in Microstrategy. He teaches courses in BI tools, high-concurrency platform management, and ETL and BI processes development. He received his MBA from the Universidad de la Rioja and his engineering degree in telecommunications from the Universitat Politècnica de Catalunya.

About the Technical Reviewer

Marcos Muñoz is a computer engineer in the Business Intelligence/Data Warehouse field specializing in Oracle and MicroStrategy technologies. He has 10 years of experience as a project manager, BI architect, and BI consultant. Marcos is particularly knowledgeable in the technical aspects of BI and DWH. He is a MicroStrategy Certified Developer and MicroStrategy Certified Engineer.

Acknowledgments

I want to take the opportunity to thank all my family and friends for being patient when writing the book, as I was sometimes unavailable to them. I also wish to thank all current and past colleagues and job employers I had along the years as most of what I know, I have learned from or with them. As a special mention, I want to thank Juan Valladares for agreeing to work on the book together, especially as he has a company to manage.

I want to thank as well, all the people at Apress that helped us through our journey. From the very beginning, our first editor Robert, as he was very fast to show interest in this book, and very helpful in the early stages through all of the process; to Rita, for all her support and encouragement and help during the writing process; as well as Laura and Susan, our latest editors, for the help and support they provided during the process of writing the book.

As a last mention, but not less important, I want to thank Marcos Muñoz for his work correcting all the mistakes we made during the chapter submission, sending all his valuable inputs, which have improved the book a lot.

—Albert Nogués

I will follow with Albert's acknowledgments, starting with him who involved me in the enterprise of writing a book; and Marcos who has helped us a lot with his comments. It has been a hard job, especially to combine it with our daily dedication to our respective jobs and lives but all three together in combination with Apress colleagues have created an effective team that has been able to finish this book on time.

Main knowledge comes from colleagues, partners, and employers, so I would like to acknowledge all the professionals I have been working with, giving special consideration to Xavi, always available when I have some doubt; Joan Carles and Rubén, great leaders that have helped me during my career; also Oriol, Juan Carlos, Laura, Jorge, Olivier, Roberto, Cristian, Claudio, Jaime... the list would fill the whole book, just add a special mention to all members of Best In BI. I hope that anybody missing here doesn't feel underestimated because it is not possible to make an exhaustive list here.

And of course to my family, Vanessa, Quim, Mama, Papa, Esther, Fredy, Javi, Dani, Lucia, Marias and the rest of them, who are the real reason to go ahead every day.

—Juan Valladares

Introduction

Welcome to "Business Intelligence Tools for Small Companies: A Guide to Free and Low-Cost Solutions." In this book we want to propose multiple options to implement a BI project based on a simple idea, implement an Agile system, easy to manage using open source tools, minimizing in this way licensing costs that usually involve these kind of solutions.

In this book that you are about to start reading we will introduce you to the main Business Intelligence concepts using user-friendly definitions; we will explore the main BI components that BI solutions must use, analyzing some of the principal open source solutions for every component; and we will also show you some best practices to manage the whole project from an Agile methodology perspective and recommended strategies for life cycle management, including the movement of all of the platform to a Cloud environment.

We will show you how to implement from the beginning a platform with all basic components required for the full solution.

This book is intended to be especially useful to the following reader groups:

- IT responsible working for Medium and Small enterprises that require implementing from zero an entire BI solution.

- IT teams in All size companies that have the requirement to implement some DataMart analysis for a single department, investing the minimum cost on licenses.

- Managers and Business Owners that are thinking of leveraging some low-cost solutions to gather some company insights.

- BI professionals that want to discover new tools and methodologies to deploy new BI solutions.

Most of our experience has occurred in big companies in different sectors such as Bank, Insurance, Grocery or Retail, which require high-capacity systems to be able to process huge amount of data, with multiple formatting and different reporting capabilities and with hundreds or thousands of users that require multiple access types and delivery options. This allows us to detect which are the most common problems in those big environments and propose a simplified, free solution with basic capabilities that most of our customers need to use.

CHAPTER 1

■ ■ ■

Business Intelligence for Everybody

When some years ago we were offered to join to our first Business Intelligence project, we thought that something in the term was redundant because at the end of the day, doing Business requires Intelligence. This is the truth, especially if you pretend to do your business correctly, because profitable business cannot be performed without intelligence.

Speaking from a computer science book perspective Business Intelligence (BI) belongs to an Analytics World. Business intelligence is a set of tools and processes that helps you to make decisions based on accurate data, saving time and effort. The main idea behind a BI tool is the possibility of easily analyzable data based on business concepts without having technical knowledge about database tools or other sources that contain the data. BI tools pretend to extract knowledge from our stored data based in three main pillars: reliability, availability, and attractive user experience.

Imagine that you are the CEO of a small company dedicated to cookie manufacturing, and based on the sales reporting by product that you are analyzing, you detect that Cream Chocolate cookies have been decreasing in monthly sales every month. The figures you are seeing are now are about half of the amount that was selling at the beginning of the year. As CEO you have different possibilities: remove the Cream Chocolate cookie from the catalog; change the Cream Chocolate formula; set a bonus for your commercial department if they sell this product; or fire the brand manager because her department is causing you a loss, as reflected in the Profit and Loss analysis. But what happens if the real problem is that this product has changed its internal code because of a change in the cream formula and the product catalog is not correctly updated in your BI system—and it's not reflecting properly the sales with the new code? Your previous decisions would be wrong because you are basing this on an incorrect data analysis. This is why reliability is a basic requirement in all IT projects but is especially relevant in BI tools because they can be used to take main decisions from strategic company management to basic operational activities. Based on that, it's mandatory that data offered by our BI system must be consistent, and every analysis dimension must ensure correct results based on the data quality.

Now imagine that you are working on a car assembly line; it's Friday at 8 p.m. and you need to make the order so as to refill your warehouse of different pieces before going home. You are launching your warehouse dashboard that suggests the quantity of every single piece that you need to ask about, and the source of this information has information from Wednesday afternoon because the refresh daily process has not finished yet. Next week you will suffer some stop on your assembly line because of missing wheels or you will have the warehouse completely swamped because you asked for 100 bumpers and they arrived last Thursday. A similar reason could cause the same result if you cannot access the system at the time required due to some maintenance activity on the platform, and you need to estimate the order based on what you think is missing.

Our system must be available to our users when our users need to use it. This seems to be an obvious condition, but there are two main factors that could cause us to fail to achieve this objective. Our system must be stable, running correctly during business hours, and data must be updated accordingly to our target consumers and their requirements.

© Albert Nogués and Juan Valladares 2017
A. Nogués and J. Valladares, *Business Intelligence Tools for Small Companies*,
DOI 10.1007/978-1-4842-2568-4_1

A last main characteristic of the system that we expect to build is that our access to the available data must be user friendly and adapted to consumer expectation and capacity. You cannot provide expert analysis of data with a tool where they cannot interact with the information and, on the other hand, you could be in trouble if your commercial staff, which has no idea about computers, is required to publish a SQL query to get the data that requires analyzing. To be able to provide a tool that is user friendly, you need first to know your user and agree with their requirements based on their needs. Also a provided solution must be adapted to user capacity.

Achieving these three points is not an easy task, but they are the basis to delivering a profitable and lasting BI solution inside your company or for your customers.

What Is Business Intelligence?

Added to the brief introduction that we have done just some lines above, and taking into consideration that we are going to talk during this book mostly about Business Intelligence, we would like to analyze with deeper detail how to answer to these two simple questions: What does BI means? How can we properly understand the BI concept?

BI has multiple definitions in multiple publications. The Wikipedia definition for BI is this: "Business intelligence is a set of theories, methodologies, architectures, and technologies that transform raw data into meaningful and useful information for business purposes." In our opinion this is a very interesting definition because it shows a full image of a BI solutions, and not only the usual focus on the front-end tools that some definitions remark. Because to have a BI solution in place implies to follow some **theories** in the definition of the process such that some specific data model applies **methodologies** that help you to achieve efficiency during your implementation project and then the later maintenance, which define the correct **architecture** that gives you an appropriate Return of Investment based on the benefit that you will obtain from the BI project, and finally choose the set of **technologies** that meets with your requirements, specifications, and economic possibilities. In Figure 1-1 you can see a diagram of main BI components. Keep it in mind because it will be useful for you to understand the whole book.

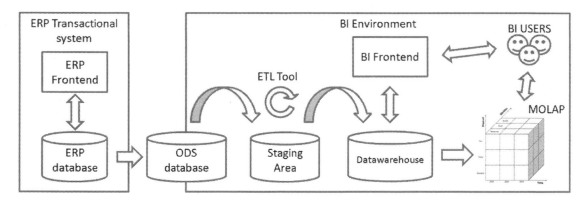

Figure 1-1. *BI system components*

In Figure 1-1 you can see that the main information source of all the system is the ERP (in spite of that, as we will see, there can be multiple other sources), then we have an ODS database that contains a direct extraction from ERP; it can be a database or some tables inside our database, but this concept usually exists, using direct extractions from ERP so as not to overload the source system. With our ETL tool we will move information from ODS to the Staging Area database where we will process the information, and finally we will insert that in the datawarehouse where we will access with our BI front-end tool. It is quite possible that we have just a database and the distinction between ODS, Staging Area, and Datawarehouse is just the tables that we use or different schemas inside the database. Finally we can have a MOLAP system that will help us to realize the budget for the next year. We will see the detail of each component along the whole book.

During this book we will analyze some theories during this introduction; we will talk about methodologies in Chapters 2, 11, and 12; we will see the full architecture of the system in Chapter 10; and we will evaluate different technologies in Chapters 3 through 9.

There are also some other interesting concepts to define BI. One of them is focusing on the advantage that a company can get from implementing this kind of system as far as you can be more efficient in administrative tasks for gathering information and use your time to analyze the information to get conclusions. Also it's important to remark that information that we are managing can come from internal and external sources to be able to analyze how we are performing our activities but also to compare with our competitors if they are publishing information or trying to analyze data from our target markets that we are going to access.

Another interesting concept is the possibility of forecasting the future. We are not talking about witches or fortune tellers; we are referring to finding the correct patterns that will allow us to anticipate what can be our sales if the conditions remain the same. What can happen if a new competitor enters on our main market and steals 20% of our market share? Or what could be the result of increasing a 25% our sales force team by 25%? In this concept, the key feature is to gain the ability to detect which variables are correlated with each other and which of them are almost independent.

Interactivity is also one of the focuses that can give you an idea about what BI is. It's really interesting that business analysts can investigate and navigate through the data to be able to discover these hidden patterns that can give you visibility of your near future.

A later element to mention with different BI definitions is knowledge, which is the result of applying BI techniques to big amounts of data stored in our databases or simplifying the formula: if you join Data + Analysis you get Knowledge.

BI Evolution

The concept Business Intelligence referred to in this book first appeared and was described by Howard Dresner in 1989. He described Business Intelligence as "concepts and methods to improve business decision making by using fact-based support systems."

From the late 1990s the usage of this term has been generalized and it's possible to find innumerable references to BI in technical books and online articles.

The appearance of BI is directly related with the consolidation of transactional systems around the world. Before transactional services were installed everywhere, the main usage of computerized tools was for high-level analysis; the amount of information saved in the systems was small enough to be analyzed directly without the need of any extra tool. When transactional systems appeared in business scenarios, the amount of data to manage increased exponentially. Think about a retail company that had monthly information about units purchased of a given product and the stock that remained in the shop; now it has information of every single ticket of any client, with the detailed products that they have purchased. They can obtain relationships among products; they can analyze payment methods; if their customers pay with cards, they can get the name of the customers and they can analyze how many times a given customer goes to our shop, the type of products they buy, and a lot of other analysis. And this is only an example; you can translate this example to your business and understand why you need BI in your own business.

■ **Note** Most of the recent references are related to the specialized software for BI capabilities, and there are many consultancy companies that have dedicated BI teams and projects just to attend to development requirements on BI tools, considering the rest of the solution as auxiliary components of the BI tool itself. So be careful if you are thinking about hiring consultancy support by ensuring that their estimation of costs for the project contains all the required stuff for your request.

From Strategic to Tactical

BI also has suffered some changes in the scope of their projects, spreading across organizations from top management reports and dashboards to daily operational analysis. Benefits of BI have been proven from top managers, and they have noticed that BI offers multiple possibilities for their organizations to get profit on BI tools implementing BI projects from the bottom to the top of their companies. Over the years we can see that implementations have moved from strategic implementations that assist top managers in the decisions that they must take to guide correctly the companies. This includes all the environments inside the organization, including the lowest ones to facilitate employees to make decisions such as which products I need to ask to refill for the warehouse or what color is the best seller to dress the mannequin at the shop I am working for. BI has moved from strategic decisions support to tactical ones.

■ **Note** Initial BI implementations were known with different acronyms that reveal the nature of the target users on those incipient deployments. One of the initial acronyms to name this kind of systems was DSS or Decisional Support System, which shows us that our target audience will be decision makers. For sure, every single person in our company will make decisions, but most important decisions are usually made by managers, leaders, and executives. Also another interesting acronym that manifests the same is EIS (Executive Information System), which in this case contains directly the names of the target users of the BI platform: executives of the company.

Big Data

Nowadays, the most important trend of BI is referred to Big Data and Data Lake concepts. Big Data itself is based on the possibility of using BI tools and analytical capabilities to extract information from the incredibly enormous amount of data that is being generated every day by our employees, customers, and platform users in many different platforms such as social networks, job networks, forums, blogs, mobile apps and resources, mobile devices, GPS information, etc., that is saved into unstructured systems and that you cannot attack with standard datawarehouse armament; and this is due to the nature of a datawarehouse, as we will analyze in the next sections. DWH is based on a structured database that contains homogenous information loaded into our system using ETL processes that ensure integrity of the data, multiple checks, and validations; and these kinds of processes are too complex to read from Big Data sources, as far as processing power required to perform that kind of analysis is too high. In a Big Data system, accuracy is not as critical than that in a DWH scenario. Using a Big Data system to analyze Facebook logs and missing a comment that could give you information from a potential customer among 1 billion users is something that we can accept, but if you miss a row in your accounting system it will generate an accounting mismatch. Big Data can be used as sourcing of our BI system: it is an extra component on the infrastructure layer, and it won't replace our sales, finance, or operation analysis that we can have in place.

In order to be able to give support to Big Data requirements, a new concept different from DWH is required. In this scenario we can locate the Data Lake. The idea behind this concept is that you don't need to process all the amount of data that you have available to create a structured data source for your BI system. Instead of that, you should access directly to your data source in order to fish the information that you require (from fishing comes the idea of the lake; ingenious, right?).

Internet of Things

It can be considered also as a source for Big Data analysis, but I would like to discuss the Internet of Things in a separate section due to the possibilities that it can offer to BI projects. Internet of Things is related to the incredible amount of information that could be extracted from incoming electronic devices that will be used in multiple elements everywhere. Now we have cars with Internet connection, fridges that can tell us what is missing inside, cleaning machines that can send us an SMS when they have finished, or cleaning robots that can be programmed to be launched from the smartphone. Imagine the amount of information that this could provide to analyze inside Big Data environments. This Internet of Things enables infinite possibilities of investigation and development where knowledge extracted from the information provides inputs that will be extremely interesting to analyze.

BI Characteristics

As part of characteristics already commented on, there are additional ones and topics related to BI where we would like to focus in deeper detail, and that they require some separate parts inside the book.

Hidden Relevant Information

Doing again an imagination exercise, imagine that you are the Sales Force manager in your country for your company and that you have five sales representatives that are under your control from different nationalities: one of them is Chinese and another one is Italian. The rest of them are natives of the country. They have regions assigned based on the amount of customers that they must visit and the region size of your country. They are making regular visits in order to visit all the customers with a minimum frequency of one visit every three months.

In this scenario, you prepare a consolidated report of all the countries and you see that every three months brings a peak in the total sum of sales for the whole country. The initial conclusion that you can arrive is that at the end of the quarter your sales representative is selling more due to some objective that they have or that your customers have some kind of quarterly purchase periodicity. You decide to change the objective process to do it monthly in order to motivate your employees to spread the sales along the whole year, but this might put too much pressure and one of them may fall ill.

Instead of the initial conclusion and reaction, to confirm that periodicity is the root cause of the sales variation, you proceed with a drill into the regions and you find that there are two regions of five that have a higher peak in the graph and the rest of the regions are stable. You decide to search the sales representative of every region and you see that they are the regions from the foreign employees. Comparing among them you can see that they are below average sales for two months in the quarter but in the third one they are over the rest. So the variability is not coming only from the periodicity but also depends on the employee.

At this moment you want to do a deeper analysis to investigate why they are more variable comparing with the rest. You can then drill into the cities visited per month and you can appreciate that in the third month of the quarter, they are always visiting the same cities based on the quarter periodicity of your sales. You can then use an analysis coming from the National Institute of Statistics with the immigration per city and correlate with the top sales cities for these two employees, detecting that the top source country of immigration in the cities where the Chinese employee has better sales results is China, and the second source country in the cities where the Italian guy is the sales leader is Italia. With these results could be an interesting testing to reorganize your customer assignment by assigning, if possible, the customer to somebody in your organization from the same nationality, because they understand better each other in the negotiations they do, instead of doing it as now by geography.

What you can learn from this example is that knowledge beside the information can be hidden among billions of rows of data, and to extract valid information from this data you need to know, not only technically how to play with it and be able to use full functionality of your BI platform, but also it's required to have knowledge about what are you looking for, in conjunction with a part of intuition.

Accuracy

When you are managing thousands, millions, or billions of rows to perform your analysis you can think that if you lose few data it won't affect the overall amount because you are more interested on trends and data evolution than in having the exact data until the third decimal. And you are possibly right, but it's also really important to have your data matched with the source of your information because you need to offer reliability to your customers. Let's explain an example in order to show you the importance of small data.

In a company that produces consumer products like cleaning products, you have a process defined with your customers to manage return orders in case of some product that is in poor condition or directly broken. This can cause small amount return orders that arrive to the transactional system and that are forwarded to the DWH. You could ignore these orders in your DWH because masterdata management is completely different, and doing an overall analysis you could see no difference; you show Gross Sales of $2,234,555.34 instead of the real $2,232,434.96. But this 0.09% of Gross Sales that has been missed has related high costs of return orders management, if you don't take them into account you can miss interesting data for your organization that could be translated in your Net Revenue in a big Margin, because some returns from important customers have some penalty related, added to the management cost.

Significant KPIs

Key Performance Indicator (KPI) is related to a measure of data that is useful to measure your company evolution. Definition of which KPIs I need to follow up to ensure that I'm not missing any significant measure of my business is one of the critical steps in the BI project implementation. You need to take into account multiple considerations when you are defining them, but the most important one is the relevance of the metric. Of course the relevance of the metric will depend on the target audience; if you are defining them with your sales department, most relevant metrics will be related to sales, Units or volume sold, Gross Sales and Net Sales mainly, because is quite feasible that these metrics are directly related to your performance and with your sales objectives, but if you are from the financial department you will be more interested in Costs and Net Revenue. In case you work for the Customer service department, you will be interested in Sales Returns, quality of goods sold, etc.

But this is not the only important parameter when you want to have a significant set of KPIs. In the case of the sales department, imagine that you have been moved from one customer to another. Your old customer was purchasing $3 Million per year and the new one is purchasing $8 Million. Is this number good or bad? Can this number tell you by itself how the performance is on this customer? For sure, it doesn't. The best BI analyses when you are measuring your evolution are based on percentages; you need to compare it with something. How is this data comparing with the last year's sales? How is this data compared with my yearly objectives? This is the kind of analysis that most of information receivers are interested on; they want to receive information that can be interpreted easily.

On Time

How many times have you heard a typical sentence like "I need you to deliver the sales report in my desktop by tomorrow morning before 09:00 AM" or "This information should have been delivered to our manager yesterday"? We have already commented on the importance of delivering the information at the right time when it's required, but we would like to develop this idea.

In order to have the information delivered on time, we first need to figure out what is the idea of on time for our customers for every set of information. In the project scope definition, temporary axis is one of the main topics to evaluate in order to validate if a project scope is realistic or not. You need to consider different elements that can affect that, such as volumetry to move, hardware capacity, process complexity, and information load dependencies. And it's also relevant in this definition to consider the future evolution of all these subjects: how it can grow up the volumetry to manage and the target volumetry (depending on the process strategy it can depend not only on the information to load in your database but also in the size of the target tables, especially if you are using update or merge strategy); how it can increase hardware capacity, if we have a scalable system or if we have state-of-the-art hardware and we cannot grow without rebuilding the full platform with new servers, to control the increase of complexity on processes and validate that prior processes won't suffer delays that could affect our ETL processes.

All these considerations are especially relevant if we are talking about daily or intraday processes, but they are not so critical when the timeline for your ETL is on a weekly or monthly basis.

Company Analytics Life Cycle and Continuous Improvement

Well, we are in the sixth month after our BI project is in place, and we are analyzing, for every month, which are the bestsellers in our product catalog and which are the worst sellers. We are seeing that, as in last month, a given product appears in the bottom of the list, and this is happening month after month for the same product, but our defined action is to follow the evolution. For next month it is quite feasible that the same product appears in the bottom of the sales list.

What is missing on our BI system? It seems to be clear... action. We need to act accordingly with the results that we have gotten from our analysis and after that, check for the results of our actions. We are inside the Company Analytics Life Cycle where our BI system can help us to understand the whole process.

- **Analyze data**: When we are talking about analyzing data, we cannot stop just in the initial analysis. In the previous example we cannot stop just by defining which are the products that have the worst ratio of sales, we need to investigate what is the main reason for that, or if there is more than one factor, that can affect to the results that we have observed. We need to try correlating information with other analysis, to try to see other visualizations of the information, to analyze the information along different dimensions trying to isolate which is the problem of this product. Maybe it is a matter of a bad design of the product, the usage of a wrong sales channel, the publicity used is not good enough or it's focused to the wrong target; maybe the sales representative of this product is producing bad results for all the products that are in his portfolio or it's a temporary product that we are trying to sell out of in its principal season – we are trying to sell ice cream in winter in our local office in the North Pole.

- **Define action plan**: Once we have isolated which is the root cause of the problem detected or which is our strength in our bestseller, we need to define a set of actions that will help us to correct/improve what is wrong. If we see that we have a really bad product in the catalog, we will choose to remove it from the catalog; if the region is not the correct one we will choose to relocate the sales representative to a better region; if we are selling ice cream in the North Pole, it is better to close our company. Our action plan must contain mainly actions, responsibility, and dates.

- **Act**: An action plan doesn't have any utility if you don't follow it. So we need to proceed accordingly with it within the scheduled calendar.

- **Check for results**: We need to measure how we are changing the KPIs that we are evaluating in order to see if our actions are giving the desired results: if they are causing the opposite effect instead of what I was expected or if my actions don't have any effect at on the result.

■ **Note**   The main result of a good analysis is a good action plan. We need to try to provide a BI tool that allows your analysts to easily elaborate an action plan and measure the results of our actions to implement the action plan.

Benefits of BI

Using BI can get you multiple benefits for your company. If your BI tool has been correctly implemented and you have achieved a correct performance on your BI system, you will get benefits that we classify as direct and indirect.

Direct Benefits

We consider direct benefits to be all that advantages that the fact of using a BI tool provides you with that come directly related with the BI implementation. For sure you will get these benefits if you are able to implement a BI solution with a good strategy and performance. One of the ways to allow that is to follow the recommendations we are trying to provide you with in this book. If you do it in this way you can get the following:

Resource optimization: If you are trying to implement a BI solution, it is quite possible that currently your analysis system is something like an Excel recompilation of different teams that is consolidated into a single workbook from where you extract different graphs, pivot tables, and high-level ratios. There is the possibility that very often you find some mistake in the format of the source Excel sheets, and that you need to modify tens of pages because you need to add a new column in your analysis or that you made a mistakes in the aggregation, requiring you to redo all of the extraction again. The BI solution should provide you directly with the information in the format that you require; you should be able to save reports in the format that you need to receive top product sellers or compare with the last period, and you should be able to schedule a job to receive these reports on a daily, weekly, or monthly basis.

Cost saving: Derived from the previously commented optimization of resources, you will get cost savings specially in human resources that are invested in the current manual analysis process. The BI project will also allow you to save multiple other costs by implementing a cost analysis reporting in your BI solution that will provide you multiple action points to optimize your processes and operations. Later in this chapter we will analyze some examples, inside the Indirect Benefits part.

▓ **Note** Try to measure the amount of time that your company is investing in gathering, aggregating, and formatting information that allows you to know the result of the current exercise, monthly closing period, or daily sales; sum all of this time multiplying by the human resources unit cost and then compare it with the total cost of your BI project. In this way you will be able to analyze the direct Return of Investment (ROI) that you will get from the BI project implementation.

SVOT: acronym of Single Version Of Truth; one of the benefits of having a reference system that contains all data required for your analysis is that everybody will use the same system so all your departments will get the same figures, instead of having multiple Excels with manual extractions, with manual modifications, and personal considerations. All the departments will agree on which total amount of sales you achieve every month or what is the last year's net revenue.

Single information responsible: Within the BI project implementation you will define a BI responsible or department that will be your single point of contact for all the information. You won't depend anymore on information arriving from different channels, Excel files filled up and formatted by multiple employees, and departments or email exchanging between all the sales controlling department. The IT department will centralize the information delivery process.

Self-service analysis: If you are the IT guy of a small company and you have the responsibility to extract and format data for business users, this is one of the most interesting benefits that the deployment of a BI solution can provide you. They will be able to generate reports by themselves. You will create an infrastructure that will allow them to do any analysis that they can think about just with the prerequisite of having available the information that they require. But they will be able by themselves to format data, drill across information available, pivot data, segment data, filter information, applying any of the modifications that you want to allow them to do.

Detailed analysis capacity: If you are able to implement a strong data model with data integrity, verified lookup tables, and complete dimensions you will get a robust system that will allow you to analyze the information at the most detailed level that you can think of. For sure, to be able to do that, you need to use a level of hardware that matches with the requirements on response time expectation, data volume, and parallelism of usage. But if hardware is powerful enough you will have the possibility to deliver an analysis system that can provide high-level aggregated analysis to the most detailed information available, always using BI capabilities to filter the amount of information that the system is returning to the user.

Indirect Benefits

Derived from the previous group of benefits but also coming from the setup of the BI solution itself, you will have the possibility of getting other intangible benefits that are not direct, but that can be considered as a consequence of your BI project. The BI project will give you the tools to make correct decisions and act following these decisions; if you do that, you will be able to achieve a set of benefits. In the following list we are showing just some examples that we consider relevant in general terms, but it can be the basis for you to define what is the focus of your actions that have been derived from your BI analysis.

Sales Increase: You will be able to analyze which are the customers that are buying your product and to analyze if you find common patterns among them, in order to focus your marketing strategies on those customers that meet these patterns. You will than have the possibility of analyzing your products through multiple product characteristics that can give you a combination of attributes that maximize or minimize sales results, and you will see which selling promotions are more effective by comparing with sales of the rest of the year. In summary, you will have within easy reach tools that will allow you to perform complex analyses easily.

Cost reduction: Derived from your BI project you can reduce your company costs based on many perspectives, as far as having a powerful analysis tool will help you to improve cost controlling process. Once you develop the BI project with the required information, you will be able to analyze all available costs in your organization, operational costs, human resources, renting and leasing, financial costs, employee expenses, departmental costs, etc.

Customer fidelization: You can analyze the response of your customers to a marketing campaign. You can also analyze the usage of fidelity cards, you will be able to validate the evolution of your customer acquisitions, and BI will allow you to join information coming from your transactional and your logistic operator to validate if they are delivering your products on time, among multiple other analysis that you can think about.

Product affinity: If you focus on your product dimension you will be able to analyze which products are related among them based on your customer preferences. You will be able to analyze which pairs of products your customers are purchasing, which are those relations that are not obvious applying basic logical thinking. You can imagine that if a customer is buying a computer monitor, it is quite possible that he is interested in a mouse and a keyboard, but maybe it's more difficult to relate which book titles or films can have a relation if it's not directly related with the category, the author, or the director.

Customer Segmentation: Segmentation allows you to group your customers based on their characteristics. There are some characteristics that are directly related to them that usually come from your customer masterdata in your transactional or your CRM (Customer Relationship Module), such as age, city, address, net income, number of children, or any preference that you can get from any information gathering tool. But also you can use segmentation based on metrics, such as the number of products acquired, which are his preferred categories, which is his payment method, which are his preferred distribution channels, gross sales coming from him, purchase period, or any other field that requires a previous calculation derived from the basic model you are implementing.

Demographical analysis: Related to customer segmentation, it is especially relevant to analyze the demography of the region in which you are interested in. You can start off from information coming from internal sources gathered with some CRM tool related to your clients or you can analyze information provided by some external source like public statistical institutions that are publishing general information about countries, regions, cities, or districts. You can cross-join the demographical analysis with the previous customer segmentation and focus on those districts where the neighborhood can be more interested in your product. Also based on demography you can decide where to open a shop or where to locate a customer information point.

Geographical analysis: This can be considered as customer segmentation or as demographical analysis but we prefer to keep it as a separate benefit because of the new powerful map visualizations that allow you to locate in a map the number of customers, sales, or whatever other metric that you want to analyze.

Production process analysis: Analyzing your production process, your assembly line, or your distribution chain, you can avoid overproduction and overstock applying just-in-time techniques with the help of your BI solution. Also you can analyze lost time, waiting time, bottlenecks, the amount of resources consumed by all steps of the production process, or which are the dependencies that can be optimized/removed from your production workflow.

Quality analysis: You can also focus your analysis on the quality of the products or services that you are offering, analyzing the number of incidents by typology, source of incidents by steps in the production line, analyze customer feedback, or validate on-time resolution and delivery of your projects.

Employee productivity analysis: It will be easier for your company to define objectives and analyze their achievement by using BI capabilities to get figures about sales force visits realized, effectivity on those visits, number of orders per employee, gross sales amount, number of pieces manufactured, number of defects caused per employee, resources consumed per employee, training realized, or working days.

Objectives

You can join and combine all these benefits and KPIs analyses to define and follow up different strategic objectives for your organization, defining in this way which is the strategy that guides your company to success. Here below you can find some usual objectives but as within the benefits list, it's only a set of examples; you should be able to define your own objectives based on your company mission.

Optimize company performance: By increasing sales, by decreasing costs, or a combination of both factors you can maximize the performance of your company. At the end of the day, what is really important for results is the net income, so any of the metrics that are included in the net income calculation can grow or decrease to maximize the performance. In order to act on any of these metrics you can use your BI solution to help you to detect the way of doing it and to follow up your actions over any concrete metric.

Detect niche markets: This is an objective that you can define based especially in external data analysis, such as demography and competitor analyses. With the combination of external analysis with the knowledge about what your company can offer to the market, you can find some niche markets for a concrete product that is under your catalog or that can be developed for this market.

Improve customer satisfaction: It seems quite obvious that monitoring the KPIs that allow you to measure how is your quality of service offered to your customers you will be able to improve them. Also your BI tool can be used to report to the customers how you are performing this quality of service or any other reporting that could be useful for them. Also it can improve the relationship with your providers as far as you can also deliver to the provider information useful for their product development and marketing. An example of this objective could be a clothes manufacturer that sells his products to different hypermarket chains; if the hypermarket sends information about sales status of the manufacturer's products, the manufacturer can evaluate if they are designing following the cool line or if their products will be returned because the hypermarket is not able to sell any dresses.

Improve employee satisfaction: Some of the KPIs and benefits shown are about employee controlling, such as productivity, sales per employee, or cost controlling, and this can cause the feeling to the employee that he is highly controlled and monitored, which is not always pleasant for them. But on the other hand you can offer many other benefits coming from BI; improvement of productivity can give the employee time to dedicate to more interesting tasks. Reduction of the time dedicated to repetitive tasks such as information gathering and formatting will allow your company employees to focus on data analysis; all the options that a BI platform offers will facilitate their work; and also they can learn about BI tool capabilities, a topic that will improve their curriculums.

Who Can Get Benefit from BI?

Based on all the benefits that we have been analyzing in the previous section, it seems obvious that almost all the company can benefit from BI in varying degrees, from the basic operator that can make decisions based on a user-friendly interface to push a defined button to the general director that can decide which is the market to invest in during the next four years; at the end, any person in our organization could be required at some point in time to make decisions and they can use BI to help him decide which options to choose. Anyway let's give some examples by department:

- General management: This was the original target team for BI tools, so they are the members on the organization who can extract higher benefit from BI tools. A good BI tool must be oriented to easily show the relevant information that requires paying attention to it, to quickly focus on the target, and to get conclusions and make decisions in a short period of time. In summary the way of life of an executive team.

- Sales department: Selling is the most important task in any company. You can produce the best product in your category but if you don't sell it your company will default. So historically these teams have been the second step in a BI implementation. Analyze trending of sales, use BI as support to define sales objectives, and also to follow which is the status on real vs objective KPIs are the main analyses that you will focus on.

- Financial team: Selling next to infinite but with a unitary cost higher than the price would cause infinite losses. So the next step in BI implementations is usually the financial analysis of the company to focus on improving the company performance.

- Purchase department: Derived from financial and costs analysis, and especially for manufacturing or reselling companies, it's also quite relevant to reduce the acquisition cost of raw materials and commodities. To help you with this hard task you can use BI capabilities, too, as always you require it to analyze information to make conclusions.

- Human resources department: Other important costs derived from financial cost analysis are salaries, diets, expenses, and other personal costs of our employees. When your financial department is pressing you to reduce human resources costs, having the support of a BI tool can help you to analyze your employees' data.

- Operators: Any person in any operations team in the company can have the possibility of making a decision: from asking for stock replenishment, to offer to the customer a product that matches with the rest of objects that he has in the cart. Again, having a good BI solution can facilitate these tasks to everybody that is in on the decision-making process.

BI Platform Components

During this section we are going to analyze which are the main components inside the whole BI platform, and this will serve as a theoretical introduction to the next chapters where you will learn in depth how to install and deploy a BI solution based on main open source tools.

Source ERP

Most common information source in a BI solution and usually the first one to be used is the Enterprise Resource Planning. This kind of transactional tool is used to control and save main operations that occur inside the company. This component cannot itself be considered as a BI component; but so far it's the source of the project so we have mentioned it.

Depending on the ERP features enabled in your company we will be able to extract main information about sales, finance, operations, human resources, or stock. In order to extract the information we have two possibilities: access through some API interface or access directly to the database that contains ERP information. Our preferred approach is the first one if the tool allows you to do it; in this way you won't be so dependent on table structure change in the database. A communication interface that allows you data extraction will be possibly kept along ERP evolution while nobody ensures you that table structure will remain the same. On the other hand, table structure can be confidential so you would need to investigate which tables and columns contain the information that you want to extract while using an API interface you will have information about with the interface functions and parameters to extract required data. The main problem that you can find in the API usage is the performance that this interface offers you, because ERP performance is usually optimized for transactional operations with low volume, but launching an ETL extraction can manage thousands of rows in a single extraction. You will need to evaluate possibilities and performances of both extraction methodologies to be able to decide what your best option is.

Database

To store data. This is the main objective of the Database. This could be enough of an introduction to databases, but let's go develop this ide a little bit more. You will probably know what a database is, so it makes no sense to start here with definitions from Wikipedia. Let's focus on the usage of the database into the BI solution.

Database is the generic name that we use to refer to data storage technologies, but in case of the BI environment we usually refer to it as Datawarehouse. We will develop this concept in the next sections. For sure it's the core component of this architecture; without a database there would be nothing. You can always do your own ETL with load procedures, your own reporting solution with some web application extracting data from the database; but without a database you wouldn't be able to do anything. The main objective of the database will be to contain the information that will be accessed from the BI tool but also it will contain auxiliary information to perform data load, and depending on the tools that we use for ETL and front end, they will require you to save internal objects into a database, which can be the same of the one containing main data or a different one.

When defining the options to configure the database you will need to take into account some considerations:

- There are parameters to optimize for datawarehouses that must be set up in a different way than a transactional database.

- You will need to define what is the availability required for your users.

- You will need to define which the load periodicity is and which the load window available to refresh the information is.

- You will need to draw the environment definition; you can have an intermediate database between the transactional one and the datawarehouse that can be used as ODS (Operational Data Storage).

- You will need to define a backup policy according to the load periodicity. It makes no sense to save daily backups if your data is changing once per month.

All these considerations and some others will be analyzed with further detail during Chapter 4.

ETL

The ETL component, an acronym of Extraction, Transformation, and Load, has a name quite descriptive. As you can imagine, its main functions are Extract, Transform, and Load. What does it mean? You won't use your transactional tables to develop the model analysis with your BI tool; you will use your datawarehouse for that purpose. So you will need to extract information from your source ERP and load it into your datawarehouse. In the middle of this process you will need to adapt the information to the datawarehouse required structure, by creating your fact tables with the desired fields, the relationship tables that must respect one to many relations between dimensions, ensuring that all the possible values of join fields are present in the table and that the fields that join with the rest of tables have no null values or that lookup tables contain all possible values of the facts and relationship tables values. As part of these basic operations, it's quite possible that you have transformations such as the calculus of aggregated tables, the process of doing a daily photo of some table to keep history of its evolution, load of security tables that allow you to do row-level security access, or any other transformation that you can require in your fill up process of the datawarehouse. We will analyze this process and some open source tools for that in Chapter 5 of this book.

Front-End Tool

Sometimes considered as the BI solution itself, you will require a front-end solution that will allow your users to interact with the data saved into the datawarehouse. This BI tool will be the main communication channel between users and data so it's highly recommended that both connections are working correctly. Your users will require enough knowledge of the BI tool and the BI tool must be user friendly. On the other side, the BI tool must be completely compatible to work with your database; you need to ensure certified interoperability between both components on vendor specifications. We will analyze later in this chapter which main capabilities of the BI tool are and also in Chapter 8 of this book we will see some open source solutions that you can use to start your BI project.

Budgeting Tool

Derived from decisions that you will make using front-end tool data, you will define actions that should be reflected in the budget for the incoming periods. Sometimes this kind of tools is not considered as part of BI, but we think that they are an important tool that finalizes the company's analytics life cycle and it also includes data analysis functionalities, so we have decided to include it in our platform definition as a BI component. As a budgeting tool you can just have a worksheet containing the objective of next year, but we are going to analyze more powerful tools that will provide you some extra functionality for that purpose, such as WhatIf analysis or data sharing. We will see more information in the MOLAP section later in this chapter and also in Chapter 9 of this book.

BI Platform Location

Once we know which components will be part of our BI platform, we need to decide where to locate them. In order to decide this we need to consider what the policy of our company is for the rest of servers that we have; in case of choosing a licensed solution, we need to consider the different licensing prices for on-premise and for cloud options taking into consideration both OS licenses and tool licenses. Also there can be some security restriction from our company to upload confidential data to the cloud and what the possible evolution on our platform is. Mainly we will have three options:

- **On-Premise**: You will have the servers located into your CPD, inside your company network, and fully dedicated to you. You will need to take care of maintenance and version upgrade for OS and software installed.

- **Cloud**: You will use a shared infrastructure of virtual servers buying only capacity and forgetting about OS and software maintenance; this will be done by the cloud company. You can have your platform always upgraded to the latest version without taking care of the upgrade processes.

- **Hybrid**: You can think in a hybrid solution by locating some servers inside your CPD and others on the cloud. In this case the most usual approach is to have the BI front end in the cloud as far as usual security restrictions are stronger for databases, so maybe your security policy doesn't allow you to have the database in the cloud but as far as your BI front-end tool is not saving any data, it can reside anywhere.

BI Concepts

Inside the BI world there are many concepts that might have different names among different BI tools, but at the end of the day they refer to the same topic. In this chapter we will review different concepts, starting with the datawarehouse concept, following with the logical model of the set of tables from the datawarehouse that are used directly for reporting, then we will see the physical model and how it's related with the logical one and which elements of the model are mapped in the BI tools and how we can use them.

Datawarehouse

One of the most important concepts that comes with a BI solution is the Datawarehouse (DWH). DWH is the basis of traditional BI systems because it is the place where the data you want to analyze resides, usually supported on a database system. The idea below a DWH is that you can collect data from multiple sources, clean it, ensure its integrity, ensure consistency, and ensure completeness to have a reliable BI implementation.

You can have multiple sources from your company, data coming from your transactional system for the standard flow of Orders/Deliveries/Invoices, data coming from your warehousing tool that controls the stock that you have inside your warehouse, data coming from the logistical company that is delivering your products, manual grouping of customers or products done by your analysts, data coming from your retail customers showing how they are selling your products, etc. You can have heterogeneous data in those different sources, but the basis of DWH techniques is to get all this data linked in some way, with complete lookup tables where you have all the possible values for your dimensions, with no gaps in your hierarchies and with unique relations among your tables. To accomplish it you will require implementing integrity processes that may require some manual action back to correct some of your transactional data or customer/product master source. Data cleansing is required because is quite usual that the transactional system allows you to introduce information without too much control. At the end it is a matter of a lack of coordination among the transactional system and the BI tool. This can be easily explained with a real example that we have found in one of our customers. In this company they use a very common transactional tool quite flexibly where you can set it among many other characteristics, which is the sales force member assigned to a given customer. This relationship, based on the process design by business rules is unique; there should be no customer with more than one sales force member assigned, but the transactional system doesn't have this constraint – it allows you to set more than one. When we receive in the BI system which is the guy assigned to the customer, we see two different possibilities, and this could cause data duplication when you analyze profit of the customers across sales force hierarchy. What we have done is implement a process that ensures that you have only one assignment; we choose one randomly, but we inform to the responsible of masterdata management to correct the assignment in the transactional to be able to publish reliable information for this customer on the next load of information.

Another real-life example that we have seen for different customers of the business rule that is fixed by the business responsible but sometimes is not followed up by business operators is related to the flow Order-Delivery-Invoice, or due to some technical problem you can have integrity issues. In this case the main problem we find is a technical one. In this customer there is an automatic system that is saving data inside the DWH database. Every time that somebody inserts an order, a delivery, or an invoice in the transactional system, it's automatically sent to the DWH. Business rule says that you cannot have an invoice without delivery related, or a delivery without order related. But if due to some technical issue you receive a delivery but you haven't received previously the related order, the BI system is rejecting the delivery to a quarantine table until we don't receive the order in our system, and it's also warning us at the end of the month to about rejected invoices or deliveries.

These two examples are only some of the near-to-infinite possibilities that you can find in the nature of the different business environments that you can find along your carrier.

▓ **Note**   When you are implementing a DWH system you need to clearly understand which are the business rules that apply to your model and to implement technical restrictions to force these rules to be accomplished, defining also the corrective actions that must be performed.

Another idea related to the DWH is the possibility of containing wider history of data versus the data that you can access in the source system, as far as one of the basic requirements for a BI system is the possibility of analyzing the past evolution of your data to estimate which can be the future evolution, and if you want to do it with consistency you will need as much information as you can get.

The DWH is usually filled up with daily, weekly or monthly processes, so it's usual that the information that you can query has some delay of at least some hours, and this is due to the usual ETL processes that are used to fill up the data, and the restrictions and checks that they must incorporate to ensure data reliability. There is also a trend in place to have intraday BI but it's only possible depending on the relation between hardware available, ETL processes optimization, and the amount of data that you want to include in your DWH.

Inside the datawarehouse you have different groups of tables:

- **Entry tables**: They contain information directly from the data source that you are trying to analyze, sometimes named ODS tables because they contain information directly extracted from the ODS database; Operational Data Storage; that usually is a clone, full or partial, of the transactional system, sometimes they are named Entry tables.

- **Temporary tables**: They contain information only during the ETL process, because they are usually used to solve some calculation process that, due to performance problems, complexity, or other technical reasons, cannot be solved in a single ETL step. These are the tables that we will consider as Staging Area. They can be located in a separate database or within the datawarehouse.

- **Final tables**: They are the set of tables that will be published in the BI tool for analysis. This group of tables will be directly related with the logical and physical models that we are going to analyze in the next sections.

DataMart

The idea behind the DataMart is to isolate information of a concrete area inside the company. While the DWH stores whole information, a DataMart will contain departmental information. Depending on the strategy, you can define the DataMart as a portion of the DWH or your DataMart can be located into a separate database and the DWH is the previous stage of your DataMart. You can isolate your DataMart from the rest of the environment by locating it in a different server, in a different database instance, in a different database, in a different database schema inside the same database or just by separating it in a logical way (by table names, with prefixes or suffixes). There is no a general recommendation to do this separation, it will depend on the amount of data that you are managing inside the DataMart compared with the whole environment, your budget to implement the DataMart, the level of isolation that you can afford, and the parametrization that your database implementation allows you to perform. All previous considerations that we have explained related to the DWH can be applied to the DataMart. The only difference is that a Datamart contains only a subset of data. Later in this chapter we will talk about projects inside the BI objects; a DataMart can be directly related to a project.

Logical Model

It's quite possible that if you are part of the IT team in your company and you want to start with the definition of the DWH structure required to support the BI analysis that you want to perform, it is quite possible that you start thinking of tables, fields, data types, primary keys, foreign keys, views, and other technical stuff, starting with sample table creation that at the end is converted in your definitive table. But then you can see missing fields, you don't know how to fill them up, how are they related with the same field in other tables, etc. What we hardly recommend is to start from a logical definition of the model that you want to implement based on entities and relations among them rather than directly start with the technical definition.

With a logical model you can see which tables are related with which others, what are the fields that can be used to join those tables, and you can easily check if the business analysis requirements meet with the model that you are proposing and its main purpose; it's used to interact and translate these business requirements to the database structure.

Inside the logical model and therefore reflected in the physical model, we can find mainly three types of tables:

- **Fact tables**: They are usually the biggest tables in the model, because they will contain business data to be analyzed, summarized, and aggregated based on desired fields. They contain the detail of sales, costs, operational steps, accounting movements, human resources data, sales force visits, or whatever data that you want to analyze. It's not required to have the maximum detailed level in all fact tables, in fact it's recommended to have some aggregated tables precalculated to have a better response time from the reporting interface.

- **Relationship tables**: These tables are used to relate multiple concepts among them. They can be based on direct relations used to define logical hierarchies, as a time table that contains the relation between day, week, month, quarter, and year, or they can be used to relate independent concepts such as product and customer based on a source fact table, as, for example, product and customers that have any sales register in the whole DWH history.

- **Lookup tables**: Also named master tables, they contain mainly the concept identifier and the description of this concept, and also it can contain the upper hierarchy identifiers in case of different attributes related. In the previous time example you can have the lookup table for Day containing Day Identifier, Date Format, and Month Identifier related, and then you can find the Month table that contains Month Identifier, Month description, and the related Year Identifier. Sometimes a lookup table can be used to define a relationship.

Relational Model

Logical models are based on a relational model as far as they are usually located into a relational database. It's called relational because the basis of the model is the relations among data, usually saved into tables. A table is defined by columns and any row of data is a relation that exists among the different fields.

Defining a simple example, you can consider a sales table where you have a product code, a customer code, and the amount of sales sold during this month. You can get from this table all the relations between product and customer that have some sale related. Also you can join tables using SQL (Structured Query Language) select statements to join information from more than one table and get derived relations. But we will see further information in Chapter 3 dedicated to SQL. There are multiple classifications of models, but in this book we are going to talk about main model types used in data warehousing, Normalized vs Denormalized models, Star and Snowflake models.

Normalized Model

A Normalized Model is intended to reduce to the minimum data redundancy, optimizing storage costs by avoiding repeating the same data multiple times. This kind of model is highly recommended for transactional solutions and also can be used in DWH models but you need to be aware of its restrictions. As you will have relationships only once, you cannot allow multiple assignments. If you change a relation it will affect all the rows in the DWH. In Figure 1-2 you can find an example of a normalized model.

Cities				Regions				Countries	
CITY_ID	CITY_DESC	REGION_ID		REGION_ID	REGION_DESC	COUNTRY_ID		COUNTRY_ID	COUNTRY_DESC
1	New York	1		1	New Jersey	1		1	USA
2	New Jersey	1		2	Massachusetts	1		2	Spain
3	Boston	2		3	Catalonia	2			
4	Barcelona	3		4	Madrid	2			
5	Girona	3							
6	Madrid	4							

Figure 1-2. *Normalized model example*

Denormalized Model

Denormalized Model wants to improve query performance by avoiding joins at the runtime. In order to do that it requires repeating data along tables in order to minimize the number of joins required to solve a user request. In Figure 1-3 you can see an example for a denormalized model.

Geography						Regions				Countries	
CITY_ID	CITY_DESC	REGION_ID	REGION_DESC	COUNTRY_ID	COUNTRY_DESC	REGION_ID	REGION_DESC	COUNTRY_ID	COUNTRY_DESC	COUNTRY_ID	COUNTRY_DESC
1	New York	1	New Jersey	1	USA	1	New Jersey	1	USA	1	USA
2	New Jersey	1	New Jersey	1	USA	2	Massachusetts	1	USA	2	Spain
3	Boston	2	Massachusetts	1	USA	3	Catalonia	2	Spain		
4	Barcelona	3	Catalonia	2	Spain	4	Madrid	2	Spain		
5	Girona	3	Catalonia	2	Spain						
6	Madrid	4	Madrid	2	Spain						

Figure 1-3. *Denormalized Model*

■ **Note**   The most usual situation in a DWH is that you need an intermediate situation between a high normalization and high denormalization. Depending on the nature of the hierarchy and attributes you will use a normalized strategy, a denormalized strategy, or an intermediate strategy; in the previous example you could have the REGION_ID and the COUNTRY_ID in the Cities table but then you could be required to join with the Regions and Country tables to get the description.

Star Model

Star model is a type of relational model widely used in datawarehousing, especially for small DataMart, where you have fact tables (we will talk about it in the next sections) that contain the information of your sales, operational or financial data, and then you have also some lookup tables linked to the fact table by some key fields. These lookup tables will contain descriptions or other concepts related to the key fields. An example of this simple model is shown in Figure 1-4.

Figure 1-4. *Star model example*

Snowflake Model

Think now about a snowflake. You will visualize a flake with a big core, then branches that split into smaller branches. In a similar way you can think of a snowflake data model where you have big fact core tables in the center, linked to them by some key tables that contain the basis to extract main attributes hierarchies, then you can find directly some lookup master table linked to these key tables or smaller key tables that relate the data with other master tables. In Figure 1-5 you can see an example of a Snowflake model.

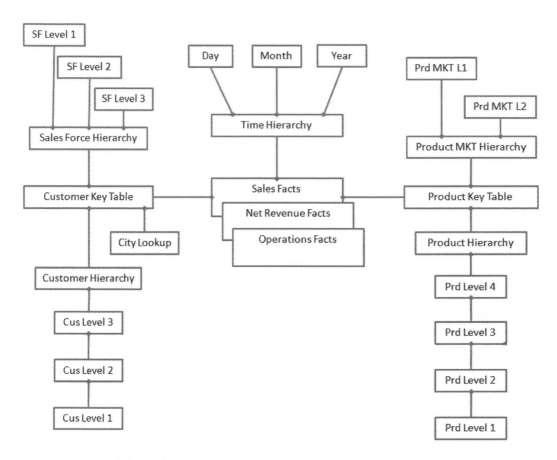

Figure 1-5. *Snowflake model example*

Physical Model

Once you have the logical model defined you will go down to the technical stuff, and you will need to define which fields contain your tables; which will be related directly with the entities of the logical model; which is the datatype for all these fields; which are the fields that will be used to join tables; which will be the unique keys of the tables; if your model will enforce data integrity with the usage of foreign keys or if you prefer to keep physical restriction out of your model and develop logical checks to facilitate the ETL process. This part will be developed in Chapter 3: SQL basics.

▓ **Note**   It's important to define a nomenclature for tables and fields. This nomenclature can be very strict but we recommend having an intermediate nomenclature level between very descriptive names and just technical ones. Fully opened names can lead into table names like SALES_AT_CUSTOMER_AND PRODUCT_LEVEL, or SALES_AT_CUSTOMER_AND PRODUCT_LEVEL_INCLUDING_PLANT. If you try to fix a very strict nomenclature you can find names like XP55GES001FT that doesn't give you any idea about what can contain this table. But with an intermediate option you can fix that your fact tables start by F_, relation tables by R_, lookup tables by L_ and then allow some characters to try to set some significant names such as F_SALES_CUS_PRD or F_SALES_CUS_PRD_PLT.

Regarding field nomenclature, it is recommended to follow these two simple rules:

1. **Do not set the same field name to different concepts**: usually BI tools recognize when you have the same field name and they try to join the information through this field, causing some data loss. You need to take into account that something that seems to be similar, as Region, can cause data mismatch if you are talking about customer region or employee region, because they can be different in some cases.

2. **Set the same field name and type to the same concept**: root cause is the same, usually BI tools will recognize the same field as the same concept but also keeping this strictly you will reduce the complexity of the model because you won't need to remember different field names in different tables to join if you are talking about the same information.

We want to remark on the relevance of these two ideas with some added real examples. Inside a customer that was analyzing the visits that their employees are doing to the stores where they sell their products, we had the information about who did the visit and we called it EMPLOYEE_ID. All customers should have one employee assigned but only one. It seems to be clear that you can have a relationship table between customer and employee, and as we were very original in that time, we called EMPLOYEE_ID the field that was related to the employee who had assigned the customer. But what happened when we were trying to do an analysis of visits per employee? The tool detected that EMPLOYEE_ID was available in both tables and it interpreted that they were the same, so it joined both tables through this field and we were losing information if the visit was done by a different employee or if the customer assignment was changed.

BI Main Objects

Across different BI tools we can find similar objects that can possibly have different names but they are related to a similar idea. In this section we will try to analyze which are these objects but you should try to focus in the idea behind the object and not in the object name itself.

Project

As a project we consider a set of information that has enough similarities to be analyzed together based on some key relation fields. In different tools you will hear about projects, universes, environments, books, deployments, but all of them are referred to the same concept, a group of tables that has sense to analyze together because they have some kind of relation. The project concept in BI is directly related to the DataMart concept in the database, which we have already defined as a set of tables related, so we will analyze in a BI project a database DataMart. Projects are usually isolated among them and many times BI tools don't allow to transport objects or join information coming from different projects, so you need to be careful when you decide if you are locating a new analysis inside the same project or you prefer to create a separate one in order to not require to redo the analysis when you realize that information must be analyzed in conjunction with the existing one.

Tables

The most usual strategy is that database tables are mapped into some kind of logical table or table definition that is an own object of the BI tool and then they are used as a basis of design for the rest of the objects. Most BI tools will have this kind of logical definition of tables inserted into their object catalog. This mapping can be from a table or from a view and sometimes the tool itself allows you to define your own views (or selects) at the BI tool level instead of doing it at the database level. All tables have at least one field that we will divide in two types in the following sections.

Facts

Facts are one of the types of fields that we can find inside our tables. They are usually numeric fields and its main characteristic is that can be aggregated. Facts are the basis of analysis and this is why they are usually named Measurement Data. With the union of the fact with the formula that we want to apply to summarize the data (we will see some example when we talk about grouping capabilities), we get measures, that also depending on the tool you can find them as metrics, indicators, or KPIs. Depending on the tool the fact object itself doesn't exist, and you map directly the field with the formula that you want to use to aggregate it. Some examples of facts could be gross sales, cost, revenue, time spent, stock volume, or space used.

Dimensions

If some of the fields are used to be aggregated, the rest of the fields will give you the level at which you want to do that aggregation. These fields are also named Descriptive data because they describe how the data is distributed and the amount of data analyzed. Nomenclature among different tools will be Axis, Attribute, Dimension, and they can be grouped into Hierarchies that will define the relationship between these attributes or dimensions. In this case we can think on date, customer, product, plant, region, country, sales office, or whatever other field that allows you to split the information.

Reports

At the end of the road the main result of a BI tool will be reports at a level that allow end users to analyze the information. Depending on the BI tool but also on the characteristics and complexity of the reports we can find different concepts like Report, Document, Dashboard, View, Analysis, Book, Workbook, etc., but all of them are referring to the final interface that will be used to analyze the information. They can go from a simple grid of sales per category to a set of visualizations that are interactive among them and that can be paged by, drilled down, pivotable, printable, or sent by email.

BI Approaches

There are different BI approaches that grant you different functionalities. Some BI tools combine more than one approach into a single platform but usually they don't cover all the functionalities. You need to define the scope of your project before you can properly choose which parts of the solution you will require.

Query and Reporting

Query and Reporting (Q&R) is the initial stage of typical BI implementations, as far as the main objective of a BI system is to deliver information in some comprehensive format for their analysis. This information is coming from a Database so to get this information we will use a Query that is a request to the datawarehouse to extract data from there. On the other hand, the information returned is formatted in a user-friendly format to be readable by the analysts.

When we talk about Q&R we are referring not only to the possibility of accessing the datawarehouse to extract information from there but also a tool that allows you to analyze it, drill into details, check for standing out figures, check for unusual behaviors out of standard trends, filter the information you are getting, and formatting it to the company standards. The main benefit of the Q&R tool is that you aren't required to know SQL language to be able to extract information from the database; this SQL is automatically generated by the BI tool. In some cases, if you have technical skills to do that, the tool can give you the possibility of modifying, tuning, or directly writing from zero the SQL that you are launching, just in case that the model in the BI tool is not fully adapted to the database below, or that you need get information from another set of tables that is not mapped into the BI catalog. This possibility will depend also on the BI tool that you are using; not all of them will allow you to create and modify queries in SQL.

Depending on the BI tool capabilities and on your organization model, they will also create their own metrics, customized dimensions, or complex filters and groupings. We will analyze BI capabilities inside the next section.

This Q&R solution is especially destined for analysts who will be able to get higher benefit of this flexibility of creating ad hoc reports than most of the components in the organization.

Information Sharing

Analysts in your company will be very pleased of having the possibility of using ad hoc reports and queries and perform complicated analysis creating complex metrics, but the vast majority of employees in your company will be very pleased if they can directly access already created reports or get their reports in their email mailboxes or in a shared location.

The standard approach in a BI environment is to have a small subset of employees with analyst capabilities that will make profit on the Q&R solution, but then to have a group of developers (they can be the same analysts group) that will create reports to share with the rest of the organization, either through the same BI tool so users can connect to the BI tool to run reports also with BI capabilities enabled, or through some distribution service such as email or a shared folder. This distribution can be usually done with a BI tool as most of them allow delivering emails automatically.

Dashboarding

Talking about BI dashboarding, you can think of an airplane dashboard. There you can find multiple indicators coming from different sources that give you an overview about the performance of the whole plane from main engines to the wing flaps. As in the airplane dashboard, a BI dashboard is especially relevant to have information about alerts, as far as you need to pay special attention to those metrics that are out of the standard working range. As commented in previous sections talking about KPI relevance, we need to focus our dashboards to those metrics that are really relevant for the analysis that we are doing and also they must be really meaningful by comparing with previous years or with objectives.

BI dashboards can also offer some features such as dynamic selectors, information panels, dependent grids, or graphs where by clicking on a part of the graph you act filtering on a dependent visualization; drill to detailed information; drill to a related dashboard; all format options that you can require regarding colors, fonts, styles, images, and shapes; multiple information layouts; tooltips; or media embedding among other features.

Data Import

One of the main trends in BI platforms is the possibility of quickly analyzing information by enabling the user access to a channel to import his own data files into a BI interface to implement quick dashboards with multiple visualizations related, giving the user self-service BI. This possibility also reduces the development time of BI projects that historically have been big projects with long delivery terms. Depending on the tool you will be able to import files in different formats (Excel, CSV, text), connect to your own datasources, to connect to web services interfaces such as Xquer, or use files that are located in shared cloud platforms.

Data Discovery

Closely related with Data Import we find the Data Discovery approach that consists in a set of visualizations specially destined to find trends and exceptions in a very easy way using a very intuitive interface. This kind of interface is the direct evolution of dashboards by simplifying controls and menus, limiting options allowed but focusing in more powerful ones. Maybe the user doesn't have a highly precise alignment grid to create perfect graphs, but they can easily relate one graph with another and filter both with simple filtering panels.

The main idea behind Data Discovery is allowing the user to create his own dashboards without having strong BI tools knowledge, just with a very intuitive interface with main functionalities.

MOLAP

Multidimensional OnLine Analytical Processing (MOLAP) databases are oriented to give you better performance in the query process. They are usually seen as a cube as far as they have multiple dimensions. Yes, strictly speaking, a cube has only three dimensions, but it's an illustrative way to draw them to distinguish them from relational tables of a relational database. Instead of multiple tables with facts, relationships, and lookup information, MOLAP databases have all the information together, precalculated across all dimensions, at all levels of the different dimensions considering the intersection with all levels of the rest of dimensions.

This precalculation has two main implications regarding performance. Information retrievals from the database are very agile, but on the other hand, information loading into the MOLAP database can take very long; in other words, if you want to have a reasonable loading time you will need to moderate the level of detail of your MOLAP cubes. You will hardly find a MOLAP database with the same, neither similar, level of detail than a datawarehouse.

MOLAP databases also usually have the possibility to save data into them by end users, and this feature opens the possibility to be used as planning and budgeting tools. It's quite usual to find a MOLAP database with a dimension named Scenario or Version that can be used to contain Actual data of current year and prevision of the next year, also with different versions of the future prevision to be able to do WhatIF analysis, as an example, what would happen if I increase my sales 15% next year by investing 10% more in publicity? Now let's try to increase 5% in this product but 15% in this other product. Now let's see if I decrease manufacturing costs by 11%, etc. You can perform and compare multiple simulations of different scenarios before publishing company objectives for next year.

Sometimes MOLAP is not considered as a classical BI functionality as far as BI was intended to be an analytical tool of reality without writing capabilities, but nowadays we think that it's quite important to have the possibility of drawing scenarios playing with different variables and once you have defined which is your objective, close the company life cycle by trying to achieve it.

Data Mining

If we have recently seen that MOLAP allows you to imagine how the future can be, we are going to see that Data Mining will allow you to predict it. Data Mining will help you to predict it based on its capability detecting hidden trends and patterns. As you can imagine, these predictive capabilities are based on past information and forecast formulas, so although you can do a very good job defining them, there will be the possibility that an unexpected fact can change fully the scenario, causing completely different results from your planning.

With Data Mining you try to find a model that justifies the relation between the input and the output of your process, so you can better understand your business. Imagine that you have a car parking business. In order to modelize your business let's start with the relation between parking lots you have and monthly invoice amount. You start with 10 parking lots and your monthly invoice amount is $ 2,000. After one year you are able to buy 10 more parking lots and your monthly invoice increases up to $4,000. Based on this data the relation between parking lots and monthly invoice amount seems to be linear. So you decide to build a full building parking of 500 parking lots but then you see that you are not able to rent more than 40, and also the rent price of these parking lots decreases. Your model was not correct because you have not considered the population density in your area, the number of existing parking lots apart from yours, etc.

It's especially interesting to find these patterns to pay attention to unusual values, null results, and data dispersion because they can give you feedback on the validity of your model.

So if you want to get most accurate results as possible you need to be very exhaustive when you are defining your Data Mining model, by including as many variables as possible to get the business model that will allow you to predict future results.

Basic forecast calculations are based on two main components: previous results and trend. Then you need to start the analysis about how they vary depending on different dimensions. Usually you can start with the time to see if there is some seasonal behavior, intra-month or multiple years' cycle. Then you can try to see if the product is affecting in some way, due to product life cycle, which is the correlation between publicity investment and net revenue, or any other variable that you want to consider.

Incoming Approaches

It's difficult to know where we can arrive in next ten years in BI approaches and options. As commented in the BI evolution section, main focus nowadays is to develop Big Data solutions to be able to access to trending to infinite amount of information that is being generated everywhere by every electronic device. This will require adapting BI tools to have power enough to analyze this amount of data and also change the philosophy of classical Datawarehouse to be able to manage this amount of information without strictness of Datawarehouse rules.

BI Capabilities

Depending on the tool that you are using you will see these capabilities with different particularities, for example, you can find a tool that allows you to drill across data just double clicking on the field header, others that also allow you to choose how to drill or others that drilling is set by administrators and you cannot go out of defined paths, but they share mainly some characteristics that we will try to analyze in this section. You are next to finish this boring introduction to start with more technical chapters, especially from Chapter 4 to Chapter 10, which will be much more interactive. Come on!

Drilling

We have already seen some examples that use drill capabilities to arrive to some conclusions but we would like to define what the meaning of drilling inside the BI environment is. Drill is the possibility to navigate across data in your model in order to get more detailed information of some singular rows. It's especially useful when you are trying to discover what is causing an unusual amount of data that stands out comparing with the rest of values that you are seeing. When you are analyzing a report by employee and you have some employee selling three times more units than the average or other employee selling less than half, you can drill into the product hierarchy to see what is the root cause of the difference, or drill across time to see if the employee has been out of the office for two weeks due to holidays, or drill across countries or regions to validate if the top seller is covering some rich area, etc.

An example using an Excel worksheet with a Pivot table to illustrate it could be the following one, where you can drill from project RUN to analyze which tasks are being done for this project, as you can see in Figure 1-6.

Sum of Hours		Column Labels ▼							
Row Labels ▼		23.05.2016	24.05.2016	25.05.2016	26.05.2016	27.05.2016	30.05.2016	31.05.2016	Grand Total
⊟ MGM				2	1		2		5
⊟ REL			2			2	3	4	11
⊟ RUN		4			3	3	3	1	14
2. Incident handling		3			3		3		9
3. Support		1				3		1	5
Grand Total		4	2	2	4	5	8	5	30

Figure 1-6. *Drilling capabilities*

Pivoting

Pivot data is referring to the possibility of moving concepts from rows to columns and vice versa, to move some concepts to the filter space, to move some concept to generate different information pages, in general, the possibility to play with fields to organize the data that you are showing in a friendly format.

As an example, it's easier to understand a report that analyzes trends if you have the time concept in columns that if you have it in rows or in case of a linear graph, it's much easier to see the evolution if the time is on the X axis. But maybe in combination of drilling, you detect a month where all your customers have an increase of sales, then you move the time attribute to the filter area, you select the month with the increase, and you move the product attribute from page to the rows area to compare them, moving the customer to columns to see the relation between customer and product for that month.

Visualizations

There are many times when seeing data graphically facilitates the correct understanding of data behind plain grids of information. This is why all BI tools offer graphical visualizations of data that, depending on the tool, can be more advanced and evolved than others. Most tools, maybe all of them, will offer you a set of graphs such as bar graphics, line graphics, or area graphics, all of them vertical and horizontal; pie charts, scatters or doughnuts are also usual in most of tools. Then you can find some specific advanced visualization that are provided only by some tools, such as word cloud, heat map, relationship graphs, waterfall graphs, sunburst, stacked bars, streamgraph, or multiple other visualizations that can be used to improve the understanding of the report.

■ **Note** Some tools allow the usage of custom visualizations that you can develop and import by yourself. If you have some specific visualization capability or any kind of customization, you should check your vendor documentation to see if they allow such kind of extra visualization.

Map Visualizations

Inside all possible visualization options there is a type of visualization that we think requires an extra section for them. Coming from the universalization of GPS capabilities in mobile devices that are used for data gathering in combination with map integration capabilities with platforms like Google Maps, ESRI Maps, or CartoDB among other multiple commercial options; you can have in your BI tool some map visualization that facilitates the correct understanding of how your sales are distributed along the country or how your sales force is covering a defined area.

Sorting

Another capability that you will need in your analysis is to have sorting options. This is also something offered by most of the tools, but also there can be some differences in the options that they allow you. You can sort ascending or descending, by one or multiple fields, based on numeric, alphanumeric or date fields, based on dimensions or metrics, and also with the possibility of sorting the pages shown.

Grouping

There are two types of grouping in the BI world. The first one is basic for BI analysis; you cannot analyze hundreds of millions of rows in detail, so you need to aggregate this information in something readable. To do that all BI tools provide grouping capabilities using SQL language that is used to query the database. In order to have grouped information you can select the detail fields that you want to analyze and use some aggregation function as sum, average, maximum, minimum, last, median, etc. – usually against numeric fields. So you get the result of the formula used over all the information available at the desired output level. This first type of grouping will be analyzed with further details in the SQL chapter.

But there is also a different grouping considered as element grouping because it's not a technical grouping based on the values of a field; instead of this, it's a group of these values to a customized aggregation. Let's explain it with an example that will show you what we are talking about. In your DWH, you have a field that classifies which is the region of your customer, and another one that is showing you where the country is. On the other hand, in your sales responsible structure, you have a team responsible for Northern regions and another team responsible for Southern regions. You can have a report of sales per country or sales per region but maybe you are interested in grouping some regions to show an aggregate value of a set of them to compare it with the rest, and the country field is not useful for you because you need some extra detail. With grouping capabilities you could group Southern and Northern regions to show information only at this detail level, nor at region neither at country level.

To do such kind of element grouping you can find different options in different tools. Some of them allow you to group the information on the fly based on the results of the report through the same report interface, and other tools allow you to create derived attributes or dimensions to show you the information grouped at this level when you use these objects; sometimes these tools allow you to group the information based on the results of a metric and also more simple BI tools will require you to move this capability to the database engine, so you need to develop extra tables/views in the database to relate different elements.

Filtering

When your base data is using hundreds of millions of rows, hundreds of fields inside thousands of tables, it's more than possible that you won't use this big amount of information in a single report. Usually reports are filtered by one or more fields. In general you want to analyze daily information, the finance closing month, compare it with the same month of the previous year, see an evolution of the last two years, but not get in a report whole history of the datawarehouse. You will filter by day, month, or year. Also you can require filtering by a product category from which you are responsible, removing from the source data that information that is related to an internal customer code that your team is using to assign free samples or limit the information to the top ten customers in sales for every region.

Again, filtering capabilities will differ among platforms. Depending on the platform, you will be able to qualify on a dimension or attribute (equal, greater, lower, between, in list, not in list, is null, is not null, contains, begins, etc.), select among different values of this dimension or filter on the result of a calculation, which can be the result at the report level or at a different level (you can filter the top ten customers to analyze their sales, or filter these top ten customers and do an analysis per product to see which are their favorite products).

Depending on the BI tool you will have the possibility to filter dynamically at runtime. You can have a fixed report with a variable filter and this can allow you to, using the same report, extract information of different customers or regions.

Conditional Expressions

Inside conditional expressions we can find many formulas such as case and if that is combined with logical ones such as and, or, and not, that will offer you the possibility to create complex calculations that can be used to customer segmentation or employee objective assignment. Imagine that coming from your transactional, you receive financial information. You receive different costs values and the sign that comes from the transactional has different meanings depending on the cost. When you are processing a sales incentive this comes in negative but when you receive an operational costs it comes in positive. If you want to have a metric that sums all the costs you will require defining a formula like this:

```
Case when cost_type = 'SI' then - Cost_value else + cost_value
```

> ■ **Note** In order to develop complex calculations you will require Conditional Expressions. Of course these types of conditions can be applied in the database but if your BI tool allows you to do that you will have the possibility to do it at BI level. There is always a balance between complexity solved on the database and complexity solved on the BI tool. Most of the times this will depend on the preferences of the developer; this can also be defined in your developer guidelines and best practices book.

Subtotals

When you are running plane reports with more than one attribute or dimension in the breakdown area, because you require doing a combined analysis of different concepts, it's possible that you want to know a part of detail of every combination of attributes, which are the subtotals of every value of one or more attributes. Also with an example coming from Excel, you can see that you will be able to know which is the total amount of hours per day including all collaborators in the project, as you can see in Figure 1-7.

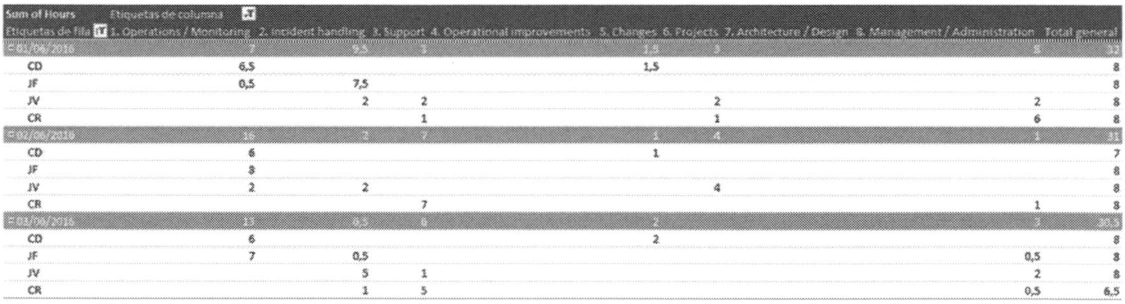

Sum of Hours	Etiquetas de columna [▼]								
Etiquetas de fila [▼]	1. Operations / Monitoring	2. Incident handling	3. Support	4. Operational improvements	5. Changes	6. Projects	7. Architecture / Design	8. Management / Administration	Total general
01/06/2016	7	9,5	3		1,5	3		8	32
CD	6,5				1,5				8
JF	0,5	7,5							8
JV		2	2			2		2	8
CR			1			1		6	8
02/06/2016	16	2	7			4		1	31
CD	6				1				7
JF	3								8
JV	2	2				4			8
CR			7					1	8
03/06/2016	13	0,5	8			2		3	30,3
CD	6					2			8
JF	7	0,5						0,5	8
JV		5	1					2	8
CR		1	5					0,5	6,5

Figure 1-7. *Subtotals example*

Administration

All BI platforms must be managed by some administration team. Again functionalities and possibilities of the platform will vary a lot depending on the tool. All of them must have in some way or another a security definition, user management, group management, role management, which will define which access is allowed to every user, which permissions and functionalities he is allowed to use, and which data the user is allowed to see. But apart from basic security management, you will have the possibility to set some limits to tune server performance, configuration parameters that affect the way of accessing the database, web configuration parameters in case of a tool with a web component, cluster configuration in case of high performance requirements, security integration with LDAP or Single Sign On authentication.

You will need to evaluate also which are your administration requirements in order to ensure that the BI tool you are thinking about is the correct one.

User Interface

Another topic that you need to evaluate is which are the interfaces that the BI tool is offering you for the end-user access and experience. In order to access to the information you will find tools that offer your users the possibility of access through a client-server installation, a stand-alone installation, a web access to connect to the system, or they offer you to receive the information with some distribution tool, mainly email or a shared network resource. Of course there are tools that offer all of them and other tools that have more limited access. This topic should also be included in your decision panel, because this could discard the usage of some tool because of your security guidelines, or because it increases the project complexity, for example, in case you need to develop a package installation in agreement with your installation Best Practices.

▓ **Note**　As summary, the best strategy to find the most suitable option is to create a decision panel in a worksheet or slide to evaluate from all capabilities that are different among platforms that are mandatory for your project, which are nice to have and which are irrelevant; set a score for every section and then choose which is the best possible tool that meets with your budget.

Conclusion

If you have arrived here you should now have an overview about what a BI solution is, what its main components and functionalities are, what its most relevant capabilities are, and the main milestones of its evolution; you should know also some concepts about BI and its main component, the datawarehouse. Now let's go to analyze, with further detail, which is our recommended methodology to implement a BI platform and we will analyze with further detail how to install and deploy main BI parts into your organization. Enjoy it!

CHAPTER 2

■ ■ ■

Agile Methodologies for BI Projects

Let's suppose that you are the leader of a BI project. Let's also suppose that you are following the typical approach to implementing a datawarehouse project with its related reporting tool, the ETL process reading from the source transactional database, and inserting data into your DWH. Following the typical approach you should gather specifications from key users, think of a very robust data model that serves to accomplish those specifications, install all components, extract all the different data that you need from different data sources, validate the integrity of all this data for all fields, define the required reporting, and then you will be able to show to the key user the result. The whole process can have taken months or maybe even years. When you are checking with the key user what has been the result, your user can have changed his mind regarding what he needs or maybe your key user has changed his mind and he has completely different ideas about what to use in his reports.

You could lose a lot of time and effort of development without getting any result and this is something that can be a real risk with the traditional approach of datawarehouse project management. To avoid this risk you can try to use what we consider what is the best of Agile methodologies applied to BI projects.

Anyway, not always using an Agile methodology is the best strategy to go ahead with a project. Traditional methodologies are based on sequential processes, high-planning efforts, and high-project management overhead, so they can be useful when you have a closed definition of project requirements, clear strategy on the project development, and long-term projects to develop. We are explaining how to work with Agile methodologies because we are quite convinced that if you are going to implement a BI project from zero, with users that have never used a BI tool, it is quite possible that you will be required to make changes to the ongoing project, collaborate with the end users to check if the output of the project is what they expect, and create an initial solution and then add new functionalities in the next releases. In these conditions the usage of Agile can help you to go ahead with your project. There is also an intermediate approach that combines some traditional features with Agile, by using it to perform some of the traditional tasks. Of course Agile also works when you have a closed definition of project requirements, so you can use Agile for solving both dynamic and static project requirements.

Also we would like to remark that Agile is not a methodology itself, it is a model to create your methodology; it is a set of rules that you need to adapt to your environment and technologies to define the exact parameters of working processes. It is important to say that maybe if you are in a small company, there are some of the recommendations inside the methodology that don't apply to you. There is some methodology that includes daily meetings, if you are the only guy for IT and you will be in charge of developing the whole solution you won't require internal developer meetings to follow up the status of the tasks, as far as you will be the only one in this meeting, unless you have multiple personalities and anyway in this case would be difficult to meet all together. In the case that your team doesn't justify implementing some things from the methodology, just ignore them and take from here the methods and activities that you consider interesting so that you will improve your productivity and be successful in your project. In order to let you decide if you are interested in using Agile methodologies let's do a brief introduction to what Agile methodologies are and how can they used to improve your customer satisfaction for your BI project.

© Albert Nogués and Juan Valladares 2017
A. Nogués and J. Valladares, *Business Intelligence Tools for Small Companies*,
DOI 10.1007/978-1-4842-2568-4_2

▓ **Note**   This chapter is a set of recommendations that we consider interesting, extracted from our experience with Agile methodologies, especially Scrum and Kanban, so do not consider them as pure Agile because what you are going to read are personal considerations about them.

Introduction to Agile Methodologies

There are some occasions where the name of the concept you are trying to define is significant enough to be auto descriptive. We are now describing one of these occasions. The main characteristic of Agile methodologies, from our point of view, is that they are agile. But, what does it mean – agile in terms of software projects? Basically there are four areas that an Agile methodology must focus on, if we look at the Agile Manifesto principles. Here are the four principles of the Agile manifesto[1] we have come to value:

- Individuals and interactions over processes and tools

- Working software over comprehensive documentation

- Customer collaboration over contract negotiation

- Responding to change over following a plan

From these values of the Agile manifesto, there are two of them that are especially relevant for us: focus on the customer satisfaction and adaptation to changes; this last concept itself is the definition of agility. Inside an Agile project we are not worried about changes that can appear in specifications, we consider that they are normal; in fact they are healthy as far as users can adapt their needs when they see what a BI tool can do. We should have in our company users with seer capacity if we expect from them that they are able to know what they want without knowing about the possibilities that your tool can offer.

To follow Agile methodologies you should focus on your team members and their knowledge instead of the defined process to develop. You will prefer that all the team members are aware on what the rest are doing, how to do it, and if there is any lock that can be solved by other members of the team, sharing experience to improve techniques instead of strong rules for development or useless control tools.

You will be interested also in having a product to show to your customers rather than in the documentation required. Documentation is a secondary task that must be done but the priority is to have a working system without failures. It is better to have a robust small project than an incongruent big system. It is better to have few reliable features rather than hundreds of features failing from time to time but should work if you check in the documentation. Here we can have a look on the concept of Incremental Delivery, as far as we will be interested in having a small initial project but with added functionalities every few weeks.

We will want to have the solution available as soon as possible in order to have customer feedback in the initial stage of the whole project, to be able to correct misunderstandings, and to change functionalities based on customer expectations. We will require a close collaboration with our customers in order to achieve that. As you will see in the next sections the customer is included in the life cycle of every iteration of development, which is the basis of Agile development.

Life is change. Nothing is static. Being agile in adapting to the change is the main advantage that Agile methodologies can offer you to achieve success in your BI project. It is good to have a plan in place, but adapt to the change will give your customer more satisfaction than following a plan. And at the end of the day, your customer will be the one who will use your system, and usage of the system is the main indicator that your project has been successful.

---

[1]Agile Manifesto is a set of principles that have been defined by the Agile Alliance to guide Agile developers to define their strategies for software development. It can be read in http://www.agilemanifesto.org/iso/en/manifesto.html

Agile is also optimization. Behind Agile we can find the intention of optimizing our resources to maximize the perceived value for our customers; they can be internal or external, the end user of the BI platform. We will try to reduce or directly avoid all those tasks that don't add value to the customer. Coming from our experience we know that a high percentage of functionalities requested in the initial requirements of a big project are never used by the customers, multiple dimensions or metrics are not used in reports, multiple reports are not executed after the initial validation, and also we have detected many times that multiple requested users who have access to the platform have never logged in on our platforms.

With Agile we will try to avoid all these useless efforts to try to focus on those functionalities that the real user really wants. I'm introducing here the concept of **real user** because we have also found project requirements defined by IT guys, by departments that will be different from who will use the tool or by very high-level managers that sometimes are not aware of the needs to perform the daily tasks of the employees, so the real user must have direct knowledge of these daily needs to define a project that can cover the most used functionalities from the beginning.

Agile is also a synonym of collaboration. There is no development team leader (it can appear implicitly based on the experience of some member of the team) but the idea is that all is discussed, estimated, and analyzed by all of the team and then the individuals do the development tasks, but the rest is done in collaboration with the whole team.

Another benefit of Agile, if you are able to implement it in a correct way, is that you will try to adjust the workload requirements with the existing taskforce in order to not stress your development team, avoiding also workload peaks. You will try to reduce the multitasking in your team in order to allow the developers to focus on their assigned task. The idea is that every member of the team is doing just serialized tasks, one after another.

Agile Approaches

There are multiple Agile methodologies that follow the Agile principles that we have just seen in the previous section such as Scrum, Kanban, Lean Software Development, Adaptive Software Development, Crystal Clear methods, Extreme Programming, Feature-driven Development, and others, but this section doesn't pretend to be an exhaustive analysis of Agile, because it would give enough content for one or two books more; so we are going to focus on two main methods, Scrum for project management and Kanban for solution maintenance. Also there is a mix known as Scrumban that has some of the characteristics of both methodologies.

Our Recommended Mix between Scrum and Kanban

Based on Agile principles, there are some components that we consider should be shared between Scrum and Kanban methodologies. The first of them is the visualization panel, as far as it is important that anybody on the team has available the status of every task. A visualization panel is mandatory for Kanban but optional for Scrum that only considers having a backlog of tasks, but we realize that is very positive to have it on a physical board. Also we consider a very interesting feature to have regular daily meetings. They are mandatory for Scrum and optional for Kanban, but we consider it a very interesting feature in any case. Both similarities are coming from the first value from Agile manifesto, to prioritize the individual evolution and interaction among them over the tools that we use, at the end the prioritization is on the knowledge that they gain.

Meetings must have a high periodicity as far as knowledge must flow among team members and it's important that they are short but frequent. The ideal period is to have a daily 15-minute meeting at the beginning of the day, and this should be mandatory. Also in Scrum you will you will have different periodic meetings as we will see in the next section. These daily meetings can be perfectly done standing up in a corner of the office, and it's especially interesting to have them in front of the other common component, the visualization panel. It is important to have a board or a wall with all the tasks located there in order to see in a very graphical way which are the ongoing topics, which of them from the backlog are the most important to pick them up, and if there is any dependence across tasks, etc. The standard way to locate the tasks is organizing per columns based on the status of the tasks and grouping per rows depending on the subject or the area. There are also some tools that allow you to organize all the tasks, so this graphical visualization can be on a board or using a projector or screen. In Figure 2-1 you can see a print screen of one of the tools that we will analyze later in this chapter; it shows the Welcome board for the Trello tool.

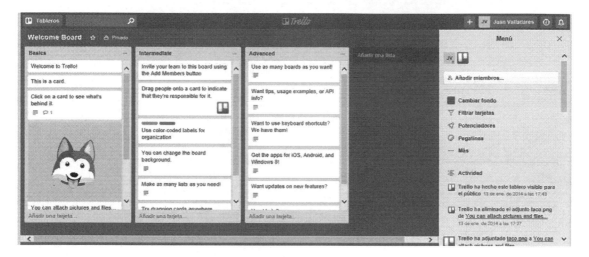

Figure 2-1. *Example of board generated with Trello*

In the daily meetings all of the team reviews the status of the tasks and moves the tasks along columns once it's advancing in the status. The number of columns and rows will be customized depending on project needs, but they should have at a minimum three columns: Backlog, In Progress, and Done. Inside Backlog we will locate all the tasks that are pending to perform, on In Progress column we will locate all that tasks that one of the members of the team or the customer are involved in, and finally we will move to Done all the finished tasks. When the new version of the software/project is released, we clean the Done column and we start again. We will see the process in detail in each Agile approach.

Developing Projects with Scrum

Scrum was one of the first methodologies that appeared to try to mitigate the problems of typical project management with sequential approach and highly organized projects in a changing environment. It was defined in 1986 by Hirotaka Takeuchi and Ikujiro Nonaka. They wanted to implement a flexible development strategy focusing on defining a common objective, so focusing on the development team.

There are several concepts inside Scrum, some of them also shared with other approaches, but as commented, we are going to focus on this methodology. To try to be clearer on the description of the methodology, which can be very abstract, we are going to illustrate the whole description of the methodology following the same fictitious example, a project developed for a hardware store company (HSC) that wants to implement a BI system to analyze their company insights. This is a medium company with 5 stores and 70 employees, 20 of them working in the headquarters, 50 of them working in the stores, in the sales force team. Every store has a store manager that coordinates all the activities required by the store, such as stock replenishment, cashier management, coordination with central teams to manage Human Resources, Commercial initiatives and offers, etc. Figure 2-2 is showing the company structure. The IT department is composed of 3 employees:

- IT manager: in charge of IT tasks coordination, contact with procurement department to purchase technical devices and software acquisitions, and contact with external providers when required.

- Field technician: in charge of maintaining and repairing computers and networks in headquarters and stores.

- Data technician: in charge of administering central servers such as ERP system, email, shared network drives, etc.

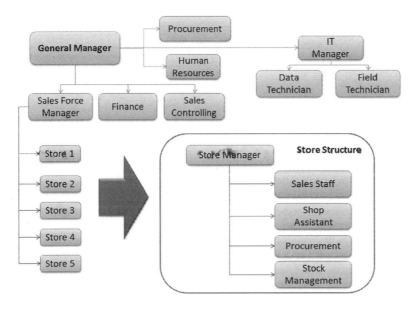

Figure 2-2. HSC organization diagram

Inside this company we will use Scrum to implement a BI system that will help the company to analyze their data, starting with sales analysis but once it's done, they will want to include stock management, procurement data, and human resources data.

Roles

In order to define what is being done by whom, let's start with the definition of roles that we have in Scrum. Also we will relate them with our example structure.

Product Owner

It is the core role from the customer side. He will be in charge of defining what they want to include in the project, which is the priority for any functionality, which are the conditions to consider the functionality as fully working and development acceptance. He will work in collaboration with the development team to define, analyze, and detail every requirement. In our example this role will be taken by the Sales Controlling manager of the company.

Scrum Master

The Scrum Master will be the person in charge of coordinating all the developments, ensuring that development team understands and follows the Agile principles; he will train the team in order to help for its auto organization; help to remove locks and impediments; and try to avoid risky interruptions that can disturb the team from its main task. He will be represented in the example by the IT manager.

Development Team

The development team will be the workforce to perform all the action required in the development. They will quote every requirement analyzing how much it could cost and how to develop it. They will organize internally who inside the team will take every task, and also they will analyze the locks that can appear during the development process to try to avoid them. At the end of the development they will analyze how to improve the development process for the next requirements. Inside this team we can define different subroles, such as the solution architect, programmers, testers, functional analysts, graphical designer, etc. They will ensure internal quality of the developments realized to avoid impacts in the future. Ideally their profile will be generalist, in terms that anyone of the team can perform any of the tasks, but this is not always possible. In the example, this role will be taken by the IT manager, the data technician, and two developers hired from an external company to help in the initial development.

Stakeholders

This is a very generic concept as far as there will be a lot of stakeholders that are interested on what we are doing. They are not involved in the development process, but they must be informed about project decisions and how the project is advancing. In our example we will consider as stakeholders all the management team, including the general manager.

Users

They will be the final consumers of the information that we are providing. In our example the number of users will be increasing, starting with the sales support team and store managers, and once new functionalities are added we will include as users the HR team, Financial team, stockers responsible for all stores, and the procurement team.

Product Owner Assistant

This role is optional; it will depend on the Agile knowledge of the Product Owner. He will be in charge of assessment to the Product Owner regarding Agile principles and rules. In our example we won't consider it.

Sprint

The sprint concept or iteration is one of the most important concepts in Scrum, also in most of Agile methodologies. It can be also considered as iteration. Sprint is the single cycle of development inside a project that will provide a fully usable version of the project available for the customers to be tested. It is a period of between 2 and 6 weeks trying, depending on the task size of the project, with the intention of being reduced to the minimum as possible. The target task size should be between 2 and 8 hours of workload, in order to detect easily tasks that have been more than one day in the "In Progress" status. In order to achieve this objective is quite possible that we need to split tasks into subtasks to fit into this maximum task size of 8 hours. We will explain how to do it in next Project Segmentation Techniques section. We will choose enough tasks to include them in the sprint to try to keep the team busy during the entire sprint but ensuring that the team won't be overloaded.

Sprint is split in three parts: **initialization** of the sprint to define what tasks are going to be included, **development** of that tasks that will provide the next version of the project, and **conclusion** of the sprint to publish the results to the customer and evaluate the performance of the team. From this sprint organization it derives into meetings explained in next section. In Figure 2-3 you can see a diagram showing the sprint organization, what are its meetings, who should assist, and the expected result of every meeting.

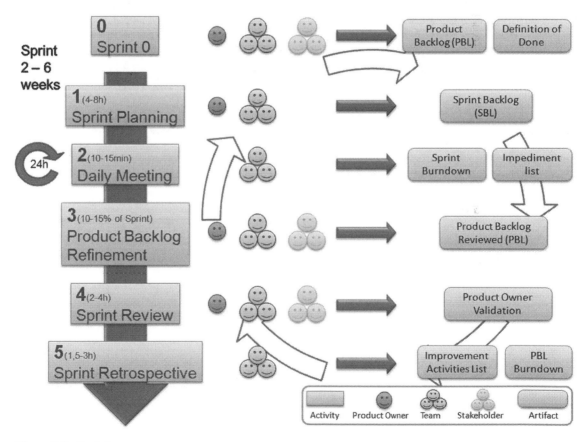

Figure 2-3. *Sprint summary*

In the example that we are following we will define the sprint duration of 4 weeks initially and we will try to reduce it during the next sprints, so initially we will have a work force of 144 hours per person as far as the first and last day of every sprint is dedicated to initiate and end the sprint. If we have 3 people in the team we will have available up to 432 development hours to manage in the sprint.

Specific Scrum Meetings

Related to the sprint we find four related meetings that are specific for Scrum, also shared with some other methodology, plus some initial meetings to define the project scope and the common daily meeting.

- **Sprint zero meeting**: The aim of this meeting or meetings is to define what the project scope is and which functionalities are expected by the user to define the overview of project requirements. From this meeting we should extract the Product Backlog, a list of requirements to develop in order to be able to deliver a working version of the project. This requirement list must contain the description of every Product Backlog Item, which is the importance of this PBI for the customer, which is the cost for the development team to implement it and if there is any risk related to the requirement. Also in this meeting we should get the **Definition of Done**, a checklist of conditions that comes from the agreement between both customer and developer of which is the expected result for all requirements. This meeting should also be attended by the product owner, the development team, and at least somebody representing all the stakeholders.

  In our example we will define in our Sprint zero meeting that we want to include the following Product Backlog included in the Table 2-1.

Table 2-1. *Product backlog derived from Sprint 0*

Product Backlog Item	Business value (1-10)	Development cost	Risks
Sales data coming from ERP system	10	200 hours	Unknown source structure
Master data hierarchies from ERP system	9	240 hours	Lack of definition of the required levels
Customized masterdata groups	7	160 hours	Extra technology to connect
Warehouse Stock information	8	320 hours	Different masterdata values
Financial information	9	400 hours	N/A
HR information	7	320 hours	Confidential restrictions

- **Sprint planning meeting**: In this meeting we choose an agreement with the Product Owner, the list of Product Backlog Items that we are going to develop in the sprint, also considered as **Sprint Backlog**. These items can be split into tasks in order to have a task size that is manageable by the team. This meeting is split in two parts: the first one including the product owner and the development team and the second one just including the development team.

- "What" meeting: The first part of this meeting should take between two and four hours to define all the PBI included and to get full understanding from the development team about what the PBI is referred to, entering into details about the exact result of the requirement or functionality.

- "How to" meeting: The second part of this meeting should take also between two and four hours and we should evaluate how to achieve the product owner requirements, which should be split into tasks of every PBI, who from the development team is in charge of every task, and how long we expect that this task can take. If some output tasks has a high effort quotation we need to try to split it again. Effort estimation must be done jointly among all the team trying to advance as much as possible, which could be the troubleshooting issues to solve.

We have a backlog workload estimated of 432 hours. Our effort valuation for first two tasks is 440 hours, so inside the initial sprint the team will try to implement the first item on the Product Backlog and try to finish the second one with the help of the IT manager. For that we will try to split every item into tasks until we arrive to a reasonable task size. In Table 2-2 we show just the initial split of the first task. We should follow with the split trying to define single subtasks from a size of between 2 and 8 hours.

Table 2-2. *Initial split for Sprint backlog table*

Product Backlog Item	Task	Development cost	Who
Sales data coming from ERP system	Analyze information source	16 hours	Data technician
	Define database model	24 hours	Data technician
	Implement ETL process	40 hours	External 1
	Define data views	16 hours	External 1
	Create logical model	16 hours	External 2
	Define BI objects	32 hours	External 2
	Create BI reports	24 hours	External 2
	Technical validation	16 hours	Data technician
	Process automation	16 hours	Data technician

- **Daily meeting**: As commented in the general approach, the development team will have a 15-minute daily meeting in the morning to comment on the ongoing tasks. In this meeting everybody in the team should answer three questions:

    - What have you done since yesterday's meeting?

    - What are you doing today?

    - Which blocking points have I found or I think can appear?

The idea of having a maximum task size of eight hours is that you can detect in the daily meeting if there is any hung task as far as you shouldn't be doing the same task in two consecutive daily meetings. From this meeting we should maintain the **list of impediments** that can appear.

Coming back again to our example, we can detect that we don't have the required credentials to access to ERP data, or that we need support from the ERP provider to understand some table structure. Maybe some component of the BI solution, such as the BI reporting tool, especially in the initial development, is pending to be delivered; we have run out of space on our development database system or many other examples of locks that can appear.

- **Product backlog refinement**: This meeting can be scheduled at any point of a sprint but should be done for all sprints, in order to review which the remaining Product Backlog is, if there is any added tasks to include it in the list, some priority change, etc. As in the sprint zero meeting, it's required that the Product Owner and the development team assist at the meeting and the expected output is the Product Backlog reviewed and updated.

- **Sprint review**: Once the sprint has finished we will review with the Product Owner and the stakeholders all the PBIs implemented. This will give our customer the possibility to review his expectatives, reprioritize tasks, change anything that he considers, or add new requirements based on a real result about what our tool can do. Then during the product backlog refinement meeting he will be able to reschedule our work.

  In the HSC example we can find that the sales support manager has detected that the product hierarchy from the transactional doesn't match with the financial required groups to extract the net revenue data, as they need to group by products with different tax rates and it's not set in the ERP tables that we have extracted. So once the sales part is working, when we try to add financial information we need to modify the masterdata load to add some added fields.

- **Sprint retrospective**: This is an internal development team meeting that will allow the team to review what we have done, how we have solved the issues that have appeared, what we can do to avoid problems to appear again, and how to avoid locks that can freeze our development. This meeting will allow the continuous improvement of the whole team improving anyone's productivity. The output of this meeting should be the **list of improvements** that we should apply to get a better performance of our development phase.

  From this meeting we could see that we would prefer to have the sprint size of three weeks and reduce the task size because we want to have a bigger interaction with the product owner. Or we could improve some method part, such as adding a new status for task validation, or we decide to have the meetings on a projector instead of a physical board.

Release

We usually will work with at least two environments: one for development and another for production, or maybe three, an added environment where the user will validate the development done with some extraction of production data. Once the sprint is done and validated we need to promote all the changes to the production environment in order to be able to take profit of the new requirements developed. We know that the output of all sprints is a set of requirements that compound a fully working solution delivered to the Product Owner for validation, but the movement to production can require some efforts such as recompilation, client download by all users, data reload, etc., and it is possible that we decide to wait and freeze some development until we have a bigger amount of requirements to move to production. The concept of release is the transport to production of a set of developments that can come from one or multiple sprints. We will deliver a release when we have been able to gather enough High-Value capabilities for the end users that they accept changing their current tool.

Again in our example we will release to production once we have developed first three items: sales data, masterdata from the ERP, and customized masterdata, as far as it is mandatory for the Sales support manager to have this requirement in place.

Artifacts Used in Scrum

From the sprint explanation, we have seen some tools that will help us to go ahead with our Scrum methodology. Also there are other tools that will help us to monitor and follow up to advance the project. All these tools listed and explained are considered artifacts by Scrum nomenclature.

User Story

It is important to have a technical definition of what is required to develop, but it is also important to have the user explanation of what he wants to get in order to be able to correctly understand what the real requirement is behind the technical one. A user story is a short definition of the requirement and it is expected to be written following a user story template, which includes the role of the requester, what the requirement is, and what the benefit of implementing it is. An example of a template could be this:

```
As <role>,
I want <requirement>
so that <benefit>
```

Moving again to our example of the hardware company, we could see these as user stories:

```
As the Product Owner I want to have the capability to drill across Customer hierarchies so
that users can analyze the detail of sales through ERP customer organization.
As the HR Manager I want to have data of salaries and bonuses available so that our
department can analyze the information by company hierarchy, analyzing HR costs, holidays,
and training.
As the Finance Manager I want to have information of Profit and Loss so that our department
can analyze PNL through main product hierarchy.
As the Sales Force Manager I want to analyze how our sales are performing by store so that I
can analyze sales per category, customer groups, and promotional campaings.
```

The Product Owner and development team will analyze and sort by priority all these user stories, including them into the Product Backlog and then the development team will analyze how to proceed technically to have this requirement in place.

There is a procedure named INVEST to validate if the user story is good enough to take it as a starting point of a development. This procedure corresponds to the initials of the words:

Independent: We should be able to develop the user story isolated from the rest of the pending user stories.

Negotiable: As the user story is not highly defined, we can negotiate with the product owner how to proceed with the development of the user story.

Valuable: The user story will give some added value to the end user.

Estimable: With the definition that the user has done, we can quote how much effort we will require to implement this user story.

Small: The result of this evaluation should be manageable inside a sprint.

Testable: There is a way to validate so that we have achieved the user expectancy for this user story.

Developer Story

A developer story is linked to a user story and it is a more detailed explanation in small sentences of what is required to do to achieve the user story requirement. It is the first step of the detailed analysis that we will require for every Product Backlog Item in order to have it clear enough to start the development (so to include it in the Sprint backlog). As per its name, a developer story is delivered by developers, and it is written in a language that both developers and the Product Owner can fully understand. We are going to explain it in a clearer way using an example of our beloved company HSC.

In this example we are going to analyze with further detail what is the developer story for the user story related to the HR manager in the previous example. First, as in the user story, we will define a developer story template.

```
The <data module>
will <Action/feature>
that will allow <requirement>
related to the <related user story>
```

With this template we are going to define our developer stories to support the HR user story

```
The HR model will receive data from the META4 system that will allow analyzing HR company
data availability related to the HR user story.
The HR model will contain the employee hierarchy that will allow drill through HR hierarchy
levels related to the HR user story.
The HR model will provide a report paged by store that will allow analyzing HR costs,
training, and holidays per store related to the HR user story.
```

Also in this case, methodology fans that like to define acronyms have defined one to see if the developer story is good enough to start the development, and its name is DILBERT'S:

Demonstrable: Once we have developed our developer story we should be able to show the Product Owner that it is working as expected.

Independent: As in the user story case we need to define developer stories that can be developed without affecting another developer story, and development should deliver a functionality working by their own.

Layered: This characteristic is very focused in Business Intelligence development, as far as most of the developments on the BI side must take into account ETL, database, and user interface developments.

Business-valued: The user must appreciate the development we are proposing, coming from the user story value.

Estimable: Again coming from the user story characteristics, the developer story must be clearly estimable. In this case the difference is that it should be done with a more precise estimation as far as the developer is writing what he is going to develop.

Refinable: The developer story must be written in a more concrete way of a user story but again it can be reviewed and adapted during the development.

Testable and **Small**: These two characteristics are directly the same as in the user story.

Product Backlog

The product backlog is a list of requirements at a high-level definition, without detailed specifications qualified by user priority and development cost that must be defined at the beginning of the project and reviewed after every sprint, so it is not a fixed list; it can evolve and change along the project. The Product Owner is in charge of creating and updating this list. In this case we refer to the example explained in the Specific Scrum meetings to understand what the Product Backlog is and in what moment it is created and maintained.

Definition of Done

The definition of done is a checklist that every item must meet in order to be considered good enough to be moved to the production environment. It is a quality validation that pretends to ensure that all the requirements have been developed with the desired quality standards. The definition of done checklist contains items such as developed code matches codification standards, tests have been passed successfully, performance tests have been successful, we have get the customer approval, or the documentation is finished.

When to Start - Ready Definition

An interesting idea added to the original Scrum methodology is the Ready definition or Definition of Ready. It is an analogy of the Definition of Done, or when our development can be considered as finished, but related to the definition that the Product Owner must do of the PBIs that he wants to develop. When we define the Product Backlog, the PBI is defined at a very high level, so it is not something that you can start to develop. In order to be a candidate to be included in the sprint backlog, this PBI must be fully defined and specified following the parameters of the Definition or Ready list. Some examples of the list of checks that a Definition or Ready can contain could be some questions such as the following:

- Can the item be finished within sprint that we are going to start?

- Has this item been estimated by the development team?

- Is this item adding value by itself? Or it must be done with other items on the list?

We need to agree with the product owner which requirements must meet a functionality definition in order to be accepted to start the development.

Product Backlog Burndown Chart

It is a graph that shows in the X axis the number of sprint and on the Y axis the number of Product Backlog Items pending to be implemented or the expected effort to develop them. The trend should be descending, but it can suffer some increase due to some PB revision that includes new PBIs to meet some new specification defined by the Product Owner. We can see an example in Figure 2-4.

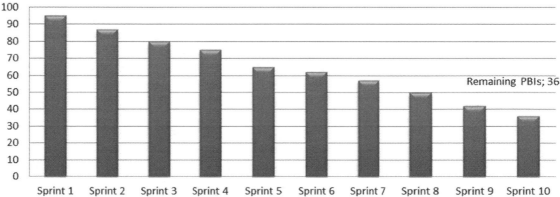

Figure 2-4. *Product Backlog Burndown*

Sprint Backlog

In the beginning of the sprint we select the list of PBIs that will be developed within the sprint. This selection is the sprint backlog. The list of items to be developed must be split into tasks that should be developed in a maximum period of 2 days with the recommendation of keeping the task between 2 and 8 hours of development effort. Sprint backlog shouldn't be modified except if there is a major lock that impedes finishing some of the PBIs. What we need to keep up to date is the information related to every task, if there is any lock, how many hours are pending to finish the development, and who has this task assigned. Also in this list we should avoid splitting too much the tasks that don't fall into micro-management.

Sprint Backlog Burndown Chart

This line graph is an estimation of the remaining effort to finish the sprint. Theoretically it should be a decreasing line that crosses the zero line at the end of the sprint. But as far as the remaining effort is reviewed every day, it can vary if when we start a task we detect that it was not correctly estimated. Usually we set two lines, one with the theoretical trend and the other with the real advance. In Figure 2-5 you can see an example of a sprint backlog burndown graph.

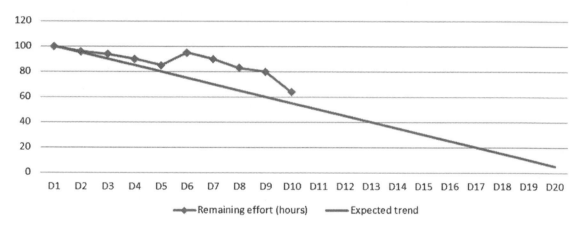

Figure 2-5. Sprint Backlog Burndown

Impediment List

During daily meetings we will detect impediments or locks that can put at risk the completion of a given tasks or a full PBI. The Scrum master must keep these locks in a list and follow up all topics to be able to unlock them. Inside this list we need to inform the lock, the related PBI, status of the lock, scheduled resolution date, and who is in charge of solving it.

Improvement List

After every sprint we will review our activity in the retrospective meeting. There the whole team should suggest any improvement that they think is relevant for the performance of the team. This list is also maintained by the Scrum master and it should contain the action item, status, scheduled finish date, and who is in charge of solving it.

Limiting Spent Time with Timeboxing

Time is limited in our life. It is also limited in our jobs and also in our projects. We need to limit the time that we dedicate to every task in order to not to lose too much time dedicated to the same task. Timebox practice defines the limits of every type of task in order to ensure that project realization doesn't hang. This timeboxing technique is applied to all levels, so we try to define how long the whole project should take, how much time we need for a release to be delivered, which is our sprint period and then set how much time must be spent in single tasks such as taking requirements, developing code, which is every type of meeting duration, documentation development, etc.

This methodology is highly time based as far as all the activities have a time constraint. Behind this strategy we have the theory of balancing three items: effort, requirements, and schedule. We can always improve two of them by balancing the other. We can use less effort if we reduce the requirements or we increase the schedule. If you want to have more requirements we can do it with more time or more resources. If you want to reduce the schedule we can do it by adding resources or decreasing requirements. What Scrum proposes with this timeboxing is to try to fix one of the parameters and maximize the others based on customer priorities. We fix that in a sprint we will spend 4 weeks. If we have 5 people assigned for development so we can invest a fixed number of hours. So our customer must prioritize the tasks that he wants to have available and we will focus on them. We will increase customer satisfaction because in spite of having some functionality missing, our customer can start validating and using the most important tasks for him.

In the same way we fix the elapsed time for meetings. We need to start by the most important topics to discuss and don't spend more time than required in the meeting. When we are searching information to implement a new functionality, doing documentation or whatever we have a maximum amount of time for that and we need to do our best to develop the task. The idea behind this attitude is that we need to get results good enough to be in accordance with quality standards but not perfect. Anything perfect would cost too much time and effort so it would be against our objectives of delivering things quickly.

Spike

Sometimes there is some task that must be done but doesn't add direct value to the customer. Based on Scrum standards, all our activities or PBIs, must provide added value by themselves, but software development, and in this case BI development, can require doing something that our customers really don't care about. In order to be able to include them into the Product Backlog/Sprint Backlog, Scrum enables the concept of spike to be able to manage it. Let's show you an example. Imagine that your user has asked you to be able to drill from one report to a different report with more detail filtering dynamically the customer that they are choosing. You have done the development of the BI solution using a BI tool that can do that kind of drilling but not in the current version that you have installed in your platform but in a new one. Upgrading the platform to the new version could be considered as a spike as something that is required to do, but you user is not requesting it directly.

Also another example could be that in order to follow up with a given requirement, your team needs to use a feature that they have never used before. If you want to succeed in meeting this requirement, it's possible that your team requires doing some training in this feature or some Proof of Concept in order to be sure that they know how to apply it to your customer need.

Maintenance with Kanban

Once you have developed main functionalities required by your customers, it is safe to say your BI system requires some maintenance, corrections, small improvements, and administration. And in order to perform this maintenance our preferred approach from Agile methodologies is Kanban, which allows you to maintain your application with a continuous improvement approach, focusing on small changes in the solution to adapt it to incoming requirements.

Kanban Concepts

There are some concepts that are interesting from Kanban theory that you will need to know in order to manage your project with agility.

Task Workflow

Kanban defines a task workflow by defining the different status that a task should take, from the initial backlog, to the development, testing, validation, and finis. This list of status can vary based on our environment and maybe based on the task type, but it is important to have them correctly defined to maximize the team performance. In Figure 2-6 you can find an example of workflow for Kanban.

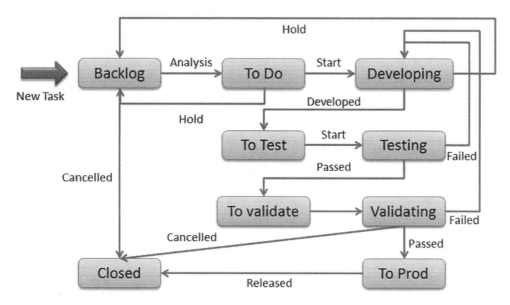

Figure 2-6. Task workflow in Kanban

Task Board and Visual Management

In order to have all the tasks controlled, to detect bottlenecks, to detect locks, and to easily communicate internally with the team which is the status of all the tasks and the advance that the team has done in a daily basis, it's important to have a board that contains all the tasks inside using a card for every task and moving it forward as far as the task is passing through the different status defined in the workflow.

The board will be organized with a column for every status and one or more rows depending on task types, priorities, incidents vs. requests, or whatever need that we have in our project. Figure 2-7 is an example of a Kanban taskboard.

	Backlog	Development		Test		Validation		To Prod	Closed
		To Do	Developing	To Test	Testing	To Validate	Validating		
New Requests									
Incidents									

Figure 2-7. *Example of Kanban Taskboard*

Visual management is a very important feature for Kanban. It is important to have a physical board, or at least a projector or screen that is permanently visible in order that all the team can see in every moment of the time how we are and which next steps are required. As an Agile methodology, Kanban pretends to have a 15-minute daily meeting with all the team to comment on the status of the tasks and what the status changes of every task are.

Work In Progress

One of the most important parameters in our Kanban system will be the Work In Progress or WIP. We will set a WIP for every status, defining the maximum number of tasks that we allow to have in a defined status. It is possible that not all the columns require a WIP and also it must be adjusted to the project that we are working on; having a high WIP will cause us to be overloaded, which is why we start with multitasking; and one of the ideas in the Agile development is that we need to concentrate on a single task in a moment of time in order to be more efficient in our development. Also we could suffer long wait times: long queues in intermediate status and loss of quality. If we define a low WIP, maybe we have people that have nothing assigned to them.

Lead Time

In a Kanban system the lead time is the amount of time that we need to develop a task, since we move it to the initial status until we close it. It doesn't take into account the backlog time. By measuring the average lead time we will be able to see how efficient the Kanban strategy is. The idea is that tasks don't stay in a status indefinitely, so if we detect that a column has more tasks than it should we need to move resources to help the ones that are having locks in order to avoid bottlenecks.

■ **Note**   It is very important to remark that Kanban sets some basic rules but that it must be adapted to every environment when we try to set up the system, and that it must be flexible and continuously adapted to improve the team performance. We will need to tune every WIP of every status, define different Kanban taskboards for different types of activities, adjust the status that a given activity type has, and all these adaptations should be done with the help of periodical retrospective meetings with all the team that will allow us to improve the way we are working.

Mix of Both Methodologies, Scrumban

Our initial consideration of mixing some characteristics of both methodologies is not something that we have invented at all; searching across Agile theory you can find the Scrumban methodology. Scrumban is applying both methods but is less strict than Scrum that pretends to have universally interexchangeable members of the team, so it allows some specialized role, that as you will see, in the next section, fits better in BI developments; and it is a little bit more organized than Kanban using iterations to organize the job as in Scrum. Scrumban fits better than Scrum in highly changing projects and also for maintenance projects or projects where we can expect multiple errors due to variability in source information: as, for example, a project based on Big Data tools.

As mentioned in the introduction of this chapter, it doesn't pretend to be an exhaustive analysis of Agile methodologies, so we encourage you to investigate them; you will be able to find hundreds of pages on the Internet with information about them. In this book we consider more interesting foci on peculiarities of Agile for BI as hopefully you will see if you continue reading this chapter.

Peculiarities of Scrum for BI

Rules and methodologies discussed above are generic for SCRUM under any software development, but there are some particularities that affect BI projects that are specific of this environment; you will easily understand what we mean when going ahead with this section.

Sprint 0 - Longer Initial Analysis

Ok, we have decided to start our datawarehouse development, we have the approval to go ahead by top management and we want to apply Scrum methodology. To do that we will gather all user stories; we will create our Product Backlog, estimating at very high level the needs and cost of every task; we will choose some of them based on the Product Owner priority; we will fill up our Sprint Backlog, analyzing in deeper detail the needs for every PBI; and we will start to develop. After four sprints we arrive at analyzing one low-priority PBI that was on the list since the beginning that had low priority but a big impact that will cause the rebuild of all work done until now.

This is because when you are implementing a datawarehouse database you need to define a logical model based on entities that must have clearly defined relations; which concepts or attributes have relationships; which kind of relations there are: relations one-to-many, many-to-many; which is the single key to relate tables; and at the end we need to define the keys to be able to launch correct SQL statements that should return consistent information. Also you need to take into consideration that you are expecting to have a long history saved there, big tables with millions of rows of information that are validated, formatted, transformed or whatever that your process requires to meet your needs. In this situation, every time that you move a change into production could require a processing of a big quantity of information if you don't have a robust solution for your table schema.

So one of the main particularities of applying Scrum to a Business Intelligence project are the needs for thinking in advance and analyzing in deeper detail the implication of the user story in the technical solution that you are trying to deliver.

Again to illustrate what could be the problem on not following these recommendations, let us show you an example from HSC Company. It is a derived example from a real one that we required to manage in our experience in a different company with the difference that in that moment we weren't using Scrum.

Inside our user stories examples, we find four different requests to add new data to the system, but let's imagine that you have 40 or 50. We follow the priority defined by the user and we start implementing the first one:

```
As Product Owner I want to have the capability of drilling across Customer hierarchies so
that users can analyze the detail of sales through the ERP customer organization.
```

We could think on a simple logical model to support this requirement based on a Star schema, with just four tables: Sales Facts and Customer, Product and Time Dimensions, something like that shown in Figure 2-8.

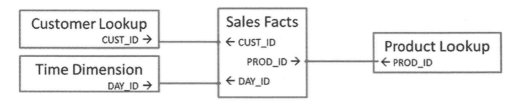

Figure 2-8. *Star example with single key fields*

As in the table model, the related dimensions in the BI tool have their corresponding ID that defines which is the key to join among tables. We follow with other user stories that add levels to every hierarchy; there is some other user story that requires improved speed and we add aggregated tables to our model, new dimensions of analysis, and more and more tables and attributes.

After some sprints we decide to include in our analysis this user story:

As Sales Force Manager I want to analyze how our sales are performing by store so that I can analyze sales per category, customer groups, and promotional campaigns.

Our initial high-level estimation was, ok, let's go to add the store to the sales fact table and let's add the promotional fact table with a lookup of stores. But then we start the detailed analysis:

- Stores don't have the same number of products; it depends on the store size and the location. Due to that, the category of some products can vary among stores because some tools are considered *Garden* for a big store with a focus in agriculture, and as *Other Tools* in another store in the center of the city. So category is store dependent. To solve this we need to add the store to the Product lookup and reprocess the whole dimension.

- Customer groups are also defined by store based on the type of customers that live around the store. Also the customer code can be repeated along stores, so to have a unique identifier of customer we need to add the store to the customer lookup and reprocess the whole dimension again.

- You think on time... time cannot vary. The relations in time hierarchy are the same for all the stores... Wrong again. To analyze the number of days that we have for a given campaign we need to count labor days. As stores are located in different cities and regions they have different local holidays. So again, we need to add the store to the time dimension and reprocess the whole time dimension.

- Your logical model also needs to be reviewed to include the field store in the Key definition for customer, customer group, product category, day, type, and multiple other concepts that you have created to analyze.

At the end the resultant model would be something like that one shown in Figure 2-9 (only focused on the initial one).

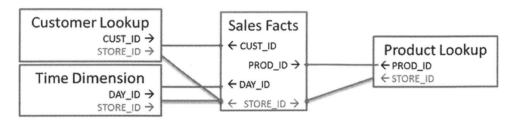

Figure 2-9. *Star example with double key fields*

■ **Note**   As a summary of this section, we will need to invest more time to analyze with further detail which user stories can require big impact modifications in our model. At the end of the day we should try to have the most complete schema definition as we can before starting the development. For sure there will be requirements that will require adaptations and changes but we should try to prevent as much of this as we can.

BI Project Segmentation

After the Sprint 0 meetings you will have multiple requests to take into account from different stakeholders, for different departments, using different tools to develop at different BI layers by different developer roles, or at least somebody with different development knowledge areas. So at the end of the evaluation of requirements done in these Sprint 0 meetings, you will possibly have a lot of work to do for a lot of teams requesting it. In that moment you have two possibilities: run far away and become a gardener in a small town or try to start with this incredibly big amount of work. If you have already arrived to this point in the book, I'm pretty sure that you will choose the second option. So, how to start?

As discussed in the previous section, you should have invested a lot of time in the analysis of all incoming user stories. From that analysis you should have get a logical schema for the database, a draft of the reports that users want to get from your BI system, and which should be the main functionality of your platform. With this overview of all the pending user stories to implement, you should be able to segment them into smaller related groups in order to be able to provide a working solution after every sprint.

In order to correctly decide how to group all the related tasks, you should analyze the requirements under three different points of view:

- **Functional area**: You need to split the tasks into smaller groups related by the related functional area that they belong to. So you can classify the pending developments as related to finance, related to sales, related to operations and also which components are cross-environment, such as customer data or product data. In order to perform the correct analysis of functional areas one helpful tool will be the logical model analysis.

- **Technical analysis**: Having an overview of all the required fields to analyze the information, you can find that there are some initially unrelated user stories that have the same technical source and/or target. Imagine that your sales manager has requested to see the information of products classified by product family and your financial manager wants to see what the free weight that you are giving to your customers to promote a given product is. Both information fields seem to be completely independent but you can realize that both are coming from the same table in your ERP system, and you are going to locate it in the same table in the target, so joining both user stories would reduce the overall cost regarding the cost of doing them separately. In order to correctly perform the technical analysis you need to focus on the physical model of the system.

- **End user experience**: In your BI tool you can have some compound reports that are mixing information from different functional areas, data coming from different systems and located into our datawarehouse into different targets. But maybe your key user wants to have all this information together into a single report, so in order to deliver a fully working screen you can decide that you should develop at the same time some separated analysis to be able to provide a finished development.

After this second analysis to group all the tasks that you can see that some of these segmented user stories are too big to be included into your sprint, and also the related tasks have an excessive effort quotation to fit into the expected maximum size of 8 hours of work. In order to be able to follow with the Agile methodology, you will need to split the segments into smaller pieces.

In order to do this you can repeat the functional analysis but focusing on selecting from the logical model the minimum parts that male sense to develop together, in order to get less tables and fields to manage, and also you can agree with the product owner that some parts of the report will appear with a "Under development" message until the next sprint. At the end of the day this would be a way of segmentation by tables, as far as you are choosing some tables that will be added to the model, with the related BI and ETL development.

Also you can split the whole project by segmenting the information loaded. You can load information for some time window (current year, current month) and let the historical data for next sprint, or you can load information country by country in a multinational environment, or company by company in a company cluster, or customer by customer, or, in the example we have been following during this entire chapter, store by store.

If these splits are not enough, you can think of selecting just some columns of your tables, specially grouping them by source tables or environments. And if this causes still too long of tasks, you can think of splitting them by type, so first you can load base metrics, then implement derived ones from these metrics, then go ahead with aggregations, etc.

Front-End vs. Back-End Activities

Another characteristic of BI projects is the differences on development in front-end activities versus back-end ones. In order to be able to show some given information in your BI front-end tool, you will require having loaded information into the back-end database. On the other hand, having information loaded into a database but without any front-end report to analyze it has no sense in a BI solution. So front-end and back-end activities are closely related; in fact most of the user stories will have developer stories from both types, as far as both activities are required, but they have a nature completely different. Back-end activities are work in the shadow, dirty work that is not directly seen by anybody while front-end reports will be accessed by everybody, the BI tool will be the user interface to access to the information on the back end.

Due to this, requirements will be completely different. The front-end tool is used directly by the end customers, so you will require taking into account that in order to define main rules. You cannot have very strict nomenclature rules and naming convention in the front end as far as your users usually won't have technical knowledge and report names, descriptions and images must be user oriented to allow the user to easily navigate through the tool. Your front end should be intuitive, and easy to use; this will be more important than the technical solution and possibilities that you offer.

Instead of this, the back-end component has stronger technical requirements, so you can set up a naming convention that forces developers to follow a nomenclature in database objects, ETL objects, programming procedures, apply best practices for performance tuning, etc. In this part you will have a stronger position to force developers to follow up the rules, and few developers to learn them, also with technical skills that will make this purpose easier. In the front end your customers have the decision power, so you need to adapt to their requirements. You will have plenty of users, maybe they are contacting you through some key user, such as the product owner, but at the end you will need to have a solution understandable by a wide variety of people, with different levels of knowledge and from different functional worlds while the back end should be accessed only by people with deeper knowledge about what they are doing.

Role Isolation - Specific Knowledge

Agile development in software development pretends to have fully flexible teams, where all the components of the team have programming knowledge and developer stories can be taken by anyone of them, but as you can suppose, based on the previous section, development tools and skills for any BI platform component are completely different. You will require technical skills to develop database structures and ETL processes, with mathematical thought, able to define structures, relationships, load procedures, and understanding of a relational model when you are talking about the back-end tools, schema definition knowledge to define BI tools models, functional and graphical design skills to define the final reports, and finding somebody that have all these different skills is difficult and expensive, so possibly you will require having different subteams inside your team focused on the different areas of development. So members on your team won't be fully interchangeable to work in the different user stories, adding complexity to the team management.

In order to be able to go ahead with the whole project you will require in your team at least somebody with any of these skills:

- **Data architect**: In charge of defining the physical model, table structures, field types, database model, etc.

- **Database developer**: This role will develop the required structures defined by the data architect inside the database.

- **ETL developer**: He will be in charge of defining the loading processes from source systems to the target database.

- **Data modeler**: This role will require knowing about how to define and implement within the BI tool the logical model that will allow the BI tool the possibility to generate the correct SQL to launch against the database. These role tasks will vary based on the tool that you choose; there are complex BI tools that allow you to create complex models with hundreds of functionalities and easier tools that won't require high data modeler capacities.

- **Front-end developer**: You will need also in your team somebody with functional knowledge to understand your customer needs and replicate them into the BI system. This role will require also having good communication skills as far as he will be the main interactor with the end customer.

- **Graphical designer**: In this case it is not mandatory to have a graphical designer in your team, but as far as reporting capabilities are becoming more and more attractive visually, it is recommended to have somebody in the team with graphical design skills that can collaborate in the definition of the reports that the user will access.

These different roles and skills can impede that you can exchange your development team from the roles that they have, if you want to have multidisciplinary teams you will need to invest in training of the team, something that on the other hand is always recommendable.

Also we are aware that if you are working in an small company trying to implement a BI solutions, it is possible that you will be the only one in charge of doing all the stuff, so we hardly recommend you to follow up with this book reading as far as we are going to try to give you the basis go ahead with the development inside all BI components. For sure, we will give you an overview of usual object types, but you will need to get deeper knowledge in the tools that you will use for every component. You can find some books dedicated entirely to each software provider, so inside this book we cannot show you all the options for all the different tools, it would be almost impossible to achieve.

Developer Story Types in BI

This section is closely related to the previous section as far as you will need different roles in your team to be able to go ahead with the development of different developer stories. We specify here developer stories and not user stories, because at the end of the day, a user will want a utility available in the front-end tool, and in order to be able to provide the user with this functionality, most of the time you will be required to split this user story into multiple developer stories for data modeling, ETL, BI tool schema, and BI tool reports or dashboards.

Not all the user stories will have all the components; maybe a user request is executing the historical load for an existing analysis and your task will be just to relaunch ETL processes and result validation, without any modification of your BI tool; maybe you have some request that only requires you to modify an existing relationship among data dimensions that affects only your BI schema, or to change a column type in the database to be able to accommodate longer descriptions for an existing concept. At the end the target of all these modifications will be to show the result in a BI report, but the modification itself maybe doesn't require developing any report.

Data Modeling Developer Stories

For the data modeling developer stories you will be required to focus on database development. It means that you will require, for sure, database skills and possibly data architect ones. Data modeling stories will define what the desired table structure is, and based on that, developers will work with database design tools to create all the required infrastructure.

In order to validate the data modeling process we will initially test with queries that the joins expected are returning correct data, we will check data integrity in the system, and we need to ensure that every column has been defined with the most suitable column type and size. We will see data modeling in detail inside the next chapters; in Chapter 3 you will see the SQL basis, and in Chapter 4 we will analyze how to install the database component, and in Chapter 5 we will talk about data modeling.

We need also to take into account how the change can be applied to the existing production environment, for example, if our transport includes changing a column type in an existing table, we will require to see if we need an auxiliary table to keep the existing information; or if we want to create a new table in production, we need to check if it must be created with the information of development or user testing environments or if it must be created empty to fill it up with the related ETL process.

ETL Developer Stories

ETL developer stories are based on the way you are going to fill up the information into the database, so in the analysis to be able to start them you will require knowing what the source of information is, how you can connect to this information source, what is the expected load periodicity, which are the target tables to fill up, how the load has been defined, incremental or entire, and which are the required checks to validate that the information has been loaded correctly.

Again, we need to validate that our development is correct. In order to validate ETL developer stories, we will require checking that the information in the source is the same as in the target, that our process doesn't duplicate information, that we don't lose information in any step of the load, that the performance of the load and the load time are correct, that we have set this load in the correct position regarding the rest of loads, and, if there are any in place, that our development fits into the best practice definition for ETL in our company. Within Chapter 6 you will find many other details about ETL tools.

BI Model Developer Stories

A developer story to modify the BI model will define a set of BI tool objects that will be required to be implemented in order to be used within the reporting interface. Object types vary a lot across the different tools; there are some tools that require big modeling efforts and that are usually more robust, targeted for big implementations, and other tools easier to manage with few modeling requirements. So depending on the tool, you are going to choose whether the efforts related to BI model stories will be big or small.

When we are modifying the definition of the BI model, we need to ensure that the modification that we do doesn't cause an undesired effect to the existing structure. So we need to validate the change we are doing but also we need to have a set of generic validations defined inside the BI tool to ensure that they don't change when we modify the existing model. The expected result of a BI model development is that when you use those objects in the tool, they generate an expected SQL query, so you will need to validate that it is using the expected fields, joining tables using the correct key fields, without any undesired table usage. In order to further analyze BI tool developments we recommend that you read Chapter 7 of this book.

Report and Dashboarding Developer Stories

As the latest type of developer stories of a standard BI system, we would like to discuss the final step that is the reports and dashboards that the user will interact with. In order to be able to develop the report we will need to know which information our user wants to see, in which order, which format, if there will be some image, drilling paths, corporate information, which is the detail of data required, if there is any grouping field, which the filters that apply to the report are, if there is any selector to interact with the information, how many panels of information are going to be shown, and any other detail that can be relevant for the report. Again it will depend on the tool that you are going to use and which capabilities it offers to you.

This will be the part that will be directly validated by the product owner and the customer, but it will have implicit the validation of the rest of the components of the BI solution, as far as running a report the user can validate if the database has been correctly modeled, if the ETL is loading correctly the information, and if our model has been correctly defined into the BI tool. Inside the validation we need to ensure also that the performance of report execution fits with the expected one.

MOLAP Developer Stories

MOLAP is out of the standard BI solution but it also can be included in the Agile development whenever we need to develop any MOLAP database. MOLAP developer stories must specify what the source of information is; which dimensions of the company are going to be included in the database; how will it aggregate all the dimensions; which interface is going to be used to load information; if the information loaded will be manual, automatic, or a mix of both; how it will be integrated with the rest of BI platform; if we will save information from MOLAP to the datawarehouse; if we will read actual information from the datawarehouse; and what is the usage period and expected load frequency for the tool.

Validation of the MOLAP development should take into account if data aggregates correctly, if metrics have been defined correctly, and that the performance for information load is acceptable. As the BI reporting, this tool will be also used by the end user so key users and product owners will also collaborate in the validation of the MOLAP developments.

Agile Management Tools

There are some tools that can help us to manage all this stuff that we have been talking about, to assign tasks, to have a board in our computer, create user profiles, groups, set delivery time to the tasks, etc. We are going to show you how to use one of them and also information about some other tools that can be interesting for you.

Trello

Trello is an open source tool that allows you to create your own working panels, where you can locate the taskboard both for Scrum or Kanban methodologies.

In the web `http://www.trello.com` you can access to the tool. Once you have been registered you will be able to access the tool. In the initial screen you will have the different boards that you have access to as shown in Figure 2-10.

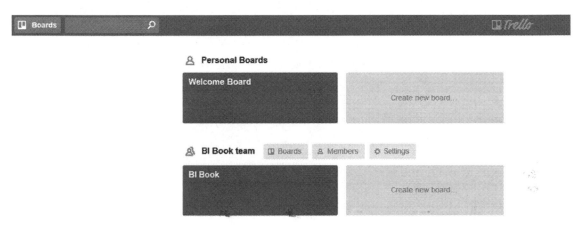

Figure 2-10. *Trello initial screen*

When accessing your board you will see different Cards that at the end of the day are columns that will allow you to organize the different statuses that a task can take. In Figure 2-11 you can see an example of board used to write this book.

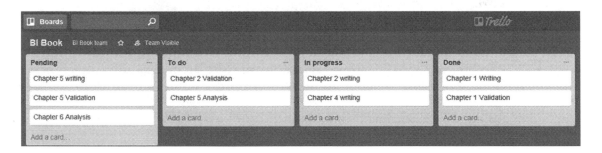

Figure 2-11. *Board example in Trello*

Inside every task you can see different fields, such as who as assigned this task (in Trello it is considered as who is member), what date it must be delivered, description of the task, comments, labels, add attachments and an interesting feature of Trello, add checklists that can help you to define steps, validations, or requirements or the Definition of Done, as you can see in Figure 2-12.

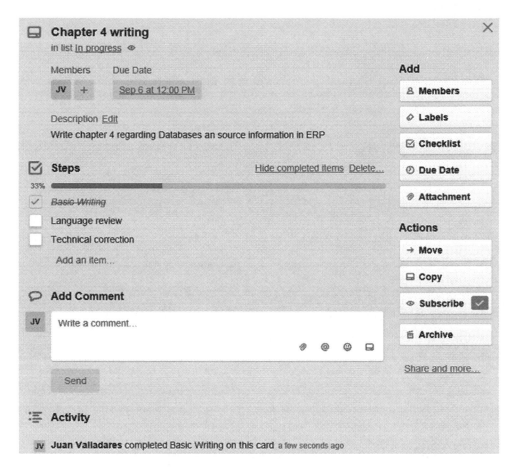

Figure 2-12. *Example of task in Trello*

Once a task is changing its status, you can simply drag and drop the task in Trello to move it to the next status. Also you can subscribe to a task in order to receive emails every time that the task is updated. On the other hand you have the possibility to create users and teams, and then you can assign the board to a team.

We think that this tool is quite simple to use but at the same time quite complete to follow up all the work done. Also you can upgrade to the paid version that will give you some advanced functionality and extra space to save attachments.

JIRA Software

It is one of the most popular software for development management from an Agile perspective. You can only evaluate the software for free but you can get it quite cheap ($10) if you want to have it installed in your server and from $10 per month if you want to have it in the cloud. We have much experience using this management tool and it is quite more complete than Trello, but as far as it allows more options, it can be a little bit more difficult to manage.

You can find it in the vendor page (Atlassian): https://www.atlassian.com/software/jira

You can customize your board to define columns and rows to locate the tasks, but you can add up to 140 fields to the tasks to be able then to group and classify, so it will be quite more flexible; but if you use a lot of these fields you will increase the overhead cost of maintaining this software. In Figure 2-13 you can see some of the fields available that can be configured for every task; you can configure more than 100 fields. Also as JIRA is oriented to Kanban usage, you can define the WIP of every column and the background of the column is marked in red when you have more tasks inside the column than the defined WIP.

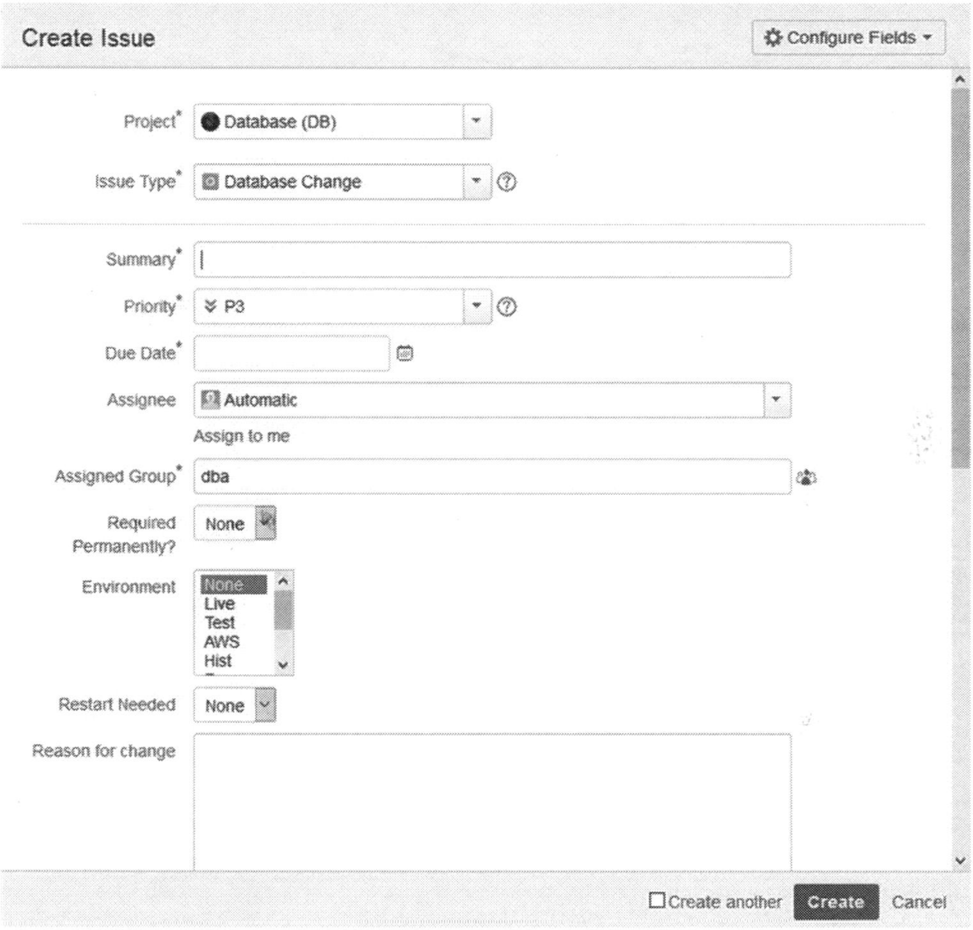

Figure 2-13. JIRA task creation options

Board is working very similar to Trello one, as you can Drag and Drop the task to change the status, assign tasks to any member of the team, add attachments, or follow a concrete task developed by other members of the team.

Apart from this you can also define your dashboards to analyze the way you are working, determine average delivery time, track recent activity in the platform, time consumed, calendar of incoming delivery dates, and multiple other gadgets that can be included on your dashboard as you can see in Figure 2-14.

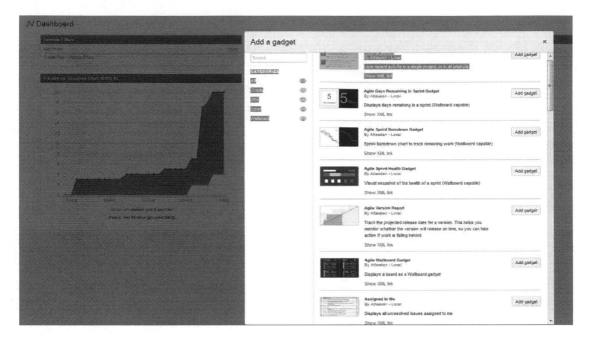

Figure 2-14. *Dashboard customization in JIRA*

There are multiple other options in JIRA, but we wouldn't want to spend too much time on this topic as far as we have multiple other components and tools to analyze. We encourage you to evaluate JIRA, Trello, or any other development management software that you find and check which of them adapts better to your needs.

■ **Note** As a summary of this chapter, we would like to recommend that you follow Agile principles but adapting them to your organization and your needs. Any management methodology should be something that facilitates our lives, not something that just adds complexity and management overhead to our projects. Keep it simple, or at least as simple as you can.

Conclusion

In this chapter we have seen how to use Agile methodologies in your project management, adding a management layer to the project that can help you to organize all the amount of tasks required for delivering a BI system with all the components. As remarked during the chapter, we don't like to be very strict on methodology rules, so we propose that you adapt them to your project using those tools that make sense for you to use. In the next chapters we will see detailed information about all BI components, starting with some SQL basics that will help you to interact with a database.

CHAPTER 3

SQL Basics

Prior to getting started with relational databases and probably, even more complicated things, you need to be familiar with the standard language used to interact with them. While you can get away without knowing SQL and working with databases (think in Microsoft Access), sooner or later you will need to learn it.

In this chapter we will see an introduction to SQL. We will concentrate only in basic topics, so you won't get lost in details that are not needed in the first stages of your project.

For the ones that already possess decent SQL skills, you may move to the following chapter, but even in that case, we strongly recommend you do a quick review.

What Is SQL?

SQL stands for Structured Query Language. As the name implies, it is a language used to query a database, but it has many other functions apart from querying as we will see throughout this chapter.

In this chapter we will see an introduction to SQL. We will concentrate only in basic topics, so you won't get lost in details that are not needed in the first stages of your project.

One thing we need to know about SQL is that it is not similar to other programing languages like Java, C# and Python. This is because it is a declarative language, instead of a procedural one. This means that instead of describing the calculations you need to do to retrieve what you want, you simply specify what you want, without specifying how to get it.

SQL was originally created having the relational model in mind. Later additions, however, added features that are not part of the relational model, so you can do things in a different manner if you need to. Many of the current database versions support both the relational and the object relational methods, which include support for object-oriented programing and more complex data types than the usual primary ones. We will see what the relational mode is and its implications shortly; do not worry.

Where this may seem a bit limiting, and probably you're right as SQL by itself is probably not enough to do every calculation you may think of, especially if you think in large enterprise applications, it is easy to use in combination with other languages, especially these procedural ones. Almost every programing language has some sort of database specific connector to interact, or uses some sort of ODBC driver, which lets you to connect to a relational database and execute queries.

Let's do a brief introduction to SQL history. SQL first appeared in 1970,[1] but the latest version is as of 2011. Not all databases implement the same version; even not all of them implement a whole version but sometimes only a subset of it. This sometimes causes problems when you try to port code that works for a specific version of a database to one from another vendor.

---

[1]http://docs.oracle.com/cd/B12037_01/server.101/b10759/intro001.htm

© Albert Nogués and Juan Valladares 2017
A. Nogués and J. Valladares, *Business Intelligence Tools for Small Companies*,
DOI 10.1007/978-1-4842-2568-4_3

■ Note There are hundreds and hundreds of books only devoted to learning SQL. Apress has two that are interesting, one called *Beginning SQL Queries, From Novice to Professional* from Clare Churcher with the second edition published in September 2016; and another focused to Oracle-specific SQL, *Beginning Oracle SQL*, by Lex de Haan, Tim Gorman, Inger Jorgensen, and Melanie Caffrey. These two can be a good follow-up to this chapter.

The Relational Model

SQL is the standard language used to interact with relational databases, but it does not implement all concepts in the relational model. Many years later, there is no fully successful implementation of the relational model yet.

The relational model basically has three concepts:

- **Tables**, which are used to represent information and hold columns and rows, whose purpose is to represent attributes and records (think in an Excel spreadsheet).

- **Keys**, to govern and implement successfully relations and identify uniquely the records in a table (more later, when we discuss about primary and foreign keys).

- **Operations**, which use the model and can produce other relations and calculations that satisfy queries.

In order to show you an example about these concepts, let's take a sample of a relational database; it may well be the typical Employee and Department tables, explaining a bit about some of these concepts using two tables from the Oracle HR schema.

■ Note HR schema and others related during this book are the test schemas you can decide to install when configuring an Oracle database. However, Oracle has released them in github and you can download them for free and port to any database you want. You can find them here: `https://github.com/oracle/db-sample-schemas`. If you want to follow up the examples that we are showing in this chapter, you should jump first to Chapter 4 to install the database.

The procedure to install them is basically cloning the github repository and run a couple of commands that will start a script. You have the full instructions in the welcome page of the repository. However, if you want to test, you may opt as well to download the Oracle Enterprise edition database from Oracle website, which is free according to the owner website, as soon as the usage is for "purpose of developing, testing, prototyping and demonstrating your application and not for any other purpose."

In Figure 3-1 you can observe that we have two tables. The first part enumerates the columns of the tables and its datatypes (we'll see them in a while). For departments we can see we have a column called *DEPARTMENT_ID*, consisting of a number of 4 digits. The "P" in front means that this column is a primary key. Then we have the *DEPARTMENT_NAME* column that is the Name of the Department; the *MANAGER_ID* that is a column to identify the ID of the Employee (in the Employees Table) who is the manager of the department, which is a Foreign key to link the Departments table to the Employees Table, and enforces that the manager of the department has to be a valid employee found in the Employees table; and the *LOCATION_ID* that is the id of the location of the department (as you may well think this may presumably be a Foreign key to a "missing" Locations table. Yes, you're right!).

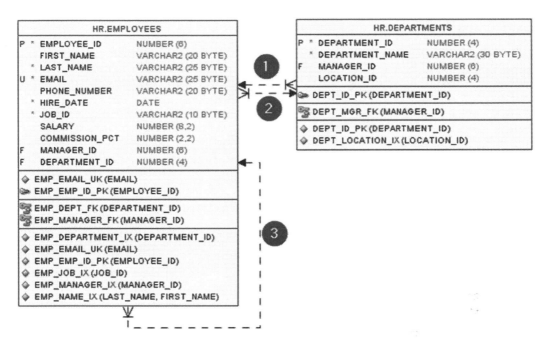

Figure 3-1. *Relation between the Employees and Departments table in the HR schema*

The columns that are used in relationships with other tables or identify the values of the table, here the DEPARTMENT_ID and MANAGER_ID are usually depicted apart in the diagram, in the form of keys with the constraint names, depicting the column name where they refer to in brackets. Also any *indexes* found in the table needed to speed queries or to ensure that relationships are properly implemented are depicted in the last part of the table definition as well.

There are still other important things to learn from this figure. These are the dotted lines between tables that define the relations. These are:

1. DEPT_MGR_FK, which is basically a relationship that tells us that the employee with MANAGER_ID number is the one and only manager from the Employees Table. That relationship is a Foreign Key and usually follows a convention of naming it with an ending "_FK."

2. EMP_DEPT_FK, which is in the opposite direction, tells us that the DEPARTMENT_ID found in the Employees table is one and only of the DEPATMENT_ID found in the departments table.

3. EMP_MANAGER_FK. This is the most complicated to understand so far, but defines the manager relationship, so that a MANAGER of an employee needs to be at the same time a member of the EMPLOYEE table, making it impossible to have a manager assigned that is not part of the table, or in our domain world, an employee from outside the company.

This is understood, which is basically all what we need to know about the relational model. Obviously the topic is much broader but for our purposes trust us that this is basically all you need to know.

Databases and Database Vendors

There are many types of Databases: Relational, Object Oriented, No SQL... but traditionally, Datawarehouses have been set up with relational databases. This is due to the fact that it is easy to learn SQL and perform analysis with it. Nowadays, many systems, especially the ones that deal with large amounts of data, tend to use another type of databases, NoSQL Databases, which are different, and usually use a key/value pair as a way to store data. While these databases can be schemaless and more practical, they do not follow the ACID principles for transactions that most relational databases follow. This is the tradeoff to pay to gain speed and be able to deliver better performance when dealing with a lot of data, and to be able to act distributedly.

▓ **Note**   Nowadays NoSQL and specially distributed databases are a trending topic with the BigData advent, and in fact, they use different languages than their relational counterparts. But even with this, you will find a lot of similarities between them and the SQL language. So, despite that you are thinking of implementing a NOSQL database for your project or any other database type, this chapter should be worth reading.

Main database vendors are headed by Oracle, which claims to have a market share of 98.5% of Fortune 500 companies. While this may seem astonishing, it is important to stress, that usually large companies have a mix of database vendors. It is likely easy to see large companies having Oracle, SQL Server and even MySQL, Postgree and open source databases like MariaDB and many others.

DB2 from IBM and SQL Server from Microsoft are also important commercial relational databases to note. You will find them in many companies although they are probably not as extended as Oracle is.

It is worth noting that all commercial databases usually have some reduced or limited version to be used for free, and even with distribution permission granted from their own firm. These versions tend to come with some sort of limitation, but as a small or mid company, it should not be a problem to you, and you can always migrate to the paid version if your business grows.

Table 3-1 compares the three versions of these commercial databases and their limitations so you can decide if any of these can fit your needs. This data is subject to change as new releases of these databases are launched, but it is correct at the time of writing.

Table 3-1. *Free versions of comercial databases*

Database Name	Latest Version	MAX CPU/CORES	MAX Ram (GB)	MAX Disk (GB)
Oracle XE	11gR2	1 core	1	11
SQL Server Express	2014	1 cpu, all cores	1	10 per database
DB2 Express-C	10.5	2 cores	16	15.360

If you move away from commercial databases, Oracle has another RDBMS, which is open source, called MySQL. While MySQL was purchased by Oracle, you can still download it and use it for free. You will see mainly this database in a lot of web projects especially, as it is easier to integrate with web languages like php and so.

There are other open source databases that are very interesting (Table 3-2). One is postgresql, which nowadays is gaining more and more popularity. Another one is the fork of MySQL, called MariaDB, which was continued by the main people that created MySQL and whose project kicked off when Oracle bought the original Database. One of the best things of MariaDB is that it is 100% compatible with MySQL. Since a few years ago, different and new features were added to MariaDB that deviate from the original MySQL. These usually have no limitation, only the ones derived from implementation and architecture limits.

Table 3-2. Open Source Databases

Database Name	Latest Version	Download Link
MySQL	5.7.2	http://dev.mysql.com/downloads
Postgres	9.5.2	http://www.postgresql.org/download
MariaDB	10.0.25	https://mariadb.org/download

All these databases can work with multiple operating systems, including Linux and Windows in the main parts of their respective versions.

■ **Note** When choosing a database to implement your project, do not focus solely on money, but also the skills of your employees. Oracle is very good and the Express version may work for you but if for any reason after some time you need to move to the paid editions, licenses aren't cheap. Migrating a database from one vendor to another can be a pain if your system is large or your applications use specific features from one vendor. In these cases, you may explore to use a cloud solution of a paid database where the price can be kept under control. Open source databases are very good alternatives, too. We will use both alternatives during the course of the book.

ACID Compliance

Relational databases tend to support what is called the ACID property set when it comes to transactions. A database that is ACID compliant means that it guarantees:

- **Atomicity**, which means that statements can group in transactions, meaning that you can control that if one fails, the transaction fails, meaning that no change is written to the database.

- **Consistency**, meaning that the database is always in a valid state, even after a failure or a not following a restriction. Usually the easiest way to think about this is a table with a constraint that checks that no employee can have a salary bigger than $10,000. If you start entering employees and salaries in this table as a part of a transaction, if you add one with a salary of $20,000, the entire transaction should fail and the table data should be restored as it was before the start of this transaction. The same applies in case of a failure during the execution of a transaction. Either the transaction was committed and made permanent, or if it was not yet finished, it will be rollbacked.

- **Isolation**, meaning that any time you query the database, what you retrieve is what it was before any transaction running started or what is after any running transaction has finished. Think, for example, you have a large table and you sent to the database an instruction to update employee salaries to set twice their salary. If the transaction has not yet finished when you query the table, you should see a previous version of the table, where the table was not yet updated. When the transaction finishes, any subsequent query to the table should see the updated table, with the employees having twice its salary, but generally speaking it is not desirable to obtain a mix, as this leads to errors.

- **Durability**, means that once a transaction has been committed, it is committed forever, even in case of a failure. So the database has to ensure, by having different mechanisms that the changes are always stored to disk, either by storing the data directly (usually inefficient) or by storing a vector file defining the changes applied to the data, so the database, once restored, is able to replay these changes.

Types of SQL Statements

At that point you are probably already wondering what you will do to query the total of money invoiced to a specific customer or a specific region. That's good, and soon you will learn how to do it. But there are other statements, perhaps, as important as queries, when working with a relational database.

Apart from querying the database, there are other types of statements that usually are used before querying, to prepare the database to hold the data that subsequently will be queried. Let's do a quick recap and review them:

- **DML or Data Manipulation Language statements**. It is important to note that this group includes the Select statements to query the tables, but it is not limited only to query because there are other set of statements that fall below this group as INSERT, used for inserting records in a table; UPDATE for updating these records; and DELETE to delete records based in one or multiple conditions. In many places you will see the MERGE statement. This is a specific statement that not all databases have implemented, and it is a mix of INSERT + UPDATE. Sometimes it is also called and UPSERT due to that reason.

- **DDL or Data Definition Language** statements, which comprise statements that are used to modify the structure of a table or either create them or drop them. These include the CREATE statement to create objects like tables and indexes but many other as well, the DROP statement to remove them, the ALTER to modify them, and a special statement called TRUNCATE used to delete all data from a table.

- **Transaction Statements** like COMMIT, to commit all changes done inside a transaction, ROLLBACK to undo all changes done. Note that if a COMMIT has not been executed at the end of a transaction, upon disconnecting, due to the consistency mechanisms implemented by almost all databases, the changes will be lost and not persisted to the database files. Some databases also implement the CHECKPOINT command that checkpoints a database, but this is not of too much use for us right now.

■ **Note**   Use caution with the Autocommit option that is enabled by default in some relational databases, like MySQL and SQL Server. This can be good in case you forget to commit at the end of your transaction but can have terrible consequences if you mess it up with an update or delete statement. Our suggestion is to turn autocommit off, and remember always to commit.

The TRUNCATE statement as all DDL statements have an implicit COMMIT. If you truncate by mistake a table there is no way to recover it, unless you restore the database to a previous checkpoint. While with some databases it is even possible to recover an update or delete issued by mistake, this is not possible with truncates. Be very careful when running a TRUNCATE statement!

SQL Datatypes

The SQL language defines a set of datatypes for column values. Not all databases implement the same datatypes, and not all the databases use the same name for compatible datatypes. However, the most common datatypes are supported in most of the databases.

In the following subchapters we review them, explain when to use them, and analyze its particularities. It is important to use the correct datatype, as choosing an incorrect one, apart from wasting space, can led to poor performance due to explicit or even implicit type conversions.

Numeric Datatypes

There are many numeric datatypes. In most of the databases these are represented by the Integer, Float, Real, Decimal, and BigInt types. There are other datatypes like the Smallint for small numbers. All these are used to store numbers and you need to choose which is the one that is suited better for each column of your table basically by taking into account the maximum value that can go, and if decimal positions are needed, and until which extent the precision is important to you.

Text Datatypes

Text is usually represented by the Char (Fixed length) or the newer Varchar (Variable length) datatypes. The String datatype and the CLOB (Character LOB) datatypes are available as well in other databases. The recommendation we give you is to stick to the Varchar datatype unless there is a very important reason to use other datatypes as usually new databases do not waste space storing a smaller value in a varchar column defined to hold more characters.

Date Datatypes

All date, time, and timestamp values fall into this category. Probably here is where most differences appear from different database vendors. It is important to check the documentation of your database to make sure that you are using the correct type. Most databases also support timestamps with time zones so if you plan to support different regions with your application and you want to use these different time zones, please check the manual before choosing the datatype.

Other Datatypes

In this area we can find other important datatypes like the binary, bool, or bit datatype used to hold binary information (true or false); the LOB or RAW datatypes used to store large information like large chunks of texts, images, or files in binary format; XML types to store XML information in databases that support it, so you can query the data efficiently; user-defined types by combining primary types; and many other data types that do not fall in any one of the previous categories.

Retrieving Data from a Table

A few paragraphs ago we presented the Employees and Departments tables. We will use these during this chapter to execute a few instructions and check for the expected results. We will start with the most basic operation you can do over a table that is accomplished by using the SELECT statement.

First, we want to show you an overview of a select statement. In subsequent subchapters we will see each block one by one and discuss its usage. The common SQL select statements are all in this form:

```
SELECT list_of_columns (separated by commas, or *)
FROM list_of_tables (separated by commas)
[WHERE set_of_conditions (separated by operators)]
[GROUP BY grouping_of_columns (separated by commas)]
[HAVING set_of_conditions_applied_to_the_grouping_of_columns (separated by operators)]
[ORDER BY ordering_conditions (separated by commas)] ;
```

If that is clear enough, then we can start by getting all the information of a table using the special SELECT * statement, and then we will see other ways of options to use with the Select statement, and we will use projection to only select specific columns from the database.

The select * Statement

The syntax of a select statement is very easy. We usually need to specify what kind of columns we want to retrieve information for and from which table or tables and then an optional predicate in the where clause to apply any filter in the data we want to use, so the rows that do not match these conditions will be filtered out from the result.

A typical select statement is as follows:

```
SELECT
      column1, column2, columnn
FROM
      schema.table1
WHERE
      column1 = 'Albert';
```

The columns that you select are part of the projection clause. The columns in the where clause are either filter or columns used to join to other tables. Let's see first the most basic statement:

```
SELECT
      *
FROM
      hr.departments;
```

```
DEPARTMENT_ID DEPARTMENT_NAME                    MANAGER_ID LOCATION_ID
------------- ------------------------------ ---------- -----------
           10 Administration                        200        1700
           20 Marketing                             201        1800
           30 Purchasing                            114        1700
           40 Human Resources                       203        2400
           50 Shipping                              121        1500
           60 IT                                    103        1400
... (output truncated)
27 rows selected
```

With that select * statement we are telling the database to retrieve all columns and all data from the departments table without applying any filter. Since we are retrieving all data a FULL SCAN of the table will be performed, meaning that the entire table will be read by the database and returned. No index, no matter that it existed, will be used to retrieve this, unless we have the uncommon situation to have all columns of the table in an index (more on this later in the performance chapter).

The select column Statement

The select column statement is a particular case of the select * statement. In that case you use a projection predicate to select only the columns you are interested in. Wherever possible, use this instead of the select *, as the latter is very dangerous. Think about what can happen if you add a column to the table. Now, the select instead of returning n columns will return n+1. The same if you remove some. During the years we have seen a lot of mistakes in processes by using the asterisk, instead of naming the columns. Also, from a performance point of view, using a select * is not a recommended solution, as you will inhibit the optimizer of using indexes if available, but this will be discussed later.

A typical select statement to retrieve the name of the departments of the departments table is as follows:

```
SELECT
      department_name
FROM
      hr.departments;

DEPARTMENT_NAME
-------------------------------
Administration
Marketing
Purchasing
Human Resources
Shipping
IT
... (output truncated)
27 rows selected
```

The select count (*) or count (column) Statement

When we are developing we might be interested in easily counting the number of rows a query returns. This can be accomplished by the select count statement. If you use the asterisk it will count all the rows from a table but with the column it will skip NULL values, that is, values that do not exist for that particular row. The sentences, in case that all the values exist, should yield the same output, but different databases implement internally differently and certain ones may have better performance than the others.

```
SELECT
      COUNT(department_name)
FROM
      hr.departments;

COUNT(DEPARTMENT_NAME)
---------------------------------------
27
```

Whereas:

```
SELECT
      COUNT(*)
FROM
      hr.departments;
```

```
COUNT(*)
----------
27
```

So we get exactly the same output.

The select distinct Clause

When we do not want to consider distinct values for a specific selection, we can use the distinct clause. This will only retrieve one per pair of columns. If we select only one column, we will retrieve only the different values for that column; but if the query consists in a projection of several columns, it will return the different tuples, meaning that will return distinct values but taking into account all the columns selected. Here is an example:

```
SELECT
      COUNT(first_name)
FROM
      hr.employees;


FIRST_NAME
----------
107
```

But, for example:

```
SELECT
      COUNT(distinct first_name)
FROM
      hr.employees;

COUNT(DISTINCTFIRST_NAME)
--------------------------------------
91
```

However, if we query the pair first_name, last_name, we will see that there are no employees with exactly the same name (first name + last name being the same):

```
SELECT DISTINCT
      first_name, last_name
FROM
      hr.employees;
FIRST_NAME           LAST_NAME
-------------------- ------------------------
Ellen                Abel
Sundar               Ande
Mozhe                Atkinson
... (output truncated)
107 rows selected
```

Sorting

Sometimes you want the result of a query to be sorted by a specific column. Sorting can be done in ascendant or descendant manner. In SQL, this is accomplished by using the ORDER BY keywords. The ORDER BY clause always goes at the end and sorts all records that have been selected and not filtered out. Let's have a look at a few examples:

If not specified, by default the order of the sort is ascending:

```
SELECT
      department_name
FROM
      hr.departments
ORDER BY
      department_name;

DEPARTMENT_NAME
------------------------------
Accounting
Administration
Benefits
Construction
... (output truncated)
27 rows selected
```

▓ **Note**   By default, sort is done numerically for number fields, alphabetically for text fields, and chronologically when you are sorting a date field.

But we can specify the direction of the sort:

```
SELECT
      department_name
FROM
      hr.departments
ORDER BY
      department_name desc;

DEPARTMENT_NAME
------------------------------
Treasury
Shipping
Shareholder Services
Sales
... (output truncated)
27 rows selected
```

We can sort the rows by multiple columns at once, so the first column in the clause will be sorted, then the second, and so on:

```
SELECT
      first_name, last_name
FROM
      hr.employees
ORDER BY
      first_name, last_name;
FIRST_NAME           LAST_NAME
-------------------- ------------------------
Adam                 Fripp
Alana                Walsh
Alberto              Errazuriz
Alexander            Hunold
Alexander            Khoo
... (output truncated)
107 rows selected
```

And we can even decide to sort some columns in ascending and others in a descending manner:

```
SELECT
      first_name, last_name
FROM
      hr.employees
ORDER BY
      first_name asc, last_name DESC;
FIRST_NAME           LAST_NAME
-------------------- ------------------------
Adam                 Fripp
Alana                Walsh
Alberto              Errazuriz
Alexander            Khoo
Alexander            Hunold
... (output truncated)
107 rows selected
```

Sometimes, however, it is easier just to specify the column position in the projection predicate, so the following will be similar to the previous example:

```
SELECT
      first_name, last_name
FROM
      hr.employees
ORDER BY 1 ASC, 2 DESC;
FIRST_NAME           LAST_NAME
-------------------- ------------------------
Adam                 Fripp
Alana                Walsh
```

```
Alberto              Errazuriz
Alexander            Khoo
Alexander            Hunold
... (output truncated)
107 rows selected
```

If we have any nulls in our columns and we want to move them to the beginning or to the end, we can use the NULLS LAST:

```
SELECT
     First_name, last_name, commission_pct
FROM
     hr.employees
ORDER BY
     3 DESC NULLS LAST,1,2;
FIRST_NAME           LAST_NAME                      COMMISSION_PCT
-------------------- ------------------------------ --------------
John                 Russell                                   ,4
Allan                McEwen                                   ,35
Janette              King                                    ,35
Patrick              Sully                                   ,35
Alberto              Errazuriz                                 ,3
... (output truncated)
Vance                Jones
William              Gietz
Winston              Taylor
107 rows selected
```

And that's all we need to know about sorting for now.

▓ **Note** Apart from ordering, relational databases usually implement some predicates to select the top X rows. This can be very useful in a datawarehousing environment if you want to display your top 5 customers, or the top 10 regions with more sales. Depending on the database this is implemented as a TOP, LIMIT, FETCH FIRST X ROWS ONLY clauses; since there is no clear standard, we won't discuss it here but we encourage you to check your vendor's manual for more information.

Filtering

So far we have seen some basic queries but these are not very useful yet. We are just selecting all the records from a table. But you usually want to select only some records that meet a condition. And that condition or set of conditions can sometimes be complex to calculate. Most of the time you will be asking your data how much have I invoiced for a specific region? Who are the employees working in the supply chain part of the business? What is the percentage of profit I have for a specific group of products? All these questions need some sort of filtering as you are targeting the calculation to a defined group.

The Where Clause

The predicate to filter rows that will be included in the result is the WHERE clause. But there is a vast set of operators to use along with. Let's review them.

If you recall from previous examples, we had 107 employees in the table, let's see how to compute the Employees that earn more than $10,000 with a query. The query will look like this, using what we have learned so far:

```
SELECT
      count(*)
FROM
      hr.employees
WHERE
      salary>10000;
15 rows selected

  COUNT(*)
----------
        15
```

So we can now deduce we have 15 employees out of 107 that earn more than $10,000$.

Operators

For our example we used a greater than operator to compare the salary value of the table to a specified number. But there are many other operators. Here a list of the most used (Table 3-3):

Table 3-3. *SQL Comparison Operators*

Operator	Meaning	Example
=	Equal	Salary = 10000
<	Less than	Salary < 10000
<=	Less or equal than	Salary <=10000
>	Greater than	Salary > 10000
>=	Greater or equal than	Salary >= 10000
<>	Different than	Salary <> 10000
IN ()	One of the values in the list	Salary in (10000, 11000, 12000)
NOT IN ()	None of the values in the list	Salary not in (10000, 11000, 12000)
BETWEEN x and y	Between that range	Salary between 10000 and 11000
LIKE	Match partial word (text). Will retrieve all employees whose first name is either Pete, Peter or PeteXXX. The wildcard % will match any character(s)	First_name like 'Pete%'
NOT LIKE	Exclude Partial Matches	First_name not like 'Pete%'
IS NULL	A specific column is null	Salary is null
IS NOT NULL	A specific column has a value	Salary is not null

▦ **Note**   Remember that if you are comparing against a text string or a char, you need to use single quotes to enclose the string or char where you are comparing against. The following is not a valid comparison: where first_name = Peter. The correct way is to enclose the text between single quotes: where first_name = 'Peter'.

There are some other operators apart of these that are similar to the IN() operator. These are operators that compare a group of rows: ANY, ALL, EXISTS, and the names are self-explanatory. We will see some use in the subquerying subchapter.

Logical Operators

There is another set of operators called logical operators. This set of operators is widely used because they act as a glue between different conditions or other operators. For the readers used to programming in any sort of language, these results will be very familiar (Table 3-4).

Table 3-4. *SQL logical operators*

Operator	Meaning	Example
AND	Both conditions need to be true to return true	Salary > 10000 and first_name like 'Pete%'
OR	At least one of the conditions need to be true to return true	Salary > 10000 or first_name like 'Pete%'
NOT	The condition needs to be false to return true	NOT (Salary > 10000)

If you remember from school, you need to be aware of the operator precedence. This means that we sometimes need to use parentheses to specify which operator comes first. For example, in the NOT operator, in Table 3-4, we have specified parentheses to tell the database to execute first the comparison of Salary > 10000 and then the NOT operator. In this case, this is not necessary, as by default, comparison operators are applied first, then the NOT logical operator. But you need to know the precedence rules. In any case we recommend you use parentheses if possible as the code is much clearer, and you can easily know which conditions should be checked first, and avoid any errors. We can guarantee you this will save you a lot of time debugging bad performing queries.

Usually arithmetic operations are executed first. So an addition, subtraction, multiplication, or division will be executed before any other condition. Later, the comparison operators are applied. After these, the LIKE, IN, and NULL ones. Then the Between statement, then the <> comparison operator, and at the very end, the NOT, AND, and OR operators in this order. It makes sense that these logical operators are left until the end because as we explained previously, these are used mainly as a glue between other operators or group of conditions.

If you made it until this point, this is very good news. You have learned probably the most important part of how to write queries. That is not enough yet, of course, as things can get much more complicated, and to write good, efficient, and clear SQL queries takes some time to master. But we are on the right path.

Grouping Data

Sometimes you don't want to retrieve individual rows from the database. You want to group by certain data and retrieve only totals, number of groups, averages, and so on. This is very typical in a datawarehouse environment. Think about a manager who wants to retrieve the sales for a specific day. Depending on the size of your company, retrieving the sales one by one will not make any sense, as you are probably more interested in a specific detail instead of revision of all of your customer purchases. For these cases, we need to be able to group records by certain groups and in SQL, this is achieved by using the GROUP BY clause.

There is a list of aggregation or set functions that you need to learn too. These specify the operation you want to calculate over the group. The most common ones used are depicted in Table 3-5.

Table 3-5. *Aggregation functions*

Operator	Meaning	Example
MAX ()	Returns the Maximum of the set.	MAX (salary)
MIN ()	Returns the Minimum of the set.	MIN (salary)
SUM ()	Returns the total or subtotal of the set.	SUM (salary)
AVG ()	Returns the average of the set.	AVG (salary)
COUNT ()	Counts records. You can specify a condition inside as well.	COUNT (salary > 10000)

To be able to include columns in the group by clause, we need to make sure that these are also in the projection part of the query. This means that we cannot add a column in the select clause, that it is not present in the group by clause. Some databases let you to do so, by modifying some parameters, but this is usually something we want to avoid, so please bear this in mind.

Knowing this we can compute, for example, the average of salaries of employees by their department, to see which is the best department of the company to work for if we only care about money, of course!

```
SELECT
      department_id, trunc(avg(salary))
FROM
      hr.employees
GROUP BY
      department_id
ORDER BY
      2 DESC;

DEPARTMENT_ID                   TRUNC(AVG(SALARY))
-------------  ------------------------------------
           90                                 19333
          110                                 10150
           70                                 10000
... (output truncated)
           10                                  4400
           30                                  4150
           50                                  3475
12 rows selected
```

So it is clear that people assigned to department with id = 90 are clearly the ones that earn more, whereas the ones assigned to department id = 50 are the ones that receive lower payslips. This is easily explainable because the department with id = 90 is the Executives department.

■ **Note**   Forget for the moment about the usage of the TRUNC() function. We added them to have a clear output and not to have results with large decimals. Depending on every database, this function has different names but basically instructs the database to get rid of the decimal part of a number by truncating the output. Caution, this does not mean the same as Round. Again; check your vendor's manual for more information.

Obviously this sometimes is not enough and you want to filter a specific group because, for example, you are interested only the ones that have at least 5 employees. This requires to use a HAVING clause as if you think, there is no way to filter that using a where clause. The key point here is to understand that any filter in the Where clause filters rows, but in this case we need to filter groups from the output, not individual rows. So the Having clause is needed. Here is an example:

First, a count of the number of employees per department:

```
SELECT
      department_id, COUNT(*)
FROM
      hr.employees
GROUP BY
      department_id
ORDER BY
      2 DESC;


DEPARTMENT_ID    COUNT(*)
-------------  ----------
           50          45
           80          34
          100           6
           30           6
           60           5
           90           3
           20           2
          110           2
           40           1
           10           1
                        1
           70           1
12 rows selected
```

And then, let's have a look at our departments having more than 5 employees assigned:

```
SELECT
      department_id, COUNT(*)
FROM
      hr.employees
GROUP BY
      department_id
HAVING
      COUNT(*) > 5
ORDER BY 2 DESC;
```

```
DEPARTMENT_ID   COUNT(*)
------------- ----------
           50         45
           80         34
           30          6
          100          6
 4 rows selected
```

It is important to understand as well that the having clause is applied after all filters in the where clause are applied. So in the previous example if we filter employees that earn more or equal than 9.000 and then remove the groups that have less than 5 employees, the result can be different, as some employees may have been already removed from the groups by the condition Salary >=9.000, affecting the total count of employees for that particular group. Let's see an example:

```
SELECT
      department_id, COUNT(*)
FROM
      hr.employees
WHERE
      salary < 9000
GROUP BY
      department_id
HAVING
      COUNT(*) >= 5
ORDER BY 2 DESC;

DEPARTMENT_ID   COUNT(*)
------------- ----------
           50         45
           80         17
           30          5
 3 rows selected
```

So as you can see, we lost two groups. This is due to the fact that department_id 100 and 60 had at least two and one employees each that earned more than or equal to $9,000. Also, departments 80 and 30, had reduced the number of members that fulfill the condition salary <$9000 altering the result of the query.

Using Subqueries

There are many times that you want to specify a filter, but this filter is calculated against some parameter that is not obvious, or cannot be set beforehand, because it depends on your own data. As you saw in previous examples, we can find for employees earning more than $10,000. That is ok, but what happens if we want to look for employees earning more than the average of the company to see if they deserve so? Clearly what we have learned so far is not enough, as the average salary of the company may change from time to time. Think about pay rises, hired or fired employees, and so on, that will alter this count. This is a very common situation and we need to introduce the concept of subquery.

To make things easier, we will be looking for the average salary of the company first, and then we will test all employees against this salary if this was something like $10,000. So the only thing that we need to do is find the way to calculate this average salary first and then the rest will be as previously shown. Let's show an example of how to calculate it:

To start, let's write a query of how to find all employees earning more than $10,000. This will be something similar to the following:

```
SELECT
      first_name, last_name, salary
FROM
      hr.employees
WHERE
      salary > 10000
ORDER BY 3 DESC;
```

```
FIRST_NAME           LAST_NAME                   SALARY
-------------------- ------------------------- ----------
Steven               King                         24000
Neena                Kochhar                      17000
Lex                  De Haan                      17000
John                 Russell                      14000
Karen                Partners                     13500
Michael              Hartstein                    13000
Shelley              Higgins                      12000
Alberto              Errazuriz                    12000
Nancy                Greenberg                    12000
Lisa                 Ozer                         11500
Gerald               Cambrault                    11000
Den                  Raphaely                     11000
Ellen                Abel                         11000
Eleni                Zlotkey                      10500
Clara                Vishney                      10500
15 rows selected
```

This is half of the work, as now we need to modify our query to instead targeting the $10,000 figure calculating the employees that earn more than the company average. To do so is easier than thinking how to calculate the latter. If you understood so far all that we have discussed, this should be easy for you:

```
SELECT
      TRUNC(AVG(salary))
FROM
      hr.employees;
```

```
                    TRUNC(AVG(SALARY))
-------------------------------------
                                 6461
```

■ **Note**   In this case we don't need a group by expression as we don't want any group, we are considering the company as a whole. If we wanted to do the same calculation by department then we would need some grouping by department_id, but we will see this later, as the subquery returns more than one row (in fact one per department), and we still need to see something else to be able to answer that query.

Ok, that's it! We have the average salary of all the company so now it is just a matter of applying the subquery concept. As you can see we have a main query here, the one to compute the employees whose salary is bigger than one amount and a subquery that is the one needed to calculate the average of the company. So it is only a matter of writing them together. We will use brackets in the where, and we will add the subquery as if it was any value:

```
SELECT
      first_name, last_name, salary
FROM
      hr.employees
WHERE
      salary > (SELECT
                      TRUNC(AVG(salary))
                FROM
                      employees)
ORDER BY 3 DESC;
```

And the result is the following:

FIRST_NAME	LAST_NAME	SALARY
Steven	King	24000
Neena	Kochhar	17000
Lex	De Haan	17000
... (output truncated)		
David	Lee	6800
Susan	Mavris	6500
Shanta	Vollman	6500

51 rows selected

CALCULATE THE NUMBER OF EMPLOYEES THAT EXCEED THE AVERAGE SALARY EXCLUDING EMPLOYEES FROM THE DEPARTMENT_ID = 90 (EXECUTIVES)

We want you to try too. How do you think we would be able to count the number of employees by department who exceed the average salary? We would like to exclude from the average the executives. Try to apply the same methodology we explained in this point to find the solution.

1. Try to think the query that will satisfy this. TIP: It should be a query containing a department_id and a count of employees.

2. Try to think in the subquery needed for it. TIP: This time we need to compute the average of the company salaries but filtering first the employees belonging to department_id = 90.

3. Link both queries and write the required query.

Don't cheat and try to think about the solution by yourself! If after a while you can't, here it is for reference:

```
SELECT
      department_id, count(*)
FROM
      hr.employees
WHERE
      salary > (SELECT
                        TRUNC(AVG(salary))
                FROM
                        hr.employees
                WHERE
                        department_id <>90)
GROUP BY
      department_id
ORDER BY 2 DESC;
DEPARTMENT_ID    COUNT(*)
-------------  ----------
           80          34
          100           6
           50           4
           90           3
          110           2
           40           1
           60           1
           20           1
                        1
           30           1
           70           1

11 rows selected
```

Joining Tables

We have learned quite a few things about how to operate with a relational database so far. However, it is quite usual that we need to operate with several tables at the same time. So what we saw so far is of little use.

Usually in a datawarehouse you will have several tables. Some of them will be lookup tables, containing master data of your customers, providers, products, and so on, while others will be fact tables, containing information about sales, costs, employees' payslips, and so on. We will see more about this in the following chapters, but start thinking about it.

For the time being we are concentrating on our two sample tables, the Employees and Departments. It is likely that at some point, you want to count for example, the number of employees in each department. At this point you may think in a solution, not too much elegant but that will work, which is selecting the DEPARTMENT_ID column from the employee table, and adding an aggregation function like count (*) and then applying a group on the DEPARTMENT_ID column. While this will work, we have some drawbacks here (and some advantages, to be honest). The most important is that this solution does not give us the department names. We only have ids, so depending on what kind of information we want to present, this is not acceptable. The advantage here is that querying only one table is always faster than retrieving data from more than one, given of course, the same filters are applied.

There are many types of joins, but the syntaxis is very similar. We will present you the ANSI 92 Join Syntax and the previous one, ANSI89. While the ANSI 92 is supported in all databases, the previous may not be, especially when it comes to outer joins. It depends entirely on which one you use, and while our recommendation is to stick to the ANSI 92, which is supported in all databases, it is also true that we are more used to the old one.

The syntaxis to write a join involving two tables in ANSI89 is the following:

```
SELECT list_of_columns (separated by commas, or *)
FROM  table1, table2
[WHERE table1.column1=table2.column1 ... ]
```

Whereas in the new syntax the format is as follows:

```
SELECT list_of_columns (separated by commas, or *)
FROM  table1
JOIN table2
ON (table1.column1=table2.column1)
```

As you can see, the two formats are similar. To generate a join, in the old syntax we use the comma, whereas in the newer syntax, we use the keyword JOIN (or similar, depending on the type of join, more later) and then adding the clause ON to specify the columns joined from the two tables. Instead of this, in the first case the joined columns are specified in the same where clause, which can cause confusion as sometimes it may be difficult to see which are the joined columns and which are the conditions applied in the where clause. Most of ANSI92 supporters use this argument as the main one in their defense of the newer syntax.

Types of Joins

As we told you previously there are many types of joins. It is very important to know what we want to select, as this will force us to use one type of join or another. We will start with the most basic one and we will see the most complicated ones using examples from the Employees and Departments tables we saw previously.

Cartesian Join

In fact this is not a join but it's also named join (cross join). It is a "no relation" relation between two tables. Due to that, this is the type of join that you, most of the time, want to avoid because usually it is reached by mistake, when specifying incorrectly the joining clauses. As the name implies, what this type of join does is a Cartesian product, which involves joining (or linking) every single row from the first table to every single row of the second table. As you will already have noticed, the total number of records is a multiplication of the number of records of the first table by the ones in the second. While it is true that you usually will want to avoid this join, it may be useful in some cases, for example, when you know that two tables have nothing in common but you want to join them anyway, or when one of the tables that has nothing in common with the other, only contains one row, and you want to append the two tables together, generating a new table with the columns of both tables.

Given a few records from the employees table (Figure 3-2), and a few records from the departments table (Figure 3-3), we can create a cross join between them (Figure 3-4):

```
SELECT
      *
FROM
      hr.employees, hr.departments;
2.889 records selected
```

Or using the newer ANSI92 syntax:

```
SELECT
      *
FROM
      hr.employees
CROSS JOIN
      hr.departments;
2.889 records selected
```

	EMPLOYEE_ID	FIRST_NAME	LAST_NAME	EMAIL	PHONE_NUMBER	HIRE_DATE	JOB_ID	SALARY	COMMISSION_PCT	MANAGER_ID	DEPARTMENT_ID
1	100 Steven	King	SKING	515.123.4567	17/06/87	AD_PRES	24000	(null)	(null)	90
2	101 Neena	Kochhar	NKOCHHAR	515.123.4568	21/09/89	AD_VP	17000	(null)	100	90
3	102 Lex	De Haan	LDEHAAN	515.123.4569	13/01/93	AD_VP	17000	(null)	100	90
.	.....	....	......	.....	... ...	.. ..	.. ....	....	. ...	...	..

Figure 3-2. *Records from the employees table*

	DEPARTMENT_ID	DEPARTMENT_NAME	MANAGER_ID	LOCATION_ID
1	10	Administration	200	1700
2	20	Marketing	201	1800
3	30	Purchasing	114	1700
4	40	Human Resources	203	2400

Figure 3-3. *Records from the departments table*

	EMPLOYEE_ID	FIRST_NAME	LAST_NAME	EMAIL	PHONE_NUMBER	HIRE_DATE	JOB_ID	SALARY	COMMISSION_PCT	MANAGER_ID	DEPARTMENT_ID	DEPARTMENT_ID_1	DEPARTMENT_NAME	MANAGER_ID_1
1	100 Steven	King	SKING	515.123.4567	17/06/87	AD_PRES	24000	(null)	(null)	90	10	Administration	200
2	101 Neena	Kochhar	NKOCHHAR	515.123.4568	21/09/89	AD_VP	17000	(null)	100	90	10	Administration	200
3	102 Lex	De Haan	LDEHAAN	515.123.4569	18/01/93	AD_VP	17000	(null)	100	90	10	Administration	200
4	103 Alexander	Hunold	AHUNOLD	590.423.4567	03/01/90	IT_PROG	9000	(null)	102	60	10	Administration	200
5	104 Bruce	Ernst	BERNST	590.423.4568	21/05/91	IT_PROG	6000	(null)	103	60	10	Administration	200

Figure 3-4. *Cross join or Cartesian join between the Employees and Departments tables*

As you can see in Figure 3-4, the column Department_ID coming from the employee table and the column Department_ID coming from the department table do not match. This is obvious because we have not added such condition, so what basically the SQL engine is doing is joining each employee, irrespective of its department to all the departments in the company. In the figure we can see that employees that are assigned to departments 90 and 60 are related to department 10 and so on.

Inner Join

The most commonly used join is the Inner join. This type of join consists in joining two tables that have at least one column in common. In the where clause or on the ON clause we add the condition table1.column1 = table2.column1 and so on to specify all columns we want to use for the join. This will create a result with the columns we selected in the projection from the two (or more) tables joined using the columns we have specified in the Where or the ON clauses. Let's see an example:

We know that the department_id is shared between the two columns, and we have been told to get a list of employees and their assigned department names. This can be accomplished with the following queries:

```
SELECT
        employees.first_name, employees.last_name, departments.department_name
FROM
        hr.employees, hr.departments
WHERE
        employees.department_id = departments.department_id;
```

or using the ANSI92 syntax:

```
SELECT
        employees.first_name, employees.last_name, departments.department_name
FROM
        hr.employees
JOIN
        hr.departments
ON
        (employees.department_id = departments.department_id);
```

And the result, as you can see this time is not doing a Cartesian product anymore as we are introducing a valid join clause, so the total number of records retrieved will be usually smaller than the result of the multiplication. If one of the tables (the smallest) has unique registers for the given column, the number of records returned will be the number of records we have in the bigger table, except if one of the tables contained a null record, or the value of the department ID doesn't exist in the departments table, as the match is impossible then. In truth we have an employee with a department unassigned in our table, so the total instead of being 107, is 106 as you can see in the output of the query.

```
FIRST_NAME            LAST_NAME                   DEPARTMENT_NAME
-------------------   -------------------------   ------------------------------
Jennifer              Whalen                      Administration
Pat                   Fay                         Marketing
Michael               Hartstein                   Marketing
Sigal                 Tobias                      Purchasing
106 rows selected
```

There is a special case of an inner join, called a Natural Join. A natural join performs an inner equijoin of two tables without specifying the join columns. You may well be wondering how the engine accomplishes this. It's easy. It uses the column names and data types of columns, so any column that shares exactly the same name, and the same data type, is automatically added to the join by the engine. Since in our example the column department_id is shared in both tables and has the same data type, we can rewrite the previous join, to use a natural join instead syntax:

```
SELECT
        *
FROM
        hr.employees
NATURAL JOIN
        hr.departments;
32 rows selected
```

However, we strongly recommend against using natural joins as you will have realized that the records number do not match, and we have carefully chosen this sample to alert you of the dangers of a natural join and unwanted side effects. If we go back to the employees table definition and the departments one we will see that apart from the department_id column, they also share the manager_id column. So since we are using a natural join, this manager_id column join is also added automatically in the join clause, and it is something that does not make any sense here, because the manager of a department does not necessarily mean that is the manager of all employees working in the same department.

■ **Note**   Using natural joins can be misleading and a source of errors. Only use it when you are sure that they can be used safely. We encourage you not to use them for the following reason: think about if we have two tables that only share one column, but at some point somebody modifies one of the columns introducing a column that previously existed in the other table: This will cause the join to add the new columns on it, causing a potential unwanted result. The same can be applied with cross joins. Use it carefully.

Outer Join

Sometimes it is useful to join tables that have partially some rows in common, and on the join result, you want to have all rows that existed in one of the source tables or both of them, no matter they do not have a counterpart in the other joined table. Specially, when we work on datawarehouses this is usually a source of problems (missing records after a join) and while our processes and ETL scripts should deal properly with this, it may be necessary in some calculations to make sure we do not lose records that do not exist in any of the involved tables in a join.

Imagine your business has sales records since the beginning of your company operations. For whatever reason, you have sales associated to products that you do not longer sell, but you used to in the past. Imagine that you have lost all the details of this product, and you don't even have an entry anymore in your ERP or transactional system. If you join the products table with the sales table, there will be no correspondence between the sales of these products and the product information. This will mean that you will lose all the sales record if you join these two tables. This will lead to problems and confusion as if you calculate sales aggregated by a product category or as a whole, or even by product, you will "forget" to count those sales. This is sometimes undesirable. As you may think, you can create a fake product entry in the product table, with the correct product id, so then an inner join will work. This can be a solution, but sometimes this is not easy due to the number of records lost or this could not be practical. In that specific cases, outer joins come to the rescue.

There are basically three types of outer joins: Left Outer joins (or left joins for brevity), Right Outer Joins (or right joins), and Full Outer joins. The three share the same syntax except for the join keywords, but they behave completely different.

■ **Note**   Outer joins are especially interesting to be used in the ETL process in order not to lose any register during the loads, but it is hardly recommended to use dummy entries for final tables in order to be completed.

Left Outer Join (or Left Join)

In a left join, records that do not match from the first table appear in the result of the join whereas the ones that do not match from the second table do not appear. The result of the join will be a composition then between an inner join (all records that appear in both tables) and the ones that appear only in the first table. To understand this more clearly, let's use our friendly employees and departments tables once again.

If you recall the Departments table, each department has a manager. Well that's not completely true. There are some departments that do not have a manager, so instead a null value is present on the manager_id column in the departments table. If the foreign key does not reference a value like in this case, if we do an inner join these departments will be filtered, as the null won't match any employee in our table (we do not have a null employee, do we?). Let's see an example of a left outer join in where we will return all departments that have managers, along with the manager name, but also all the departments that do not have a manager. So the result of the join should be the same 27 departments, along with the manager name for the ones that have one.

```
SELECT
      d.department_name, e.First_Name || ' ' || e.Last_Name as Manager
FROM
      hr.departments d
LEFT JOIN
      hr.employees e
ON (d.manager_id=e.employee_id);
```

```
DEPARTMENT_NAME                       MANAGER
-----------------------------         -----------------------------------------------
Executive                             Steven King
IT                                    Alexander Hunold
Finance                               Nancy Greenberg
Purchasing                            Den Raphaely
Shipping                              Adam Fripp
... (output truncated)
Benefits
Shareholder Services
Control And Credit
Corporate Tax
Treasury

27 rows selected
```

As you can see from the previous excerpt, we have 27 records as we have 27 departments and some of them show a null Manager. This is because the Outer Join adds the records that have an entry in the department table but not in the employee table (manager).

Right Outer Join (or Right Join)

In the same way we have a Left Join, we have a Right Join. The idea is the same but this time, the records that will be added to the result of the inner join are the ones that exist in the right table (second table) of the join whereas the ones that are only present in the left table (first table of the join) will be lost.

Let's imagine that now we want to know for which department an employee is a manager. As you think, not all the employees are managers, so if we repeat the same join that we did in the previous example, this time we will have all the records from the right part of the join (employees table), and which department they manage:

```
SELECT
      d.Department_name, e.First_Name || ' ' || e.Last_Name as Manager
FROM
      hr.departments d
RIGHT JOIN
      hr.employees e
ON (d.manager_id=e.employee_id);
```

```
DEPARTMENT_NAME                 MANAGER
------------------------------  -------------------------------------------------
Administration                  Jennifer Whalen
Marketing                       Michael Hartstein
Purchasing                      Den Raphaely
Human Resources                 Susan Mavris
Shipping                        Adam Fripp
IT                              Alexander Hunold
Public Relations                Hermann Baer
Sales                           John Russell
Executive                       Steven King
Finance                         Nancy Greenberg
Accounting                      Shelley Higgins
... (output truncated)

                                Mozhe Atkinson
                                Alberto Errazuriz
                                Allan McEwen
                                Douglas Grant
```

    107 rows selected.

As you can see this time, there are employees that do not manage anything.

Full Outer Join (or Full Join)

Imagine that you want to have both, a Left Join and a Right Join performed at the same time. Then a Full Outer Join or Full Join comes to the rescue. Imagine that you want the list of departments and its managers, but you want also at the same time all employees and the department they manage. Clearly you need a combination of both. Let's see an example:

```
SELECT
      d.Department_name, e.First_Name || ' ' || e.Last_Name as Manager
FROM
      hr.departments d
FULL OUTER JOIN
      hr.employees e
ON (d.manager_id=e.employee_id);
```

```
DEPARTMENT_NAME                 MANAGER
------------------------------  -------------------------------------------------
Executive                       Steven King
                                Neena Kochhar
                                Lex De Haan
IT                              Alexander Hunold
                                Bruce Ernst
... (output truncated)
Payroll
Recruiting
Retail Sales
```

123 rows selected.

As you can see, now we have Departments with a Manager, Employees that do not manage any department, and departments without a manager.

■ **Note**   Left and Right Joins are much more used than Full Outer Joins, especially in a datawarehouse environment. But it is important also to know that there is always the option to combine both into one single statement.

Table Aliases

Sometimes, we, as humans, are a bit lazy. Referring table names all the time by their names are difficult, and furthermore, if the same table is used several times in a select statement, it is confusing. Fortunately the SQL language solves that by letting coders give nicknames or alias to tables. The following statements are similar:

```
SELECT
      first_name, last_name, department_name
FROM
      hr.employees, hr.departments
WHERE
      employees.department_id = departments.department_id;
```

   and

```
SELECT
      first_name, last_name, department_name
FROM
      hr.employees e, hr.departments d
WHERE
      e.department_id = d.department_id;
```

The only difference is that we added two table aliases to refer their original tables using an alias or nickname. The alias need to be added following the table name, and then can be used in the where clause and subsequent clauses to refer the original tables.

Correlated subqueries

We saw previously how subquerying works. But we advised that some subqueries need to use joins so the external table can compute specific calculations. Imagine we want to retrieve the employees whose salary is above the average of the salary of all people in their departments. We can't do it directly with a subquery only as we need to compute the average salary for each department and then compare each employee against this average salary. But these two queries are linked, as the department where the employee resides has to be the same than the one we are calculating for the salary, so effectively we need a join. This is one of the most common examples. Let's see how to solve it.

   The query will consist, as mentioned, in two parts. One, called the outer query will select the employees that meet the condition, and the other query, called the inner query will be the query that computes the average salary per department. The relationship between the inner and the outer query will be specified inside the where clause of the inner query, as the outer query cannot reference columns of the inner query, unless these are in the FROM clause, which is not the case (these are in the WHERE or filtering clause). The query will look something like this. Note the aliases for the inner and outer table, as explained in the previous paragraph:

```
SELECT
      first_name || ' ' || last_name EMP_NAME, salary
FROM
      hr.employees emp_outer
WHERE
      salary > (SELECT
                        AVG(salary)
                FROM
                        employees emp_inner
                WHERE
                        emp_inner.department_id = emp_outer.department_id)
ORDER BY 2 DESC;

EMP_NAME                                         SALARY
---------------------------------------------  ----------
Steven King                                       24000
John Russell                                      14000
Karen Partners                                    13500
Michael Hartstein                                 13000
... (output truncated)
Renske Ladwig                                      3600
Jennifer Dilly                                     3600
Trenna Rajs                                        3500

 38 rows selected
```

Set Operations

Joining data from different tables is very useful, but there are other operations that need to be learned. What happens if you want to concatenate data from two tables that have an identical layout? Think, for example, in two different tables containing the sales data from 2016 and 2017. There has to be a way to "concatenate" them and use it as a single table. Fortunately, there is one.

We are going to introduce a few set operators that make these and other tasks easy. Let's start with the union operator.

Union and Union All Operators

The Union operator, as introduced previously, concatenates the result of one query, with the result of another. It is important to know however, that there are some requisites. The two tables, or queries, need to return the identical number of columns as well as have the same datatypes per each column. This ensures that the result can be a concatenation of the results of both queries or tables, and that the data will be aligned and placed in the column that has to be.

The sales example introduced in the previous paragraph, is the clearest one to understand how a UNION or UNION ALL statement behaves.

To show an example for this, we need to do some work on the HR data. Let's go back to our employees table, and we will create a new two tables based on the salary an employee is paid. We will create one table for employees earning less or equal than $6,000 and another one for employees earning more than $6,000.

Let's run the following two statements:

```
CREATE TABLE
      hr.employeesLTE6000
AS SELECT
      *
FROM
      hr.employees
WHERE
      salary <=6000;
```

and

```
CREATE TABLE
      hr.employeesGT6000
AS SELECT
      *
FROM
      hr.employees
WHERE
      salary >6000;
```

the output should be something like that:

```
Table HR.EMPLOYEESLTE6000 created.

Table HR.EMPLOYEESGT6000 created.
```

So now we have in our HR schema, two new tables, one for employees earning more than $6,000 and another one for employees earning less than or equal to $6,000. Since both tables have the same number of columns and same datatypes we can use a union statement to concatenate them back:

```
SELECT
      *
FROM
      hr.employeesLTE6000
UNION
SELECT
      *
FROM
      hr.employeesGT6000;
```

And the result of the selection is the 107 employees we have in our original employee table.

Imagine now the following new set of tables:

- HR.EMPLOYEESLTE6000, which contains all employees earning Less than or equal to $6,000.

- And a new table called HR.EMPLOYEESGTE6000, which contain employees earning more or equal to $6,000.

Let's create the missing table:

```
CREATE TABLE
      hr.employeesGTE6000
AS SELECT
      *
FROM
      hr.employees
WHERE
      salary >=6000;
Table HR.EMPLOYEESGTE6000 created.
```

And amend slightly our previous query to use the new table:

```
SELECT
      *
FROM
      hr.employeesLTE6000
UNION
SELECT
      *
FROM
      hr.employeesGTE6000;
```

The result is the same. That's fine. But what happens when we use the same query with the UNION ALL operator instead? Let's see it:

```
SELECT
      *
FROM
      hr.employeesLTE6000
UNION ALL
SELECT
      *
FROM
      hr.employeesGTE6000;
109 rows selected;
```

Oops! We have a problem. We have two records more; we have duplicated data! It may be the case that you really wanted to do this, but likely that won't be the case. We just duplicated the data for the employees that earn exactly $6,000, as they are in both tables. So UNION removes duplicates whereas UNION ALL does not. So you may be wondering, let's use UNION always instead of UNION ALL. That's partially true. Most people do, but it is usually not a good decision. Since UNION removes duplicates, this is a process most expensive to perform for the database engine that UNION ALL. So use them wisely. If you don't care about duplicates, use always UNION ALL as it will always be faster. If you care about duplicated records, use UNION, which will always discard repeated rows.

▓ **Note** The Union statement does not guarantee that rows will be appended from one table after the other or sorted depending on the table they came from. Rows from the first table may appear at the beginning, end, or mixed with the ones from the second table. If you want to instruct a specific order by whatever reason, you need to use the ORDER BY clause at the end of the statement.

The Intersect Operator

The intersect operator works in the same manner as the logical operator with the same name. Basically it reads data from both queries or tables and only keeps the rows that appear exactly in both tables. Again same preconditions are needed to be met than in the case of the Union and Union All statements. Let's see an example using the tables we created previously:

```
SELECT
        *
FROM
        hr.employeesLTE6000
INTERSECT
SELECT
        *
FROM
        hr.employeesGTE6000;
```

As you may see, two employees are returned, which are exactly the two that appear in both tables, having $6,000 as a salary.

```
EMPLOYEE_ID FIRST_NAME           LAST_NAME
----------- -------------------- ------------------------
EMAIL                    PHONE_NUMBER         HIRE_DATE JOB_ID          SALARY
------------------------ -------------------- --------- ---------- -----------
COMMISSION_PCT MANAGER_ID DEPARTMENT_ID
-------------- ---------- -------------
        104 Bruce                Ernst
BERNST                   590.423.4568         21/05/07  IT_PROG           6000
               103           60

        202 Pat                  Fay
PFAY                     603.123.6666         17/08/05  MK_REP            6000
               201           20
```

Intersect can be used to find coincidences in different tables or in different columns of the same table as far as you can compare selects with a single field; it is not required to use star in the select statement. Intersect is also worthy when trying to find duplicates before mixing tables.

The Minus Operator

The minus operator, also known as except in some database engines (but caution as the syntaxis may change a little bit), is an operator that subtracts from the first table the records that are also contained in the second table. Let's see an example to illustrate it:

We will be subtracting from one of our previous tables, the EMPLOYEESGTE600, which contains all employees earning 6000 or more, the ones from the EMPLOYEESGT6000 table which contained Employees earning **more** than 6000. So the result of the subtraction operation should be a new set of rows containing only the employees that earn exactly 6000, which are the ones that will be only present in the first table. Let's check it:

```
SELECT
     *
FROM
     employeesGTE6000
MINUS
SELECT
     *
FROM
     employeesGT6000;
```

and the result are the two employees that exactly earn 6000:

| EMPLOYEE_ID | FIRST_NAME | | LAST_NAME | | EMAIL | | | |
PHONE_NUMBER		HIRE_DATE	JOB_ID	SALARY	COMMISSION_PCT	MANAGER_ID	DEPARTMENT_ID
------------	---------	----------	----------	---------------	----------	-------------
104 Bruce			Ernst		BERNST			
590.423.4568		21/05/07	IT_PROG	6000		103	60
202 Pat			Fay		PFAY			
603.123.6666		17/08/05	MK_REP	6000		201	20

Working with Dates

Dates are important in datawarehouses. While we can work with them and store them as if they were arbitrary text strings, this is not good for performance because at some point we might need to operate with them and we will have to convert them to the date format.

We saw that all databases have a support for different datatypes in column definitions. Here, in this paragraph, we will see how to operate with date-related functions. Depending on your database vendor the functions may have a different name, but they essentially do the same.

The SYSDATE, NOW(), CURRENT_DATE and GETDATE() Functions

All Relational Database engines have a function to retrieve the current date of the system. There are different functions usually to retrieve only the date part, the date, and time part, the milliseconds since a specific point in time and the like. Depending on the database, the function may be called differently but all operate in the same manner.

Knowing the current date is usually important when we want to perform date operations like addition, subtraction, filtering, and even more important when we want to refresh or reload data based on date criteria. Imagine we want to refresh the data only for the current month, or the last 30 days. We need to compare the dates in our datawarehouses against the real current date. So first we need to see a way to retrieve the current data from our database. In Oracle, for example, this is accomplished using the sysdate reserved word. In SQL Server, this can be done using the GETDATE() function while in mysql we can use the NOW() function and in Postgres we can use the ANSI CURRENT_DATE expression. Let's look at a couple of examples:

```
SELECT
      SYSDATE
FROM
      DUAL; --(This is Oracle's)

SYSDATE
--------
18/06/16

SELECT
      NOW(); --(This is MYSQL's, but you can also do select SYSDATE();)
2016-06-18 12:46:42

SELECT
      CURRENT_DATE --(This is Postgres)
date
-----
2016-06-18
```

Adding Days to a Date

In the same manner we can get the current date, we can operate with date fields. We can add days, months, and years, subtract them and remove parts of a date. Again, this depends on each database implementation so we encourage you to check your database vendor manual or to check it in Internet for more detailed information about how to operate with them. In this chapter we will show you some samples using Oracle express edition and mysql.

For Oracle we are running a query like this one:

```
SELECT
      SYSDATE, SYSDATE + INTERVAL '1' DAY
FROM
      dual;
SYSDATE   SYSDATE+INTERVAL'1'DAY
--------  ---------------------
18/06/16  19/06/16
```

Whereas for mysql, we need to amend it slightly:

```
SELECT
      SYSDATE(), SYSDATE() + INTERVAL '1' DAY ;

SYSDATE()             SYSDATE() + INTERVAL '1' DAY
------------------    -------------------
2016-06-18 15:22:33   2016-06-19 15:22:33
```

With the same procedure we can subtract any value. Note that the INTERVAL keyword is very convenient as we can change the DAY interval by any time interval we need: MONTH, YEAR, HOUR, MINUTE, SECOND ...

Conditional Expressions

The SQL language has some conditional expressions built in. Conditional expressions test a condition and execute action or another depending on whether the result of the test evaluates as true or false. As with any programing language, the condition of the test needs to be a Boolean condition, so the result of evaluating it will always be true or false.

The Case Expression

The Case expression is a very useful control statement to execute partially a calculation or gather data from a specific column based on a condition. The syntaxis of a case statement is as follows:

```
CASE
WHEN BooleanExpression1 THEN Branch1
...
WHEN BooleanExpressionN THEN BranchN
ELSE BranchN+1
END (alias for the column)
```

Let's imagine now that we want to retrieve in a column with a value depending on the date. We will query the month number and we will translate it to the month Name. Obviously there are better approaches to do this, but for illustration purposes, this will be a good one.

```
SELECT
     CASE
            WHEN to_char(sysdate,'mm')=06 THEN 'June' ELSE 'Another Month'
     END CurrentMonth
FROM
     dual;
CURRENTMONTH
-------------
June
```

This is Oracle syntaxis using sysdate and the dual pseudotable but the same can be written in for example, mysql/mariadb:

```
SELECT
     CASE
            WHEN month(now())=06 THEN 'June' ELSE 'Another Month'
     END CurrentMonth;
CurrentMonth
-------------
June
```

The Decode() or IF() Expressions

The decode expression is also used as a conditional statement. We do not encourage you to use them unless you are used to it or you are used to programming with languages that have the if/else clauses as sometimes it can be a bit cryptic, making debugging somewhat difficult. It can be used in exchange of the case and where it is more compact, it is usually difficult to understand if there are many of them nested.

The syntaxis is: DECODE (Statement,result1,branch1, … resultn, branchn, [else_branch]). Let's see an example, the equivalent to the one we did for the previous CASE command:

```
SELECT
      DECODE(to_char(sysdate,'mm'), 06, 'June', 'Another Month') CurrentMonth FROM
      dual;

CURRENTMONTH
-------------
June
```

And in mysql/mariadb instead of decode we will be using IF(BooleanExpression,if_case,else_case):

```
SELECT
      IF(month(now())=06, 'June', 'Another Month') CurrentMonth;
CurrentMonth
-------------
June
```

Conclusion

Only one chapter to learn SQL is not enough. However, we wanted you to introduce the widely used language for interacting with relational databases, so you can start thinking about writing your own queries. In this chapter we saw a brief introduction to relational databases, the type of statements available for us, data types, how to retrieve and count data, sorting, filtering, and grouping, and then more advanced statements including nested statements or correlated subqueries.

We recommend that you to read a little bit more about SQL, or purchase one of the many available books that will give you an edge on the first-needed step to successfully build a BI solution. If your code is good and well written and performs the calculations that it is entitled to do, then the following steps will be far easier, and will run much faster than a poor design or a faulty code.

In the following chapters we will start to see the components that will conform our datawarehouse, how to logically define entities, and then we will translate this into requirements to build the schema we need to store our data. We know that this chapter has been a bit dense but we promise you that the following ones will be more practical as we will start working in our solution from scratch, and we are sure that you will learn a lot from it and that they will help you to build your very own BI solution.

■ ■ ■

## CHAPTER 4

Project Initialization – Database and Source ERP Installation

In the first chapters of this book we have seen many much theoretical aspects of the BI. We started with a general introduction, a chapter dedicated to project management, and also another chapter dedicated to SQL introduction. But at the end of the day, until now we haven't done anything practical so far, so it's time to start getting our hands on the development of the solution. Before starting, however, we need to install our source ERP, where the data will come from. After installing and configuring it, we will mess with it around for a while, and then we will select our database to store the datawarehouse.

The Need for Data

All BI systems need some sourcing of data. This data can be either structured or unstructured. Also, this data can come from multiple sources and have multiple shapes. Also, we may face some data quality issues, as we may have missing and/or redundant data. But there is always a starting place, where the source data is generated.

This book will teach you how to extract, transform, load, and report your data. But there is a first preliminary step that consists in sourcing this data. For this book, we decided to use Odoo (formerly known as OpenERP) as a source of our data. Yes, it may have been easier using any of the free-to-use available sample database schemas that are on the Internet or that come bundled with almost each database. And yes again, we will use some to show you some advanced points later on the book, such as indexing and so on. But we want to show you a real application and the complete data pipeline. And, in real life, this usually means connecting to an ERP.

Most of the BI projects start sourcing some data from SAP. However, SAP is not targeted to small/mid business. The complexity of the system is overwhelming. and to make it worse license prices are high. So, it is likely that you will have to target any other ERP's. There are many of them affordable, most of them even run in the cloud, and you pay as you go (mostly by the number of users using it or by the modules you purchase, or based on the usage). But we feel that there are good open source projects, available for free, that may suit your company's needs.

Setting the System Up with Odoo ERP

If you already have an ERP installed in your company, or using some sort of program to manage customers, do the invoicing and so on, Congratulations! you already have your data source. Your data source will be your actual ERP or actual transactional database where the program or programs you are using are accumulating the data.

© Albert Nogués and Juan Valladares 2017

A. Nogués and J. Valladares, *Business Intelligence Tools for Small Companies*,
DOI 10.1007/978-1-4842-2568-4_4

For the ones that are still thinking about one, we are presenting in this chapter how to install and play with Odoo to gather data to develop our BI system.

As explained in the introduction section, we will use a Linux machine to install Odoo. We decided to use Ubuntu server 16.04 LTS version. This is because it is a Long Term Support (LTS) release, so it means that it will be supported for more time and ought to be more stable.

Installing Odoo may be a bit complicated due to different prerequisites it has, so it is more a task for your sysadmin. If you don't have one, or you can't afford to hire one, we recommend you use the simplified version that we will present in the following paragraphs.

In case you want to install it from scratch, before installing it make sure you install a PostgreSQL database, an Apache web server, and that the Python version matches the required one by the Odoo version you're installing.

The Bitnami Odoo Package

For our deployment we will use a self-contained Bitnami installer, which will install and set up Odoo for us. This package includes PostgreSQL database, the Apache server, and the program files. For the ones that are more interested in trying it first and don't want to deploy the package on an existing machine, Bitnami also offers a virtual machine with everything already installed, including an Ubuntu sever.

To download the Bitnami package, open a browser and check the following url:

```
https://bitnami.com/stack/odoo/installer
```

If we want to install the package by ourselves, then we need to choose the installers option in the menu bar as seen in Figure 4-1.

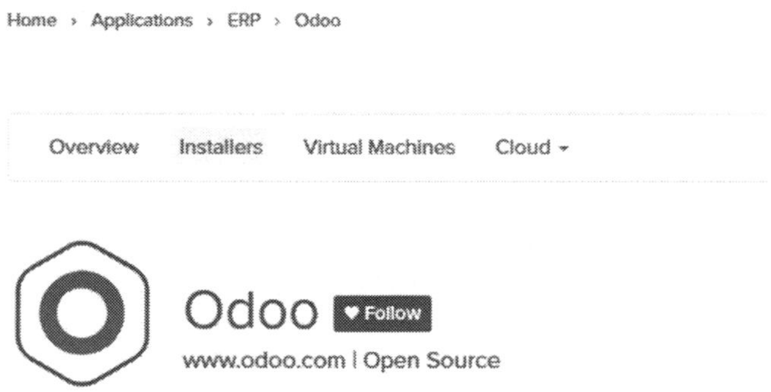

Home › Applications › ERP › Odoo

Overview   Installers   Virtual Machines   Cloud ▾

Odoo ♥ Follow
www.odoo.com | Open Source

Figure 4-1. *Finding the appropiate Bitnami installer*

And then select the one suitable for our operative system. At the time of writing the one available for Linux is Odoo 9.0-3 (64-bit) and it is the recommended download. For the ones wishing to try it first, instead of choosing the Installers option in the menu, choose the Virtual Machines and download the Ubuntu VM image containing the installed program. For the user and password bundled to log in the machine, read the page carefully.

Downloading and Installing Odoo

From here we will cover the basics of the package installation. Once we have found the suitable version, we will obtain the URL. Again, at the time of writing the download link is this:

```
https://downloads.bitnami.com/files/stacks/Odoo/9.0.20160620-1/bitnami-Odoo-9.0.20160620-1-
linux-x64-installer.run
```

But this may have changed if a new release is available. With that URL, we will move to our Linux box and use wget or curl to retrieve the installer. First we change to the /tmp directory to download the package there:

```
cd /tmp
```

And then we call wget to download the installer:

```
wget https://downloads.bitnami.com/files/stacks/Odoo/9.0.20160620-1/bitnami-Odoo-
9.0.20160620-1-linux-x64-installer.run
```

And the download will start. After a few seconds or minutes (depending on your network connection), the file will be downloaded in the machine. See Figure 4-2.

```
3-linux-x64-installer.run
--2016-04-22 11:15:24--  https://bitnami.com/redirect/to/96854/bitnami-odoo-9.0-
3-linux-x64-installer.run
Resolving bitnami.com (bitnami.com)... 50.17.235.25
Connecting to bitnami.com (bitnami.com)|50.17.235.25|:443... connected.
HTTP request sent, awaiting response... 302 Found
Location: https://downloads.bitnami.com/files/stacks/odoo/9.0-3/bitnami-odoo-9.0
-3-linux-x64-installer.run [following]
--2016-04-22 11:15:25--  https://downloads.bitnami.com/files/stacks/odoo/9.0-3/b
itnami-odoo-9.0-3-linux-x64-installer.run
Resolving downloads.bitnami.com (downloads.bitnami.com)... 54.192.28.239, 54.192
.28.7, 54.192.28.43, ...
Connecting to downloads.bitnami.com (downloads.bitnami.com)|54.192.28.239|:443..
. connected.
HTTP request sent, awaiting response... 200 OK
Length: 146937506 (140M) [binary/octet-stream]
Saving to: 'bitnami-odoo-9.0-3-linux-x64-installer.run'

bitnami-odoo-9.0-3- 100%[===================>] 140.13M 23.7MB/s    in 9.7s

2016-04-22 11:15:35 (14.5 MB/s) - 'bitnami-odoo-9.0-3-linux-x64-installer.run' s
aved [146937506/146937506]

bibook@bibook:/tmp$
```

Figure 4-2. *Downloading the installer in the machine*

Before running the file, we need to make it executable, so we do a chmod +x:

```
bibook@bibook:/tmp$ chmod +x bitnami-Odoo-9.0-3-linux-x64-installer.run
```

Once there, we will start the installer by running the following, using the sudo to run it as a root, as we will need it to install it at a different directory

```
bibook@bibook:/tmp$ sudo ./bitnami-Odoo-9.0-3-linux-x64-installer.run
```

And we follow the steps on screen. After being launched, the installer will suggest a path in the home directory. We are using /opt to install it, so amend it in case you don't have this one. See Figure 4-3.

```
Welcome to the Bitnami Odoo Stack Setup Wizard.

----------------------------------------------------------------------
Installation folder

Please, choose a folder to install Bitnami Odoo Stack

Select a folder [/opt/odoo-9.0-3]:

----------------------------------------------------------------------
Create Admin account

Bitnami Odoo Stack admin user creation

Email Address [user@example.com]: anogues@albertnogues.com

Password :
Please confirm your password :
Do you want to configure mail support? [y/N]: N

----------------------------------------------------------------------
Setup is now ready to begin installing Bitnami Odoo Stack on your computer.

Do you want to continue? [Y/n]: y
```

Figure 4-3. *Installing Odoo in the machine*

At the end of the installation process the installer will ask us if we want to start Odoo services and components. We choose yes (Y).

Bitnami and Odoo Configuration Files

In this subchapter we will review the place for all configuration files and the startup scripts so we can control our Odoo installation.

In case we want to stop or start later in the stack, there is a script to rule all the installation called ctlscript.sh located in the top of the install path, in our case /opt/odoo-9.0-3. Launching the script as root user lets us to start and stop associated services:

```
bibook@bibook:/opt/odoo-9.0-3$ sudo ./ctlscript.sh
usage: ./ctlscript.sh help
       ./ctlscript.sh (start|stop|restart|status)
       ./ctlscript.sh (start|stop|restart|status) Postgresql
       ./ctlscript.sh (start|stop|restart|status) Apache
       ./ctlscript.sh (start|stop|restart|status) openerp_background_worker
       ./ctlscript.sh (start|stop|restart|status) openerp_gevent
```

```
help        - this screen
start       - start the service(s)
stop        - stop  the service(s)
restart     - restart or start the service(s)
status      - show the status of the service(s)
```

But at this point we already have the services started so we do not need to do anything.

Apart from the services, there is another important file to know, which the default configuration for the services is, including ports, URLs, local paths, and other important things. The file is called properties.ini and again we need root access to see it. It can be found in the root directory.

By default, the ports in use by the application are the standard ports used by the services installed. In our case, the Apache server listens in port 80, whereas the PostgreSQL database is listening into the default 5432 port.

Apart from these service configuration files, there are other important files. Perhaps the most important one is the application configuration file, which can be found in /opt/odoo-9.0-3/apps/odoo/conf/openerp-server.conf

This is an important file because it contains some randomly generated passwords necessary to connect to the application. In that file, we will be able to find the database username and password, the database name, and the password for the admin account. The values we have to note from this file, along with their meanings are the ones found in Table 4-1.

Table 4-1. *Most important Odoo configuration parameters*

Value	Description
admin_passwd	The default admin password to log on in the web interface
db_host	The IP address of the database server, which will be localhost
db_name	The database name, bitnami_openerp by default
db_password	The default database password
db_port	The default port, 5432
db_user	The database user, bn_openerp by default

Installing psql and Checking the Database Connection

The last task to do before connecting to the program is installing the psql client, so we can access the PostgreSQL database from the command line. If we are using Ubuntu this is accomplished by running the following commands:

```
sudo apt-get install postgresql-client-common postgresql-client
```

After this we can test the connection and check everything is working fine by connecting with the psql client:

```
bibook@bibook:/opt/Odoo-9.0-3/apps/Odoo/conf$ psql -U bn_openerp -h localhost -d
bitnami_openerp
```

We will be prompted by the db_password value; we input it and we should see the psql client prompt. If we get to it, then all is ok.

```
Password for user bn_openerp:
psql (9.5.2, server 9.4.6)
Type "help" for help.

bitnami_openerp=#
```

Now we are done. We are ready to go to visit our newly installed ERP!

■ **Note**   If you are trying to access the database from an external machine, it is necessary to enable remote access first. In this case, you need to modify the PostgreSQL configuration file by making the database listen under all interfaces (*) and allow the remote login of the users. This stackoverflow article may be useful: http://stackoverflow.com/questions/18580066/how-to-allow-remote-access-to-Postgresql-database

---

Accessing the Application

Once all configuration steps are done, we can log in into the application. By default, we can log in using the localhost URL:

```
http://localhost
```

The welcome screen will be shown asking for our credentials. At this point we need to use the username chosen during the installation and use the admin_passwd value from the openrp-server.conf configuration file.

We can see an example of the login screen in Figure 4-4.

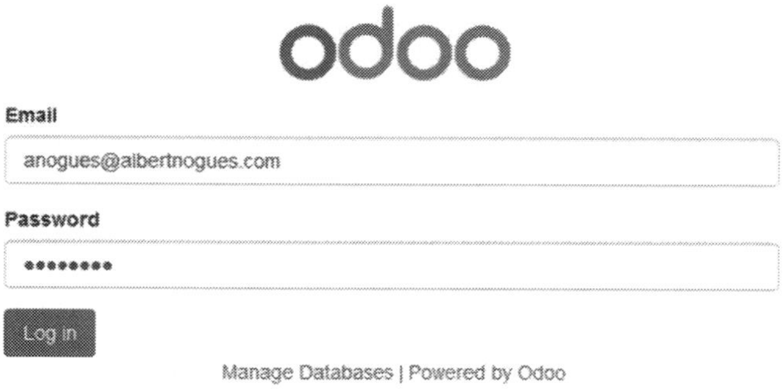

Email

anogues@albertnogues.com

Password

••••••••

Log in

Manage Databases | Powered by Odoo

Figure 4-4. Odoo login screen

Once entering, we just logged in using the administrator account.

Configuring and Installing Modules

By default, the installation sets up our Odoo with several modules enabled, but it may be the case we want to customize them, choosing which modules to install, choosing if we want to install any sample data, and so on. As you see in the lower part of the picture, there is an option called Manage Databases, which lets us start a new Odoo project from scratch.

We will choose this option as it lets us set up a new Odoo database, and we will install the Sales module for using it as an example during the book. At the same time, in a new installation we can instruct Odoo to load some test data. This is interesting as we can use it for illustration purposes, trough unfortunately, it inserts very few records. An example of how to fill the database information is found in Figure 4-5.

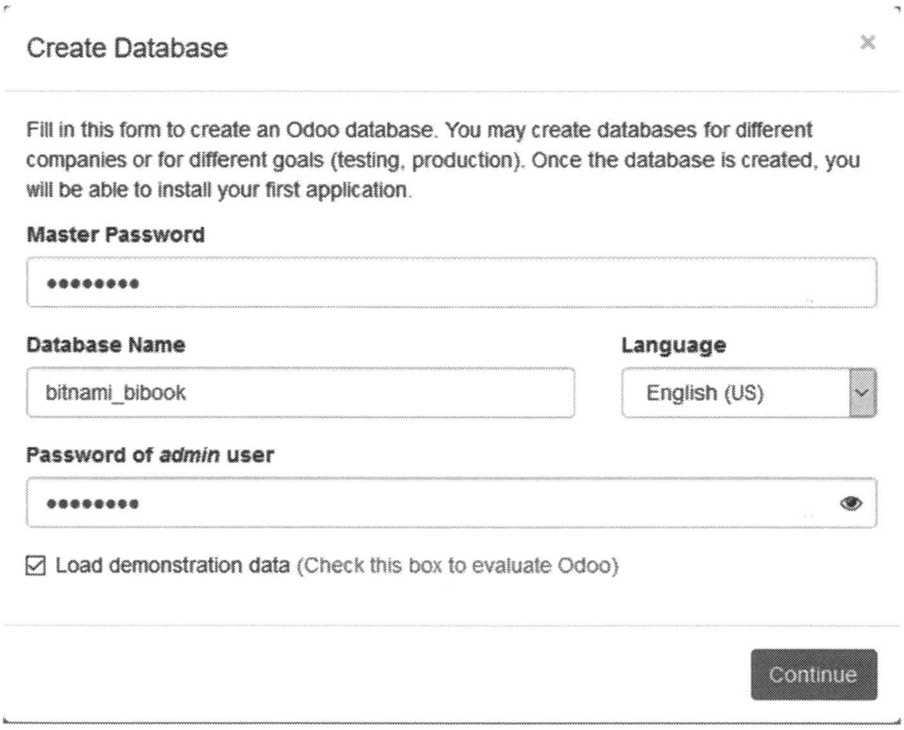

Figure 4-5. *Create a new Odoo application*

The create database dialog asks us the master password, the new database name, the language of the database, and the password of the admin user. We also have the check box to load some sample data, so we will be checking it. The master password is ***admin_passwd*** whereas the password of the admin user can be chosen by us.

This process takes a while, as the program creates and fills a new database, so please be patient and wait until the end of it.

After the database has been installed, we can go back to the main screen and ***log in with the username: admin and password, the one we put for the admin user in the new database dialog***. Once logged, the system will show you the list of applications available to be installed. For this book project, we will use mainly the sales module, so let's install and configure it by pressing Install in the Sales Management module it from the main screen as can be seen in Figure 4-6.

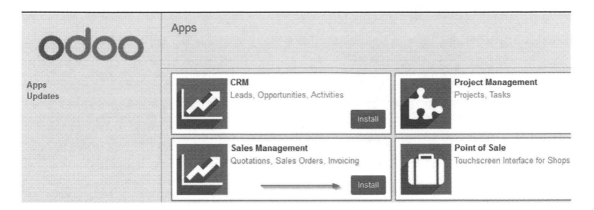

Figure 4-6. *Installing the Sales Management module*

Again, we will need to wait a while, as some scripts need to create the tables needed for the module and prepopulate some of them with the test data. Time for a coffee break!

After a while, the installation will be finished, and we will see that by default the ***Sales Management*** module has installed also as a prerequisite the ***Discuss*** and ***the Invoicing module***. That's perfect, as we will work with invoices.

Now it is time to check the test data and see the customers we have and the invoices linked to them.

If now we refresh the main screen, we will see that we have new options on the top menu, including Sales, so we click on it and go to Customers, to examine our customer base. This new menu can be seen in Figure 4-7.

Figure 4-7. *Checking our new customers*

As you will see we have some customers available now. By clicking on each company, a new screen appears showing us the contacts of the company and several other interesting options, like the details of the customer, the total invoiced, and the number of invoices and so on.

Once we have installed and tested that the application is working fine, we can start learning from the data model. Odoo has a complex data model, but fortunately the source code is clean, the access to the database is not complicated, and there are some useful resources on the Internet that detail the model a bit. With these bits and pieces we should be able to understand the basics of the base tables and the two modules that we will work on: Sales Management and Invoicing. In the following paragraphs we will detail the important tables and the model where the tables belong; we will inspect the columns and the rows of these tables and compare what it is stored on the database against what we see through the application.

There is one remaining thing to do though. The sample data we installed is a bit poor and does not cover all possible scenarios. For example, there are no invoices paid: all are in pending status. We need a bit more richness of data to be able to perform some reporting so we will create a few records and append them to the current test data. For this purpose, we will have to create some test data, but we will take care of it later on, when we start extracting the data from Odoo.

Selecting our Datawarehouse Database

There are many options for selecting a RDBMS to become our datawarehouse database. Either that or we can use a NoSQL database. But this is not yet the trend, and most of reporting tools, including the ones we will use, are designed to work with relational databases. While this paradigm may change in the future, almost all currently datawarehouses are based in relational databases.

There is a good website to track the most used databases, so you can see for yourself an which trends are in the market right now, and options to consider. This information can be accessed at the following website:

http://db-engines.com/en/ranking

Once we are aware that we need a relational database, it is the right time to select one. If we had a decent amount of budget and a lot of data to process, we will be probably looking at a commercial database. These tend to be powerful and despite the fact that there are some open source databases that also scale well, it would make sense to stick to a privative version. Think specially in terms of support, updates, and troubleshooting, and also the number of experts available in the marketplace, which is a very important factor you have to consider when you decide to use one technology or application over the other.

If we opt for a small deployment, it may make sense to use PostgreSQL as we saw before; it is the one that is used to store the metadata of Odoo, and it is directly bundled with the Bitnami package. If we opted for a manual installation of Odoo, we also had the chance to supply a MySQL database for the metadata.

Most of the time this decision is all about what kind of technologies we already have in the company (or licenses) and what is the main expertise we have in the company in terms of databases. It does not make any sense to set up a PostgreSQL datawarehouse if we do not have anybody in the company that can take care of it, when problems arise (trust me, they will!).

Our recommendation is to keep it simple and deploy a database that we are used to. If we work in an Oracle database environment then installing an Oracle database makes sense, but this is sometimes not feasible. As a small company, we can install one of the free (and limited) versions of a commercial database that we saw in Chapter 3. Oracle XE can be a good choice, but use caution, because as we saw in Chapter 3, it has some memory, CPU, and space limitations, so thinking about the future depends on our business case to decide if it is enough or not. The same applies for a Microsoft SQL Server Express.

Assuming we are a small company and we don't want to spend money in any license, the options reduce to either one of the free versions of the commercial tools or an open source or free database. In our case we will be installing either a MySQL/MariaDB for our deployment. There are a couple of good reasons to do it:

- It is a free database (or open source in case of MariaDB).

- It is the most extended one, and it has easy to administer and maintain. There are a lot of people that know how to do it and plenty of information on the Web.

- It scales well.[1]

But as always, choose what suits you the most. In our case it will be MariaDB, as it is a truly open source database, and the best thing is that it is MySQL compatible, so using the same tools or drivers you have to connect to MySQL can be used for MariaDB as well.

We decided against using the PostgreSQL bundled with the Bitnami Odoo package as a datawarehouse database because we want to have the source ERP database isolated from the datawarehouse database. In a very small deployment this is probably not very important, but thinking in a bigger deployment it could cause the failure of the whole project. The PostgreSQL database will be accessed several times a second probably by the ERP (transactional) tool. So there is some pressure on this database. If we also connect our BI tools and ETL processes to the same database, this may have an impact on the transactional tool. This is an undesired consequence of using the same database, because we want to ensure the highest availability possible, especially for our operational programs.

Since datawarehouses tend to be less sensitive than our operational programs, we will install a dedicated database to act as a datawarehouse for our platform. This is a fairly common scenario in BI deployments.

Sourcing and Installing MariaDB

There are many ways to install MariaDB. For Linux, several distributions already have it as the default database in the software manager tools. Others still have MySQL, and will miss the MariaDB repositories, and others may force you to install the packages manually, either using a precompiled package or compiling the source by yourself.

Despite what we said in the previous subchapters, when we explained that you should really install the datawarehouse database on a separate machine, as we have a little test environment, we will be installing it on the same machine. We can do it as we don't have any other MySQL server nor MariaDB running on that machine, as the Bitnami package uses a PostgreSQL database instead of a MySQL/MariaDB one, and they use different ports so we don't even need to bother about them. So for the sake of simplicity, we will be using the same machine this time but please bear in mind previous considerations.

Windows Installation

If you are using Windows, you can directly download the binaries from the MariaDB website and install them as usual, for example, for the current version at the time of writing this book it was the following:

```
https://downloads.mariadb.org/mariadb/
```

And you will have to download the msi package for windows (32- and 64-bit, whatever your flavor is).

---

[1]http://www.theregister.co.uk/2013/09/12/google_mariadb_mysql_migration

Linux Installation

Although you can download the source files for Linux there as well, we will be downloading the already compiled packages for our distribution. We are explaining here how to add the MariaDB repositories to your current software manager package for a couple of distributions, the most common ones, but look for the instructions on MariaDB website for any other type of deployment.

For both distros we will be using the repository configuration tool provided in the MariaDB website and available under the following URL:

```
https://downloads.mariadb.org/mariadb/repositories
```

Installation in Ubuntu

Installation in Ubuntu is straightforward, as we will be using the apt-get package manager. However, by default, in older Ubuntu installations, we need to add the MariaDB repositories as these are not available by default.

If we are using a current version or newer than 14.04 included, then we may skip this part as these binaries are already loaded in the package distribution list. For prior versions we need to add them first. We need to go to the repository configuration tool and in the distribution list, select Ubuntu, choose the version we have, the version of the MariaDB, and the mirror we will be using as in Figure 4-8.

Downloads Setting up MariaDB Repositories

To generate the entries select an item from each of the boxes below. Once an item is selected in each box, your customized repository configuration will appear below.

1. Choose a Distro
- openSUSE
- Arch Linux
- Mageia
- Fedora
- CentOS
- RedHat
- Mint
- Ubuntu
- Debian

2. Choose a Release
- 16.04 LTS "xenial"
- 15.10 "wily"
- 14.04 LTS "trusty"
- 12.04 LTS "precise"

3. Choose a Version
- 10.1 [Stable]
- 10.2 [Alpha]
- 10.0 [Stable]

4. Choose a Mirror
- TEDECO Facultad de Informática. Universidad Politécnica de Madrid.
- Klaus-Uwe Mitterer
- Mirror.ba
- Cu.be Solutions
- Nucleus.be
- Show All Mirrors

Here are the commands to run to install MariaDB on your Ubuntu system:

```
sudo apt-get install software-properties-common
sudo apt-key adv --recv-keys --keyserver hkp://keyserver.ubuntu.com:80 0xF1656F24C74CD1D8
sudo add-apt-repository 'deb [arch=amd64,i386,ppc64el] http://tedeco.fi.upm.es/mirror/mariadb/repo/10.1/ubuntu xenial main'
```

Once the key is imported and the repository added you can install MariaDB with:

```
sudo apt update
sudo apt install mariadb-server
```

See Installing MariaDB .deb Files for more information and for instructions on installing MariaDB Galera Cluster.

You can also create a custom MariaDB sources.list file. To do so, after importing the signing key as outlined above, copy and paste the following into a file under /etc/apt/sources.list.d/ (we suggest naming the file MariaDB.list or something similar), or add it to the bottom of your /etc/apt/sources.list file.

```
# MariaDB 10.1 repository list - created 2016-08-07 15:52 UTC
# http://downloads.mariadb.org/mariadb/repositories/
deb [arch=amd64,i386] http://tedeco.fi.upm.es/mirror/mariadb/repo/10.1/ubuntu xenial main
deb-src http://tedeco.fi.upm.es/mirror/mariadb/repo/10.1/ubuntu xenial main
```

Figure 4-8. *Adding the repositories for old installations*

So the first step will consist of executing the following commands to add the repository in our apt cache:

```
sudo apt-get install software-properties-common
sudo apt-key adv --recv-keys --keyserver hkp://keyserver.ubuntu.com:80 0xF1656F24C74CD1D8
sudo add-apt-repository 'deb [arch=amd64,i386,ppc64el] http://tedeco.fi.upm.es/mirror/
mariadb/repo/10.1/ubuntu xenial main'
```

Once done we can proceed to update our apt cache and install the program, as we will do with a current version of the distribution:

```
sudo apt update
sudo apt install mariadb-server
```

Once we run the following, a new screen will appear asking you to download new software as in Figure 4-9, and we confirm with "Y." After a couple of seconds (or minutes) depending how powerful your machine is, all should be installed.

```
bitnami@ubuntu:~$ sudo apt-get install mariadb-server
[sudo] password for bitnami:
Reading package lists... Done
Building dependency tree
Reading state information... Done
The following extra packages will be installed:
  libaio1 libdbd-mysql-perl libdbi-perl libhtml-template-perl
  libmariadbclient18 libmysqlclient18 libreadline5 libterm-readkey-perl
  mariadb-client-5.5 mariadb-client-core-5.5 mariadb-common mariadb-server-5.5
  mariadb-server-core-5.5 mysql-common
Suggested packages:
  libclone-perl libmldbm-perl libnet-daemon-perl libplrpc-perl
  libsql-statement-perl libipc-sharedcache-perl mailx mariadb-test tinyca
The following NEW packages will be installed:
  libaio1 libdbd-mysql-perl libdbi-perl libhtml-template-perl
  libmariadbclient18 libmysqlclient18 libreadline5 libterm-readkey-perl
  mariadb-client-5.5 mariadb-client-core-5.5 mariadb-common mariadb-server
  mariadb-server-5.5 mariadb-server-core-5.5 mysql-common
0 upgraded, 15 newly installed, 0 to remove and 2 not upgraded.
Need to get 11.7 MB of archives.
After this operation, 118 MB of additional disk space will be used.
Do you want to continue? [Y/n] _
```

Figure 4-9. *Installing the binnaries with apt-get*

After the installation has finished we have our MariaDB server operative, but there still are a couple of steps that need to be completed.

If the installation is in an old release, and you're installing MariaDB 5.5, there is an extra step to be done. After the installation you need to launch the (same) script than in MySQL to secure the installation. The script is called mysql_secure_installation (call it with sudo mysql_secure_installation) and should be available from your path. This script allows you to change the root password and remove some unneeded users. You may have been asked previously to set the password; if that is the case, you can skip the first part of the script. The remaining things that will be asked are the following:

```
Change the root password? [Y/n]
Remove anonymous users? [Y/n]
Disallow root login remotely? [Y/n]
Remove test database and access to it? [Y/n]
Reload privilege tables now? [Y/n]
```

Our suggestions would be to answer Y to change the root password unless you set one during the installation steps, to remove all anonymous users, to disable remote logins UNLESS the Odoo installation has been done on a separate machine, in this case answer No; otherwise you will need to tweak the permissions to the users later, to remove the test databases as they are not needed and to reload the privileges tables after the operations.

After that it is time to try to login into the server, and we can use the command tool to test the connectivity and the setup. It is not necessary but we recommend restarting the services. That can be accomplished by the following commands:

```
sudo service mysql stop && sudo service mysql start
```

After that you can review the service status by executing:

```
sudo service mysql status
```

And connect to the database with the client:

```
mysql -u root -p
```

And the program will ask the root password you just set up in the previous steps. If you see something like the MYSQL banner "Welcome to the MariaDB monitor…" then all is ok and ready for us to create out first database!

■ **Note** In newer versions of MariaDB, it may not be necessary to launch the securitization script after the installation has finished, as some of these tasks, like changing the root password, are already built into the main installer. If that is the case, simply skip that section of the document and go straight to test the connectivity. If there is any problem with the password or you are not able to log in after following the steps listed in this chapter, check this useful thread: http://askubuntu.com/questions/766900/mysql-doesnt-ask-for-root-password-when-installing

---

Installation of MariaDB in centos

We already covered the installation of MariaDB in Windows and Ubuntu, but for other people using Centos, Fedora, or Redhat Enterprise, here we cover the basics of its installation

By default, Centos7 comes shipped with MariaDB 5.5 installation (same one we installed in Ubuntu) so the process will be similar, but in this case using the yum package manager. Time to start!

```
yum install mariadb-server mariadb
```

Package	Architecture	Version	Repository	Size
Installing:				
mariadb	x86_64	1:5.5.47-1.el7_2	updates	8.9 M
mariadb-server	x86_64	1:5.5.47-1.el7_2	updates	11 M
Instalando para las dependencias:				
perl-Compress-Raw-Bzip2	x86_64	2.061-3.el7	base	32 k
perl-Compress-Raw-Zlib	x86_64	1:2.061-4.el7	base	57 k
perl-DBD-MySQL	x86_64	4.023-5.el7	base	140 k

```
perl-DBI               x86_64      1.627-4.el7       base       802 k
perl-IO-Compress       noarch      2.061-2.el7       base       260 k
perl-Net-Daemon        noarch      0.48-5.el7        base        51 k
perl-PlRPC             noarch      0.2020-14.el7     base        36 k

Transaction summary
===========================================================================
Install 2 Packages (+7 Dependant packages)

Total size to download: 21 M
Total size installed: 107 M
Is this ok [y/d/N]:y
```

After that, the installation will be finished, and then we can start the server with:

```
systemctl start mariadb
```

And now we can secure our installation in the same manner as we did in the Ubuntu installation.

```
mysql_secure_installation
```

Answering the same that in the Ubuntu version and after that we are ready to run the client and test the connection

```
mysql -u root -p
```

And again we should see the famous "Welcome to the MariaDB monitor..." If not, keep reading a little bit more, for troubleshooting actions.

■ **Note**   Again note that if we decided to install MariaDB from one of the official repositories, this will install version 10.1 with slight changes in the installation procedure.

Solving Connectivity Issues

It may be possible that by default we cannot connect from an external machine to our new MySQL installation. Some packages disable this by specifying the loopback address as the bind address for the MySQL engine.

If that is the case, the solution is fairly simple, we need to edit the /etc/mysql/my.cnf file with our desired editor (vi, nano ...) and look for the address that defines to which IP the database listens to, which will be the loopback IP, something like this:

```
bind-address: 127.0.0.1
```

And change it by binding it to all interfaces or to the IP of the interface you want to bind too (usually eth0) but may be named differently, hence the file now should read

```
bind-address: 0.0.0.0
```

Or

```
bind-address: (put here the ip of your network interface you want to bind to)
```

After that we only need to restart the service...

```
sudo service mysql restart
```

And we are good to go by specifying the IP address where we want to connect

```
mysql -u root -p(your password here with NO space)-h (your ip here with a space)
```

i.e.

```
mysql -u root -pp4ssw0rd -h 192.168.2.10
```

Creating Our First Database

It is now time to connect to our MariaDB installation and start playing with it. The goal is to convert this database in our new datawarehouse, where we will be downloading data from multiple sources. The transactional system, in our case our brand new Odoo application, will have details of customers, products, orders, invoices, and so on that we want to analyze. These along with other pieces of external information will be extracted and loaded into our datawarehouse. We will see in future chapters how to extract that information, and we will be developing some processes that will take care of them throughout the book, but first we need to prepare our database.

For our implementation, we decided to create two databases. We do not have an ODS database so we will be calling ours staging, even there are none/few transformations from the source database. For this, of the two databases to be created, one will be the so-called staging database that will emulate an ODS database, where usually raw data is placed, whereas the other will contain the current datawarehouse. This is a common setup, as usually we need a place to put our data first, which will come from multiple sources, before transforming to the final shape. In some cases, it is feasible to write an ETL script that does all this work and can directly place the data in the datawarehouse, but this is hardly ever a good solution. There are many arguments against it:

- Sometimes data from multiple sources is needed to be mixed and this makes things difficult, as the same tool cannot always mix every single piece of data.

- Performance tends to be worse if we work with external sources of data, as well as the availability can be compromised too.

- It is easier to recover from mistakes. Usually source data is in raw format, and as we seen in previous chapters, it may not be correct. We need to do some cleaning, and inserting directly into our final tables can compromise already formatted data and make recoveries much more difficult or even cause data loss.

With that in mind we can proceed to create our two databases and for that we won't be very original; we will call them staging and dwh, which stand for DataWareHouse. The commands to accomplish that are the following:

First, we need to connect again to our database in case we were disconnected:

```
mysql -u root -p
```

We will be creating two users with permission for these same databases only. Each user will only have permission on its own database. Later, in following chapters we will create users that will have permission to interact between the two databases, but right now we don't need them.

Show Databases and Tables Commands

Let's start by creating the two databases. As we are logged as root, we won't have any problems in creating the two databases. The commands we need to run are as follows:

```
CREATE DATABASE IF NOT EXISTS dwh;
CREATE DATABASE IF NOT EXISTS staging;
```

These two commands will create us the two databases, but still nothing else. No user, as we already haven't created them will have access (except our root user), and even if the users already exist, would not have any access, as we did not give them the permissions. Let's add the two users, and then the permissions to each one to access its own database.

Creating Users for the Two New Databases

We have created the databases, so now we need the users. Creating users in MariaDB is fairly simple, but we need to specify if the user will be local (it will only access the database from the same server) or it will be a remote user. Depending on the setup we choose we need to execute one of the two options:

If our user will access the database locally we can create a user locally. The statements are as follows:

```
CREATE USER 'dwh'@'localhost' IDENTIFIED BY 'p4ssw0rd';
CREATE USER 'staging'@'localhost' IDENTIFIED BY 'p4ssw0rd';
```

If our users need external access, which is likely the case, we can add a wildcard (%), and they will be able to connect from everywhere. Both users can coexist on the same database.

```
CREATE USER 'dwh'@'%' IDENTIFIED BY 'p4ssw0rd';
CREATE USER 'staging'@'%' IDENTIFIED BY 'p4ssw0rd';
```

With that, we have our users ready to access. But at this point, they will not be able to operate on the databases. Let's assign each one the required permission over their databases.

▓ **Note**   Later on, when we move to the ETL and Reporting chapters, these users will be used, so it is high likely the users need external access, so take this into account. In any case, this can be changed when needed so don't worry about it very much right now, but just have it present, in case of connection issues.

Grant Permission to Databases

We created the databases and the users, so now it is time to link all two concepts. For this we need to give permission to the users, to manipulate the databases. The permission structure in MariaDB/MySQL is quite easy to understand. It follows a well-defined format:

```
GRANT [type of permission] ON [database name].[table name] TO '[username]'@'%';
```

And to revoke permissions, the syntax is very much similar, only changing the keywords:

```
REVOKE [type of permission] ON [database name].[table name] FROM '[username]'@'%';
```

Note the wildcard % at the end, which has to be replaced by localhost if we are working with local access users.

For our case we will grant privileges to our local and external users:

```
MariaDB [(none)]> GRANT ALL PRIVILEGES ON dwh.* TO 'dwh'@'localhost';
Query OK, 0 rows affected (0.00 sec)

MariaDB [(none)]> GRANT ALL PRIVILEGES ON dwh.* TO 'dwh'@'%';
Query OK, 0 rows affected (0.00 sec)

MariaDB [(none)]> GRANT ALL PRIVILEGES ON staging.* TO 'staging'@'localhost';
Query OK, 0 rows affected (0.00 sec)

MariaDB [(none)]> GRANT ALL PRIVILEGES ON staging.* TO 'staging'@'%';
Query OK, 0 rows affected (0.00 sec)
```

Once this is accomplished, we can proceed to test our users, for this, we exit our root session and test both users. Now the connection should be done with the new user instead:

```
mysql -u dwh -pp4ssw0rd
```

Everything is going well and we should see our client logged in:

```
Welcome to the MariaDB monitor.  Commands end with ; or \g.
Your MariaDB connection id is 14
Server version: 5.5.47-MariaDB MariaDB Server

Copyright (c) 2000, 2015, Oracle, MariaDB Corporation Ab and others.

Type 'help;' or '\h' for help. Type '\c' to clear the current input statement.
```

Then we can check which databases the user can see

```
MariaDB [(none)]> show databases;
+--------------------+
| Database           |
+--------------------+
| information_schema |
| dwh                |
+--------------------+
2 rows in set (0.00 sec)
```

And then change to our database:

```
MariaDB [(none)]> use dwh;
Database changed
```

And then check that no tables are present yet:

```
MariaDB [dwh]> show tables;
Empty set (0.00 sec)
```

Then we can exit and test the other user, with exactly the same procedure:

```
MariaDB [dwh]> exit;
Bye
```

We will connect as the staging user now, and check this only after one sees the staging database:

```
[root@localhost anogues]# mysql -u staging -pp4ssw0rd
Welcome to the MariaDB monitor.  Commands end with ; or \g.
Your MariaDB connection id is 15
Server version: 5.5.47-MariaDB MariaDB Server

Copyright (c) 2000, 2015, Oracle, MariaDB Corporation Ab and others.

Type 'help;' or '\h' for help. Type '\c' to clear the current input statement.

MariaDB [(none)]> show databases;
+--------------------+
| Database           |
+--------------------+
| information_schema |
| staging            |
+--------------------+
2 rows in set (0.00 sec)

MariaDB [(none)]> use staging;
Database changed
MariaDB [staging]> show tables;
Empty set (0.00 sec)

MariaDB [staging]> exit;
Bye
```

▧ **Note** We have presented all the commands in this chapter using the command-line client interaction. For inexperienced users, it may well pay to try some GUI tool to interact with MySQL/MariaDB databases. There are many of them free, being the official MySQL workbench probably the most known: `https://www.mysql.com/products/workbench/` that works in several operative systems, but there are other good programs like HeidiSQL and dbForge for windows systems that are free or at least have a free edition.

Analyzing the Data Source

We have now our datawarehouse ready to go. So it is now time to start inspecting a little bit more the Odoo software and its database structure as it would be the source of most of our data in the ETL chapter.

If you recall from the beginning of this chapter, we installed the Odoo software using PostgreSQL as the database to hold the application metadata. Here, we will see a small overview to the PostgreSQL client, so we can inspect the source database and Odoo tables. Here, we will only be presenting an overview at what tables we will consider later for our system, and explain the basic relation between them and the module where they belong.

Inspecting our Model

We can start by connecting to the database directly and start browsing the tables. However, depending on the applications we have installed we will find many and many tables. Fortunately, there is a very good Internet resource of Odoo modules in the following webpage:

`http://useopenerp.com/v8`

While the documentation available is for version 8.0 and we will use 9.0, as it is not a very important thing in our case as differences are minimal.

As you can see on the Web, Odoo application is split in a set of categories, or modules, which are similar to the applications in Odoo. While they do not translate directly, as there are some common categories that are used by many applications, and are considered the core of Odoo, some of them have similarities. For example, we can find a Sales Management category as well as a Sales Management application in Odoo, but for example, we cannot find an Invoicing category, whereas we have an invoicing app in Odoo. This is because Invoicing is included in the Sales Management category as seen in Figure 4-10:

Sales Management	
Automated Action Rules	base_action_rule
Contracts Management	account_analytic_analysis
Dates on Sales Order	sale_order_dates
Delivery Costs	delivery
Invoice on Timesheets	hr_timesheet_invoice
Invoicing Journals	sale_journal
Jobs on Contracts	analytic_user_function
Margins by Products	product_margin
Margins in Sales Orders	sale_margin
Prices Visible Discounts	product_visible_discount
Products & Pricelists	product
Sales Analytic Distribution	sale_analytic_plans
Sales Management — Quotations, Sales Orders, Invoicing	sale

Figure 4-10. *The sales management category in Odoo metadata*

If we click further in the Sales management link, we can observe the following module URL:

`http://useopenerp.com/v8/module/sales-management`

111

And we can also see the details about the Sales Management module in Odoo. We can see a small description of the applications and flows supported by this module alongside a few screenshots of the menus and options covered by this module, all the Python classes involved in an UML diagram of classes, in case we want to customize this module by modifying the code directly in the application, and a few things more, especially interesting for our project, the Models section.

A model is one functionality implemented in Odoo. We have, for example, the sale.order model, which contains all information for the orders' functionality of the application. For example, for this module, we can see all the columns that are mapped to different tables of the database, and will contain the required logic to implement this functionality. It is also important to note that a model can be used by many applications.

The diagram of the Model is important as well, as it gives us information of the relationships between other models and classes, but at this point we are more interested in the columns and types of these columns; so by looking at this section we can start to elaborate a list of the field we will be able to extract and copy to our staging area.

For example, based on the business needs, we are interested in extracting information about our orders. To support our analysis, we are not interested in extracting the entire orders model, as probably there will be information that we don't need. The first step will be extract bulk information for Odoo model tables to our staging area. For example, we might consider copying the following fields from the order model:

Table 4-2. *Columns in the sale.order model that we want to extract*

Column Name	Description
name	Name of the order
state	Status of the order: draft, sent, cancel, progress, done …
Date_order	Date when the order was executed
User_id	A key to the Salesman who took the order in the res.users model
Partner_id	A key to identify our customer in the res.partner model
Order_line	One or more lines of product orders that reference the sale.order.line model

By getting these fields, we will already be able to calculate in our datawarehouse, for example, the total of orders by customer, the total orders generated by salesman, the number of items per order as an average, the total orders per month, and for example, the percentage of orders that actually finish in an accepted invoice, just to name a few possible analyses.

Probably to get more meaningful information we need to extract some more information from other models. For example, it is likely that we will want our employee or customer names instead of their internal id's, right? So this will require us to do a bit more of exercise and look for the columns we need in the res. users (for our employees) and res.partners (for our customers) to extract all their names, addresses, and whatever relevant information we need for them.

■ **Note**   Trying to understand a data model, especially if it is as big as this one, without having too much knowledge of UML, entity relationship, and databases may be a bit intimidating at first. But the samples we will do through the book are easy to understand and all will be guided, so don't get nervous if you are a bit overwhelmed now.

Having a look at the res_users we can quickly see that the employees' names aren't there. See Figure 4-11.

🔑	1	id
	2	active
🔑	3	login
	4	password
	5	company_id
	6	partner_id
	7	create_date
	8	create_uid
	9	share
	10	write_uid
	11	write_date
	12	signature
	13	action_id
	14	password_crypt
	15	alias_id
	16	chatter_needa...
	17	sale_team_id

Figure 4-11. *The fields in the res_users show that we do not have the user name anywhere*

This poses a problem for our sample, as we need them. So we need to look elsewhere. Fortunately, again the model comes to our rescue, and we are able to locate them. Following the information from the model webpage we can see the following section, after the column definition of all the available columns for the res_users:"Columns inherited from add-on base > res > res_partner.py > class res_partner".

At the time being we just need to focus on the last step of the sentence. This gives us the clue of where to go and find the data we need. In this case the system tells us exactly we need to look the data up in the res_partner table.

And having a look at the model we see that the res_partner table contains a name column, which will identify our employee. But at the same time this table is also used for the customers you may be wondering, as we explained previously. You are correct. For Odoo, all users that form part of our system, and even organizations, are considered members of the res_partner model. This means that all the companies, including our very own company, all of our employees, all of our customers no matter where they are companies or individuals, will have at least one entry in the res_partner table.

This fact will imply, that when we work on the extraction, we will need to consider doing multiple joins to this table, obviously using different keys and columns, depending what kind of information we want to extract. But we will only care about this later on.

Setting up a PostgreSQL Connection to the Server

If you recall from the first part of this chapter, when we installed the Bitnami package, we opted to use the bundled PostgreSQL database. This means that the metadata of all Odoo applications and modules will be stored in a PostgreSQL database. This is a good exercise as in later chapters, when we work on the ETL part, we will see how to connect to different databases, as our datawarehouse will be in MariaDB.

For this chapter however, we are more interested in being able to connect to the application metadata and browse our database. Let's start.

There are many applications for this purpose, and in different flavors. We can opt to use again the command line, or go for a GUI application. While the command line is sometimes useful, if we are basically interested in browsing relationships and data content, tuning a console client for a nice display is very tricky. For this reason, this time we will use a GUI tool to inspect our Odoo metadata.

There are many GUI tools that are free and will fit the bill. In fact, PostgreSQL comes with a bundled one. We, however, prefer using HeidiSQL. The good thing is that while HeidiSQL is designed for MySQL and MariaDB databases, it supports PostgreSQL databases as well. And it is open source so it's free to use!

■ **Note**   You can download HeidiSQL freely from its own website: `http://www.heidisql.com/`

For the installation of HeidiSQL, we can click on Next several times. If you want to change the path or any other configuration, feel free to do it. If you have followed the default steps during the installation, it is likely that the parameters to connect would be the same as the ones shown in Figure 4-12. This is the splash screen that appears when you first open HeidiSQL, and is the place to define any new connections you want to create. Our connection would be to the PostgreSQL engine, so make sure it is the one selected in the Network Type drop-down box.

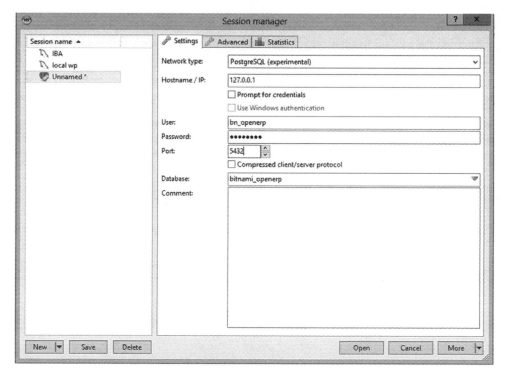

Figure 4-12. *Connecting to the PostgreSQL Odoo metadata*

Once connected we will see a couple of databases, most of them internal to the database, but one called public stands out. We need to click on this one, and the list of tables will appear. One of these is the res_partner. If you double-click it, we will enter in the table mode. At this point we do not want to edit the table, so do not save any change you made by accident (if any!).

In this view (Figure 4-13), we are able to see the columns, the relationships with other tables, and even browse the data, all with a few mouse clicks:

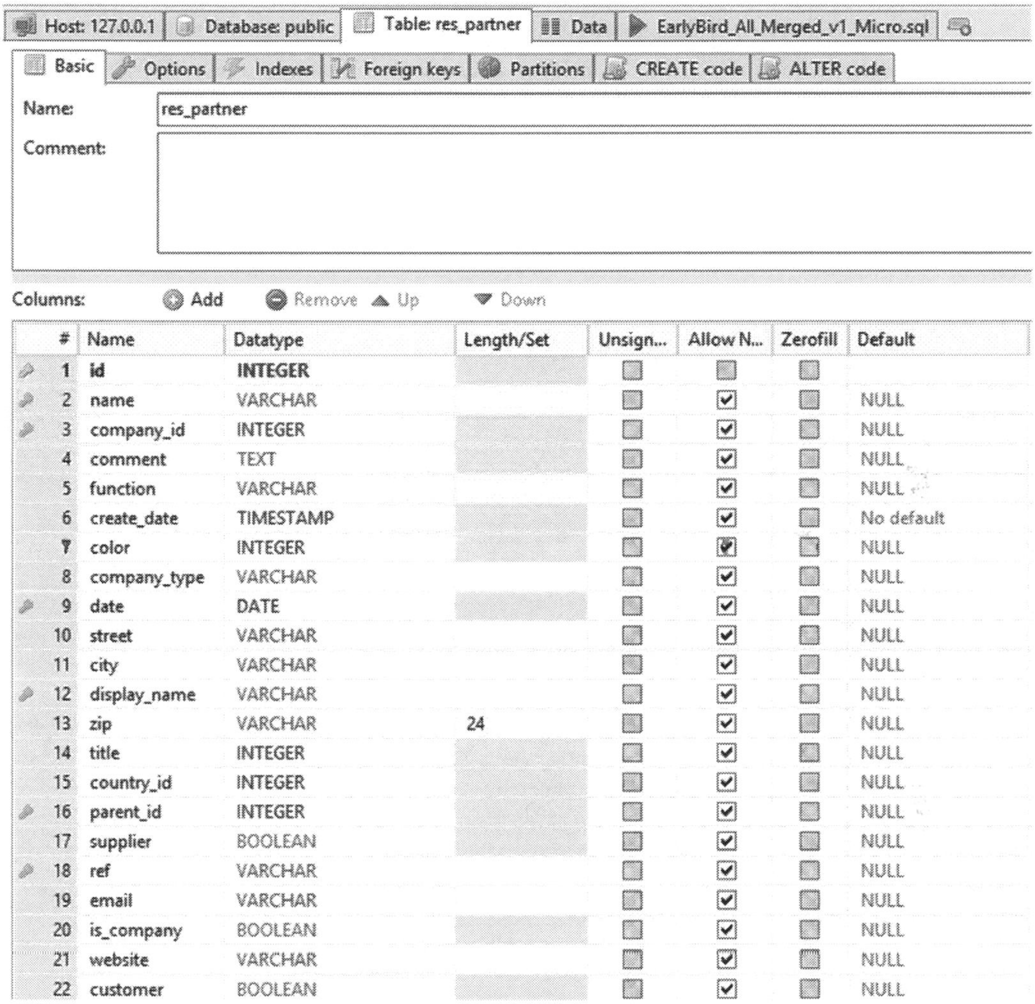

#	Name	Datatype	Length/Set	Unsign...	Allow N...	Zerofill	Default
1	id	INTEGER		☐	☐	☐	
2	name	VARCHAR		☐	☑	☐	NULL
3	company_id	INTEGER		☐	☑	☐	NULL
4	comment	TEXT		☐	☑	☐	NULL
5	function	VARCHAR		☐	☑	☐	NULL
6	create_date	TIMESTAMP		☐	☑	☐	No default
7	color	INTEGER		☐	☑	☐	NULL
8	company_type	VARCHAR		☐	☑	☐	NULL
9	date	DATE		☐	☑	☐	NULL
10	street	VARCHAR		☐	☑	☐	NULL
11	city	VARCHAR		☐	☑	☐	NULL
12	display_name	VARCHAR		☐	☑	☐	NULL
13	zip	VARCHAR	24	☐	☑	☐	NULL
14	title	INTEGER		☐	☑	☐	NULL
15	country_id	INTEGER		☐	☑	☐	NULL
16	parent_id	INTEGER		☐	☑	☐	NULL
17	supplier	BOOLEAN		☐	☑	☐	NULL
18	ref	VARCHAR		☐	☑	☐	NULL
19	email	VARCHAR		☐	☑	☐	NULL
20	is_company	BOOLEAN		☐	☑	☐	NULL
21	website	VARCHAR		☐	☑	☐	NULL
22	customer	BOOLEAN		☐	☑	☐	NULL

Figure 4-13. *Browsing the PostgreSQL table columns and data*

The data tab will be very useful, later on, to see if the data we are extracting matches with the one in the source system. The column list gives us clues about columns we might need to export and the keys in yellow and green in front of the column numbers show whether a column is a primary key (yellow) or a foreign key (green) referencing a column in another table. This information completes the one we can find in the model webpage and will help us all the time to know exactly what information we need to retrieve and from where to obtain it.

Conclusion

In this chapter, we saw how to install the operational system, in our case, the Odoo suite. We also see how to do a basic configuration of the application. Following that, we installed MariaDB database to hold our datawarehouse, and we created the two required databases: the ODS database, which we named staging; and the datawarehouse database.

Following our datawarehouse installation, we had a look at how Odoo is structured, where it stores the information, and how to access that information from a GUI tool. If we completed all the steps in this chapter we are ready to go to move to more advanced concepts and finally, start playing with our data and analyzing it. For that reason, it is important that you followed all the steps in this chapter, as subsequent chapters will assume that the installations and configurations discussed in this chapter are already implemented.

CHAPTER 5

■ ■ ■

Data Modeling for BI Solutions

You have received the request from your boss to implement a datawarehouse inside your database server that you have recently installed. You are the chosen person to lead and maybe develop, but this always will depend on the size and resources of your company, a solution that must allow your company the analysis of its data. So, after doing a previous analysis based on your user requirements, having the database available to go ahead and information enough to feed your system you should start with logical and physical development of the database solution that will be accessed from the BI tool.

If you have followed the instructions from the previous chapter you will have the ERP available to use as information source; the chosen database installed available to allocate company data; and in this chapter, we are going to propose to you a set of techniques and solutions that will allow you to analyze the sample data that comes in the Odoo system for testing purposes.

The model that we are going to propose in this chapter is a basic analysis in snowflake layout that will allow you to fill up a base fact table, some dimension tables and derived lookups to be able to drill into the information from top-level analysis to the highest level of detail available, something similar to the model of a snowflake seen in Chapter 1. Once we have this simple initial model we will analyze some configurations and structures that will allow you to offer more functionality to your customers (as always, they can be internal or external customers). We will see all this points with a combination of the explanation about the proposed solution and a deeper analysis about data modeling rather than theory as explained in the introduction chapter.

Naming Convention and Nomenclature

Before starting with the design of the database we propose that you define a basic naming convention to make easier the management of the database. When you start with the development of a model, you can think that the most intuitive way of naming tables is just to set a name for the information that it contains. And it is true, but when your model starts growing up, you can find some troubles. Imagine that we have a company that is selling tools to wholesalers and also we have a retail channel to sell directly to people. Analyzing our sales environment coming from the ERP solution, that we use to manage sales to wholesalers, we can have a table that contains our different customers that we have inside this ERP sales module and we can just think on to name it CUSTOMERS. We develop the whole analysis of the module including multiple dimensions, fact tables, ETL related processes, the BI tool over that accessing to the database and when we have the entire sales module implemented we want to analyze data coming from our CRM where we have information for retail customers.

In this moment we will have a different concept of customers. What we consider customers in sales are wholesalers for CRM and customers in CRM didn't exist in sales. The nature of customers is not the same, neither the source nor the fields that you have available so it makes no sense to mix all customers into a single table. You can name the table CRM_CUSTOMERS or RETAIL_CUSTOMERS, or any other invention but then you will start to complicate the names. On the other hand you can think that it is better

© Albert Nogués and Juan Valladares 2017
A. Nogués and J. Valladares, *Business Intelligence Tools for Small Companies*,
DOI 10.1007/978-1-4842-2568-4_5

to have different tables for customers and apply a naming convention but, oh! You need to modify ETL processes, database references, BI tool interface, and any other dependence on the table CUSTOMERS. So, believe us, invest some time before starting the development in defining a naming convention will have benefits in the future.

During our experience we have seen different naming conventions in different customers and we recommend using an intermediate approach, following a strategy that allows you to identify tables and environments but letting the possibility of giving a significant name to database objects. If you force yourself to follow a very strict nomenclature you will need a dictionary to translate what is the usage of every table and query it every time you want to write a new SQL query.

Here below we are going to propose a naming convention based on this consideration, but of course you can adapt it to your preferences or needs. In our naming convention we propose to use a name with different parts separated by "_". Names of database objects would follow this nomenclature:

OT_ST_ENV_DESCRIPTIVE_NAME_SUFFIX

Where each part matches with the following explanation. You can see an example at the end of the explanation itself:

First part, Object Type (OT), will have 2 letters maximum that should indicate which the object type is:

- **T**: Tables

- **V**: Views

- **PR**: Procedures

- **PK**: Packages

- **MV**: Materialized Views

- **F**: Functions

- **TR**: Triggers

Next part, Subtype (ST), sized in a single letter, is an optional part specially used in tables and views, which indicates which is the table or view type. In this exercise we consider three main types, related with the table types defined in Chapter 1:

- **F**: Fact table

- **L**: Lookup table

- **R**: Relationship table

ENV part is related to the environment that the table belongs to. It is between 2 and 4 characters and is variable depending on your needs. As example we propose that you use:

- **MD**: Masterdata related to customers, products, plants, stores, time, or any other data that is cross-functional.

- **SAL**: Sales data related to your invoicing system.

- **FIN**: Financial data related to the accounting system.

- **OP**: Operational data related with your production process.

- **HR**: Human Resources information subset.

- **STG**: Staging tables coming from the ERP or other sources.

DESCRIPTIVE_NAME part: it is so descriptive to need much more explanation, it should contain a name that allows you to easily identify which information is contained in the table with words (or acronyms) separated by "_".

SUFFIX part is also an optional part that we recommend especially to use it to mark temporary tables. In case you need to load some given table you need three steps, and we recommend that all these steps have the same name that the final table but adding at the end TMP1, TMP2, and TMP3 as suffixes.

So if we have a table (T) that is a lookup (L) of the billing month from sales environment (SAL) we would name this table: T_L_SAL_BILLING_MONTH, without any suffix in this case.

■ **Note**   When naming objects in your database, you need to validate what is the maximum name size that your database allows. As an example, in Oracle the maximum table name size is 30 characters.

Modeling Steps

In Chapter 1 we saw an overview of data modeling considering mainly two steps in the process, defining the logical model by setting a set of entities, relationship types among them, and attributes that exist in any table, and then going ahead with the physical model when we already defined all field names, types, precision, and other physical considerations for table creation and location. Now we are going to add two steps to the process of defining our model, business modeling, and dimensional modeling. You can see the whole modeling workflow in Figure 5-1.

Figure 5-1. *Modeling Workflow*

Going back to Chapter 2, this analysis can be considered as part of the initial zero sprint that we suggested to do at the beginning of the project. Let's describe each of the steps in deeper detail.

In order to help you with the modeling phase we strongly recommend that you use some of the modeling tools available for that purpose; there are also some free software or free editions of commercial software that can help you with the definition of logical and physical models. We will evaluate later in this chapter some of these free software modeling products to show you how to use them in a helpful way.

Business Model

Business modeling is directly related to business people, key users, product owners, or whoever is the interlocutor from the business side (maybe yourself, dear reader). It is a description of what your BI project must provide. In relation with Chapter 2 concepts, it would be mapped to user stories and developer stories, so it is the definition of what you want to analyze, which dimensions you want to use for the analysis, and which granularity you want to apply to analyze it. It can be a descriptive text, a graphical definition, or a combination of both in a language that business users must understand. A useful tool that can be helpful for this analysis is a granularity matrix, where you can cross which dimensions are used in each analysis and at which level; you can see an example in Figure 5-2. In order to get this information your first step will be to always talk with business teams to get their requirements and then have a look at the available information to see if you can meet the defined requirements.

	Sales	Finance	Customer Service	Stocks
Customer	Ship to code	Level 3	Ship to code	
Product	Single Product	Family	Single Product	Single Product
Time	Day	Month	Day	Day
Plant			Plant	Plant
Employee	Employee		Employee	
Currency	Currency	Currency		

Figure 5-2. *Granularity Matrix*

In this granularity matrix you can see that we will have six dimensions in our model: Customer, Product, Time, Plant, Employee, and Currency; and we will have four analysis areas: Sales, Finance, Customer Service, and Stock. Inside the matrix we locate for dimensions available that will be the level of the dimension that we will have for each analysis area.

Logical Model

Once we have collected all the requirements and put them together in our business analysis, we will be able to define which entities must be taken into account into our system. For that we will need to analyze in detail which are the concepts that our key users expect to have available to analyze, just validating that we have the possibility of obtain them from our source system. It seems to be quite logical that if they have some requirement of analysis it is because they are using it on the source ERP, but you shouldn't accept this as truth until you don't validate it. Sometimes you won't have the possibility to access the source ERP directly but to an intermediate staging environment where you have just extraction of some tables and some fields and you don't have access to the required ones to implement the related functionality. Sometimes your user wants to analyze the information at an aggregated level that they consolidate into an Excel file that you should include into your source systems.

In our example we will define different entities related to each dimension as you can see in Figure 5-3. Here we can see the sales entity that is referred to the information directly related to the sales process and different dimensions defined in the previous step with their related entities. In each of them the main entity will contain the join relation with the central sales entity, and the rest of them will be liked to this main entity of each dimension. As an example, Time entity will be related with Sales entity and Month, Quarter, and Year entities will be related with the Time one.

Dimension	Entities
Sales	Sales
	Status
Time	Time
	Month
	Quarter
	Year

Dimension	Entities
Customer	Customer
	Customer Country
Product	Product
	Category
Employee	Employee
	Employee Country
Currency	Currency

Figure 5-3. *Entities definition*

Also in the logical model definition we will define how those entities are related, if it is a one to one, one to many, or many-to-many relationship. In the initial proposal of the model that we are defining all relationships are one to many. When analyzing some advanced options we will see some exception for the relationship between sales and currency that will be many to many. The rest of the main dimension entity will be related with the sales entity with a one-to-many relationship, in other words, we will have multiple rows for a customer in the sales entity but only one row in the customer dimension entity. Similar to this, secondary entities of the dimension will have a one-to-many relationship with the main entity. Following with the example, we will have multiple customers in one country, so multiple rows for a given country in the customer entity but a single entry for a country in the Customer Country entity.

Finally, in a logical model we need to define which attributes are placed in each entity, in other words, which characteristics of the different business concepts will be available for each entity. Speaking about products we could think of color, size, category, manufacturing plant, or whatever other concept related to the product. Speaking about customers we can think of region, city, address, country, email, etc. For time dimension we can have day, quarter, week, semester, year, day of the week, month of the year, or other attributes related with time.

Dimensional Model

During the logical modeling process you will need to identify which facts we will need to take into account and which are our dimensions that can be considered as groups of related attributes with a given relationship among them. Sometimes this is considered as part of the logical model, but sometimes, especially in cases of big complexity, you can do it in a separate analysis getting as result the dimensional model for your data.

When defining the relationship types between attributes you will see that you can have the same relationship types than for entities, one to one, one to many, and many to many, but in this case we are taking into account the expected granularity for fields inside tables rather than granularity of tables. The clearest output for the dimensional analysis is a hierarchical schema as seen in Figure 5-4.

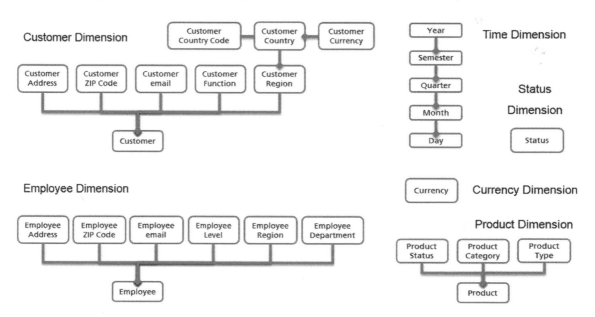

Figure 5-4. *Dimensional Model*

In this case the different relationship types have a similar meaning but with a conceptual difference. When you have a one-to-many relationship between two attributes, it means that, for example, in the case that Customer Region vs. Customer, you can have multiple customers that are located in a region but only a region assigned to each customer. In case of Customer Country Code vs. Customer Country, the relationship defined is one to one, so a country can have only a country code, and a country code can belong only to a country, a real relationship as far as this country code corresponds to the international phone code. Finally, we have the Currency and Status dimensions that in this initial version are defined as a single attribute dimension.

Also, added to this graphical analysis we should get the list of facts that we want to analyze in our project, with a description about source and conditions. We should get something similar to the one shown in Figure 5-5.

Entity	Fact	Description
Sales	Quotation Quantity	Quantity ordered when order is in Quotation status
	Ordered Quantity	Quantity ordered when order is in Confirmed status
	Delivered Quantity	Quantity delivered to the customer
	Invoiced Quantity	Quantity invoiced to the customer
	Invoiced Amount	Amount invoiced to the customer
	TAX	TAX applied to the invoiced amount
	Total Invoiced	Sum of amount and tax
Product	Product price	Product Price from catalog

Figure 5-5. *Fact list for Sales environment*

■ **Note**   Thinking of the next chapter, related to the implementation of the ETL process, we can advance as much information as we can when developing the data model. In this case we can specify more technical information if we have it available, such as source table name and field.

Physical Model

Finally we arrive to the physical model. This is the last step of this part. We have been talking with key users to understand and document their needs, we have defined which logical entities will be involved in our model and which is the relationship among them, we have been analyzing which dimensions and facts we will have in our models and the attributes used inside our dimensions; now we are going to perform the last step of data modeling before going ahead with the creation of objects in the database.

Especially useful in this step is the usage of a data modeling tool because after defining the structure of tables, columns, column types, primary keys, foreign keys, and so on in a graphical interface, we will have in most of them the possibility of generating creation scripts for all these objects to be executed into the database or directly the possibility of creating the related objects.

Inside our physical model we will define as main objects of the model which will be **tables** used to locate the information, and they will be directly related with **entities**, and each entity will be mapped to a physical table. Depending on the options available in our database we will have the possibility to define some parameters related to each table, such as partitioning, location, schema or owner, compression and multiple other features that will vary depending on your database software and also on the modeling software that you choose.

The next step will be defining which fields will contain our tables that we have defined on our system. They will be closely related with **attributes** of the logical and dimensional model. For fields we will define the name following our naming conventions; the field type, usually they will be numeric; character or date, with different subtypes depending on the database; we will define also the size and precision of the field and other parameters like if they can be null or not, among others, that also depend on the database and the data modeling tool.

With tables and fields we have the basis to define a working project but there are some other elements that can help us to improve performance and data integrity in our system. They are **indexes, primary keys, and foreign keys**. An index is a sorted structure that contains different values of a field and a pointer where this data is located, improving the response time of the system when we search for a given value. A primary key is an index that also defines the set of fields whose value combinations identify a single row in the table, it cannot exist the same repeated combination value of fields of the primary key in more than one row. In a lookup table of an attribute where we have an identifier and a description, the primary key would be the identifier as far as we expect to have only one row per attribute value, what will ensure you that you don't have any duplicated information if you do a select joining any table with this lookup. On the other hand in a fact table it is possible that we cannot define a primary key, it will depend on our design, but usually in datawarehouse environments you will have aggregated information at the desired level what will cause implicitly that your key is the set of columns that you have included in the group by clause.

Also we can have in our system **foreign keys**, which are the physical equivalence of **relationships** among entities. The usage of foreign keys has its advantages and disadvantages, so if you want to implement them you need to know in advance which problems you can suffer. The main advantage is that they ensure data integrity across tables. If in the sales table you have a foreign key by product identifier on the product dimension table, you will ensure that you don't have data related to a nonexisting product code, so joining with product table won't cause you any loss of data. A main disadvantage is that they can complicate the ETL process and also cause some slowness during the data load. When you have a foreign key enabled you cannot truncate the referenced table, neither delete registers that have information related in the dependent tables, something that seems to be logical but as we will see in the next chapter, sometimes the ETL process is easier to manage by truncating and reloading fully some lookup and dimension tables. Instead of that we will require the usage of insert/update statements or merge statements if your database and your ETL tool have this possibility. Also there are some database vendors that allow you to disable primary keys constraints to reload reference tables and enable them again once the load has finished. But again we will see these details within next chapter.

Data Load Strategies

Before going ahead with the physical model that we are proposing based on our example, let's comment something about data load strategies for our tables that we should have clear at this point, especially relevant if we mix it with the commented restriction that the usage of foreign keys could cause. We will have mainly three types of tables based on how are they loaded:

- **Manual/fixed tables**: They won't be included in any ETL process; they will have fixed information until we manually update it. They are useful for static or very low changing attributes, such as company, country, or in our model, currency lookup tables.

- **Incremental tables**: In this case we will upload new registers that appear in every load. Maybe we have only available new information to load, or maybe we have the whole history but we prefer just to load new rows instead of reloading the table from zero every time. In order to be able to have reliable information inside we will need to have a primary key of the table so we can detect which information is new and which is already inserted into our table. For existing registers we will have two options: update the old registers with new information for non-key fields or keep them as they were. The first one is the most usual option in this case.

- **Full load tables**: This is usually the quickest and simplest way to develop an ETL, and it is especially useful to reload lookup or dimension tables. It also can be required to use a full reload process for some aggregated fact tables if we can receive past information or if the aggregation is based on changing attribute relationships. If we have an aggregated fact table at the customer group and category level, it is possible that we need to fully reload this table if customers can change from one customer group to another or if the product can be reclassified across different categories.

In our sample model we will use a mix of strategies for loading tables, depending on the dimension. We will analyze them in deeper detail within the next sections, but as a general rule, we will have an incremental load for our fact tables, incremental load for most of the dimension tables, but cleaning some data and full load for lookup tables just keeping those keys that already exist in the dimension tables, in order to not have values with no related information.

Defining our Model

In our initial model we are going to implement a snowflake design with a low level of normalization. But, what does it mean? As seen in Chapter 1, the snowflake approach is the idea of having central fact tables accessed through some relationship tables that are linked to lookup tables as well or to smaller relationship tables. Table size is decreasing as I'm separated from the central fact table so it remembers to the fractal structure of a snowflake. As commented also in Chapter 1, normalization in a database model is used to minimize the space used in the database by avoiding repeating descriptions and relationship fields. We are going to propose a model that has some normalization but is not excessive.

For example, in the case of masterdata hierarchies, we are going to use a single relationship table to relate the key field with the rest of concepts instead of using every concept lookup to relate it with its parent. In case of a time hierarchy we will have the concept day that will be the relation key between the fact table and relationship one. Then in the relationship table we will have the relation between Day and Month, Quarter and Year. We could have the relationship between Month and Year inside Month table but we prefer to use a relationship table in order to reduce the number of joins required in the resolution of queries from the BI tool. In Figure 5-6 we can see the initial model defined to support basic analysis. Usually you will need to analyze all your user requirements to define a model that is able to solve all user queries, but in this case, as far as we act as our own users, we have done a reverse process. We have analyzed the information available on the Odoo model and based on that we have proposed an example model that we think can be useful, but this is not the usual procedure, as normally you will investigate in the source to gather the information that your user is requesting.

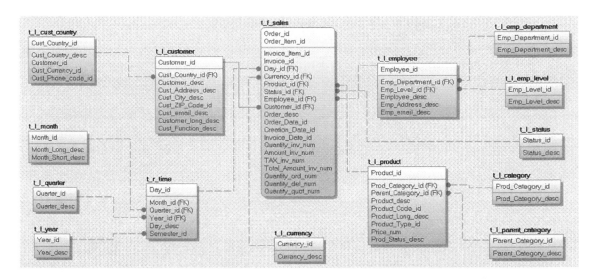

Figure 5-6. *Initial data model*

Defining a snowflake model will have some particularities that our users must accept and we need to have it very clear. When we aggregate through defined dimensions we will have a potential view of data, we mean, based on the current situation of the dimension, not in the situation that was in the moment of the fact. Based on our example, a product has related a single category; you cannot have a product belonging to more than one category. Based on this premise, product category changes will affect all the history of facts of the product. In the example of hardware stores, imagine that you sell to your customers poison for snails. In the initial classification it had *Chemical Products* category, but your company has decided to create a new category for *Garden* and they move the poison for snails from Chemical Products to Garden. The whole history of poison sales will be reclassified and the total amount of Chemical Products will decrease. This characteristic must be accepted and signed by the key user and the product owner in order to avoid discussions years later. You can think that it is not logical, your sales shouldn't change but one of the functionalities of your BI system is to have statistical information, KPIs evolution, and analyze trends that allow us to do forecasting and if you want to do an accurate forecast, you need to have the vision of how would be your historical data if you have the current situation of hierarchies; here we have the "potential vision" concept. But, why do we want to have a potential vision? Imagine that you are a food manufacturer with 20 food brands and you are thinking of selling a whole brand to get financial resources. In order to achieve the right decision you will analyze your historical data comparing your different brand results. Then, during the last six months, your company has been moving the different products that you have: chocolate with cookies from MyCookies brand to MyChocolate brand. If you analyze the historical trend by brand as it was, you will see that during the last six months your sales have been decreasing so you can decide to sell MyCookies brand. But if you are analyzing it with the potential view, you will see how the historical trend is using current hierarchies, so you can have a comparable view across brands and you will be able to make more accurate decisions.

Let's analyze with deeper detail the data model proposed to allocate our BI system and the reasons for every related decision.

Sales Dimension

As a central table for our model we have chosen Sales one as far as it is the most usual analysis to start when you are starting to implement BI analysis. In this central table we will have most of the fact fields that will be mapped to metrics and KPIs, related to amounts that can be summarized, such as quantity (number of units) and amount of the sale. Information will come from orders and invoices tables from our transactional, which we will consider generically as Sales Documents. We will define different statuses of every sales document, based on the sales status as shown in Figure 5-7. Also we will have different fields for quantities based on this status of the document that will be defined at line level as far as we can have some quotation that has some line that is not finally ordered, or the quantity of the order is lower than the quoted one, some of the lines of the order maybe are not delivered due to lack of stock or whatever other reason and maybe some of the delivered items cannot be invoiced because it has some defect. So quantities can differ for a given line of a document.

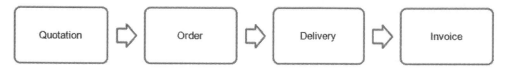

Figure 5-7. *Sales documents status*

The source of this table will be some ERP tables that will contain information about orders and invoices and also the detail of order and invoice items; detail about the ETL process will be analyzed in the next chapter. We will have some common fields informed since the first stage of a sales document, the quotation state, and other fields that will be filled up only when the sales document goes through the workflow. Also there will be some field that will have a different source depending on the status of the document as the Day identifier.

As common fields we will have the customer identifier, employee identifier, product identifier, and currency. Then from the initial status, quotation, that at the end it is an order that has not been confirmed, will have available the order and order line identifiers, order description, and the creation date. We will be able to inform also the number of products related to this quotation. In this status of the sales document the day identifier will be informed with the creation date, which is the only one that can be considered reliable in this status. Status identifier as you can imagine will be set up to *Quoted*.

Once the order has been confirmed, we will be able to inform the order date and the ordered quantity. Also we will update the day identifier with the value of the order date and we will change the status to Ordered.

Delivery status is optional in the Odoo platform, so for this example we won't take it into account in this table; we will just load the delivery quantity with the order quantity when we receive the order. In case that you have a workflow in your company that requires the management of the delivery document and take into account the real delivery quantity, you could just add required columns to this analysis.

The latest status that we will take into account into this process is the *Invoiced* one. When an order is invoiced, we will receive from the transactional system the invoice and invoice item identifiers, the invoice date, and the required numeric fields for our analysis, invoiced quantity, invoiced amount, taxes, and total amount including taxes. In this point we will update the status to Invoiced and the day identifier with the value of the invoice date.

Maybe the strangest topic that you see in this process is the combined day concept depending on the status. And possibly you will be right, it is a complication of the analysis but we have included it mainly due to two reasons. First to show you that sometimes we can have calculated fields that are not a direct extraction from our ERP, we have processing capabilities to implement derived concepts from the initial source. The second point is that we have the possibility to play with different dates and we want to do a complete analysis using a single field. Usually the most important date concept to take into account is the invoice date because it is the one affecting the official results of the company, but when we have only the *Ordered* status we won't have informed the invoice date, so if we filter through invoice date we will lose these rows in our analysis. We can use double or triple filtering using two or three fields, but we prefer this strategy to have the day field always informed with the most meaningful option in every status. But there are also some technical reasons below this. As invoice date is usually the most important field, most of the reports will be filtered by invoice date. In order to optimize the performance of the queries, one of the options, as we will see in Chapter 7, is the possibility to use indexes or partitioning using this date field, and in both cases it must be informed. For sure we can inform it with a dummy value in order to avoid nulls, but it makes more sense to use an alternative date field that will show us related information, and it can be used to extract the total product quantity invoiced in this month including also that orders that have been received but are still not invoiced by filtering only by one date field.

Regarding quantity fields, they will contain what the initial amount of the quotation was, which quantity has been finally ordered, which quantity we have delivered, and also which has been invoiced. In a perfect system all of them should be the same, but if we have all these quantities available we can analyze our selling process, by comparing discrepancies among all these fields.

Status Table

Different statuses of the sales document will be stored into a small manual table that we will fill manually. We consider it inside the sales dimension as far as it is an attribute directly related to the sales itself. Inside this table we will save just the status identifier and the status description. It will be mainly used to drill across available status inside the reporting tool in order to avoid the necessity of going to the fact table to reach them.

Currency Dimension

Currency dimension will consist just in a very simple table, just with a currency identifier and the currency description in order to show it with a long name in case it is needed. We will keep manually this table as far as it is expected to be a very slow changing dimension. In this case we prefer to keep it as a separate dimension, in spite of this it could be considered inside the sales one, because we will propose some evolution of this dimension when exploring other data modeling possibilities in the next sections.

Customer Dimension

In case of customers we will have two different tables that will allow us to define different customer concepts. The first relationship table will allow us to join with the fact table through the customer identifier and then we will have some fields related to the customer that will allow us to save some customer-related characteristics such as customer description, customer address, customer city, ZIP code, country identifier, email, a long description, and the function that this customer contact has inside his company. Also we will have available another table that will be the lookup of countries that will have the country code that will be the international code for countries, country name, the country code (two letters international code), currency of this country, and its phone number prefix.

The information source for this dimension will be also the ERP system where we have identified some tables that contain the customer and country relevant information. We will need to save information about all active customers and also historical ones that can be inactive but that have sales in our system, if not we could lose information once the customer is deactivated. We need to ensure data integrity, that all possible values for customer are included into the customer table, and also that we don't have any duplicated customer identifiers, so if there is some field that changes in the customer table, we need to modify the existing register, we cannot add new rows related to an existing customer identifier. Also in the next sections we will talk about adding geolocation information that will allow us to locate our sales information across a map.

Employee Dimension

In order to locate our employees' data the main dimension table will follow a simpler structure than our customer dimension; it will contain employee identifier, description, email, and home address and also the department to which he belongs to and the level that he has, which can be from Trainee to Chief Executive Officer in the sample data that we have in Odoo.

As part of the main dimension table we will have two small tables that will be lookup tables, one for department and the other for level, as far as in the previous table we will have only an identifier of both department and level.

When loading the dimension table we will do it in an incremental way, updating existing rows and inserting new ones and for department and level tables we will truncate and insert data based only on existing values of the t_l_employee table.

Product Dimension

Our product dimension will contain all the information related to the products that we are selling in our business. In fact, as explained for the customer table, it must contain all the products that we have in our catalog and also the ones that are out of catalog but have some historical sales. So we need to ensure that all the products that we have in our facts table are inside the product relationship table to ensure data integrity and that there are no duplicated rows by product identifier.

Regarding fields of the product hierarchy we will have available the product identifier; that it is an internal ERP numeric field; as key of the table, we will have the product short and long descriptions, the product code used to identify it in the ERP interface, the category identifier that will allow us to join with the category lookup table, the product type and status and the price of the product. We have defined also another table as category lookup that will contain the category and its description. We would like to remark in this dimension that we can have a potential view about some concepts like the category, as explained in the introduction, but also you can have a potential view of some facts, such as the amount of sales applying the current price to the historical data.

Time Dimension

One basic dimension present in almost all models for data analysis is the time dimension. Depending on the BI tools you can find that this dimension is not created physically in the database but calculated with own BI functionalities or database functions based on the date field present in the fact table. In this case we have preferred to create it physically in the database as far as we don't fix any specific BI tool or database; we just suggest some and maybe the one you choose doesn't allow you to implement this feature so we will go ahead with this physical definition just in case.

In this example we have defined a basic time dimension with five attributes: day, month, quarter, semester and year, but we could use also other attributes like week, day of the week, week of the year, month of the year, and others. We will use numeric fields to locate date attributes because it facilitates the maintenance at the database level, but we will have available the date description in a date format in order to be used in the BI tool that usually has date transformation capabilities.

Apart from these topics commented, in a time dimension we can have multiple particularities, such as the way of defining the week and the relation of the week with year and month; or your company could use a parallel fiscal period different from the natural year and natural month and you should require implementing an alternate time hierarchy based on this fiscal period concept.

Exploring Data Modeling Possibilities

Until this point you have seen how to model a simple snowflake structure with direct relations among tables that will allow you solving aggregation queries across some attributes. Nothing that you couldn't have defined into an Excel file using some formulas to link related information. Now it is time to see some other features and possibilities that could help you to do more advanced analysis in your BI system giving you powerful options and flexibility in the analysis.

View Layer

Based on our recently seen physical model, let's suppose that we want to analyze sales based on Order date field, defined as a numeric field. But in our BI model we have the possibility of using a lot of functions based on date fields such as operate with dates, get the first day of the month, or create a dynamic filter that gives you the accumulated sales in the last three months. We could use a database function to transform the field into a date type in order to get the whole functionality inside the BI tool, or create a field that retunes the year or the month. And we can do that using a database view, without saving the data calculated into the database, just using some function at runtime. In Figure 5-8 you can see how from a single numeric field in a table containing dates in format YYYYMMDD we can create multiple date-related attributes.

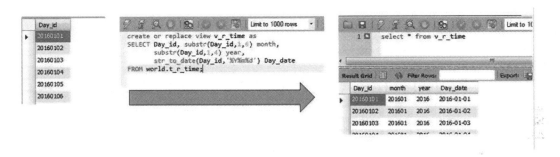

Figure 5-8. *Using views to add nonexisting fields*

Using a view layer between the physical table and the map done in our BI tool will allow us to have some flexibility to add some extra logic below our BI system. We can force a join with a parameter table to improve join performance, we can use it to adapt some field name or precision to align it with other tables in the environment, we can apply a filter in the code to remove some undesired data from all the analysis that are based on that table. Possibilities that offer the usage of an intermediate view layer are multiple but you should use them carefully in order to not affect the system with some undesired effects. You can use views, for example, to filter the returned values applying any condition that you can think of, but we have seen the example of a lookup view for an attribute that returned three rows from nine that existed in the table because they were the used values, but to see only the values that had related data in the fact table, they were joining with the whole fact table that contained millions of rows. So to show only three of nine rows the view was querying millions, with the logical loss of performance.

Also by using views you can add complexity to the whole solution, having another layer where you can add conditions, calculations, and formulas it is possible that you feel lost in case that you are require to analyze some incident. Also based in a real case, we inherited management responsibility of a running BI platform and we were required to analyze a mismatch between source system and the BI report. We analyzed the whole ETL process searching in all the steps were we could lose the information, until we saw that the information was properly loaded in the final table. However, the information was not matching so we analyzed a very complex report that was doing multiple steps validating so we didn't lose any register in any SQL join. Finally we realized that somebody had set a wrong filter in the view code so data returned was different between the table and the related view.

Materialized Views

Talking about views, most of the databases have specific types of views named materialized views. These types of views are currently a table that stores information but based on a select statement that depends on other tables. So they are similar to a physical table where we insert the result of the related select to have it precalculated, at the end of the day they can just be considered as a final ETL step that must be executed when the dependent table contents have changed. The main advantage of these kind of objects is that they have related some functionalities at the database level that allow you to refresh the information with simple statements, depending on the database system you can do just partial refresh to improve refresh performance, and you have the flexibility of creating or modifying them based just with a select statement like standard views, so you have some advantages of both object types, views, and tables.

Data Split

Sometimes we can be in the situation that our users require some data that doesn't exist at the desired level in the source system and we are forced to "invent" this information. This seems to be something out of Business Intelligence and more related with magic but we can ensure you that you can be in that situation. Let us explain it with some examples.

When we are talking about inventing we don't refer to just adding random data to a table, we refer to create some model that allows us to analyze information at a different, more detailed level than in the source of information. Imagine that we have a budgeting process defined in our company that it defines a set of sales objectives that the company should achieve to get a reliable state for the future. In these kinds of budgeting processes it is normal that you don't have the information defined at the lowest possible level, taking as example the product hierarchy, you won't have the target defined for any single product; you will have the expected sales by category for the next year. But then, during next year you will want to compare how are you advancing in your target, so you can split the overall target for the category across the different products under this category based on previous year sales, precalculating a split ratio that allows you to see data at the desired level. In Figure 5-9 you can see an example showing the required structure to implement this model and some sample data showing how it should be defined. In this example we show a structure that allows us to split the sales objective per category to sales objective per product, based on a company that has four categories and for simplification reasons, we analyze two of them assigning just two products inside each category. In order to calculate the ratio, we have created a view that calculates the sales at product level for the whole previous year and assigns it to this year. Also this view does an aggregation at category level in order to be able to calculate the ratio that we will use to split the objective. As result our BI tool should be able to join the target table and the ratio view and get the expected sales per product for the current year.

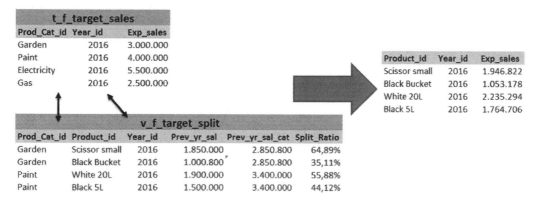

t_f_target_sales

Prod_Cat_id	Year_id	Exp_sales
Garden	2016	3.000.000
Paint	2016	4.000.000
Electricity	2016	5.500.000
Gas	2016	2.500.000

Product_id	Year_id	Exp_sales
Scissor small	2016	1.946.822
Black Bucket	2016	1.053.178
White 20L	2016	2.235.294
Black 5L	2016	1.764.706

v_f_target_split

Prod_Cat_id	Product_id	Year_id	Prev_yr_sal	Prev_yr_sal_cat	Split_Ratio
Garden	Scissor small	2016	1.850.000	2.850.800	64,89%
Garden	Black Bucket	2016	1.000.800	2.850.800	35,11%
Paint	White 20L	2016	1.900.000	3.400.000	55,88%
Paint	Black 5L	2016	1.500.000	3.400.000	44,12%

Figure 5-9. *Data split example*

Fact Normalization and Denormalization

We have already commented that our model is following a snowflake structure with some level of normalization. In the introduction chapter we already saw the difference between a normalized and a denormalized structure, and we have already seen that normalization helps us to keep integrity while denormalization provides a better performance for SQL queries. But now we want to talk about fact normalization and denormalization. In case of normalization we change the structure of the table to have as few numeric columns as possible, if possible with only one fact column, and then we define different attributes that allow us to split the information changing columns per rows. As usual it is easier to show it with an example.

In this example that we can see in Figure 5-10 we are showing two levels of normalization: the first one defining a new attribute named Sales Document status that can be Quoted, Ordered, or Invoiced and assigning to Quoted Status columns quoted quantity and quoted amount, to Ordered status columns ordered quantity and ordered amount and to Invoiced status columns Invoiced quantity and invoiced amount. In the second step shown in Figure 5-10 we have created the concept metric and moved it to rows, keeping just a column with the value of the metric.

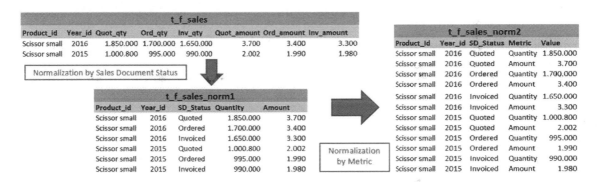

Figure 5-10. *Normalization example*

Now you can be thinking, why do I need to normalize? There can be different reasons that can be justified depending on your business requirements. When using Business Intelligence tools, sometimes it is easier to manage the attribute metric than multiple metrics. In the next section we will see an example of phase transformation where it is useful to have the attribute metric to use it as a selector.

Another benefit of using normalized structures is that it is more flexible to add new values of the normalized attributes. Imagine that you want to add a new metric with the cost of the product in the table t_f_sales of Figure 5-10. You would need to modify the table structure adding the new column and all the required dependent objects, such as views, BI model, ETL process, etc. If you want to add more data to the table t_f_sales_norm2 of the same example you just need to add more rows and no modifications in the model will be required. Just adding more rows in the ETL process and you will have your data available.

Also it is interesting to use normalized structures if you have fact columns that are sparsely filled up and you want to avoid nulls in the visualization. A typical scenario of this situation is in highest detail financial analysis where you have a lot of cost types related to invoices. You will see that a single invoice has related some standard costs derived from your product manufacturing but then some of them will have promotional discounts, others will have coupons, sales incentives, rappel agreements, etc. So in a single row you will have only few fact columns informed, but from a relationship point of view, you will have the register that relates the invoice with the cost, in spite of it is empty, what can cause some invalid analysis if you try to analyze the number of customers with rappel discounts (yes, you can filter by cost <> null, but you need to remember it). If in this example you change the structure creating the cost type attribute and only a column for value, you will have only registers related to rappel for those invoices that have rappel informed, not to all invoices.

You can also denormalize a table by doing the reverse process. The main question that you can be thinking of could be, why? There are some reasons that you can justify apply a denormalization. First is reducing the number of rows of a table; what is especially interesting is if you have a high density of the facts. If you have the same table keys for an attribute you can move the attributes to columns creating a column per each relationship fact-attribute value. Imagine that you have four product categories available and you are selling all four categories to all your customers. If you create a column for each category you will have a quarter of rows in the final table.

Another interesting feature of a denormalized strategy is that you can do comparisons between different attribute values. Following the example of categories, you can have a column that summarizes all categories and a percentage with the value of each category over the sum of all categories. Of course, if normalizing your model provides you with more flexibility, denormalizing it causes stiffness, so in this example if you require adding a new category you will require modifying the model to add new columns and the related dependencies in your BI system and ETL process.

Time Transformations

Drilling through modeling possibilities we arrive to the transformations time. The main objective of a time transformation is to show information related to a given month against other months' information. As it is becoming usual in this section, again the question is, why? Having a transformation attribute allows you to easily compare different periods of time in a single row or column. In Figure 5-11 you can see the transformation table that is joining the fact table using source_mth_id column and the month_id column of the fact table. The reporting tool is getting the month attribute from Month_id column of the transformation table and adding the invoice_qty column from the sales table. In the transformation table we are showing just data for February 2016 but we should fill it up for all months of the year.

t_r_time_transformations				t_f_sales		
Month_id	Source_mth_id	Transformation_id		Month_id	Category_id	Invoiced_qty
201602	201602	Current Month		201501	Garden	30
201602	201601	Previous Month		201502	Garden	35
201602	201512	Previous Month - 2		201503	Garden	30
201602	201502	Previous Year		201511	Garden	40
201602	201602	Year to Date		201512	Garden	50
201602	201601	Year to Date		201601	Garden	32
201602	201502	Previous Year to Date		201602	Garden	38
201602	201501	Previous Year to Date		201603	Garden	33

Invoiced Quantity		Transformation					
Month	Category	Current Month	Previous Month	Previous Month - 2	Previous Year	Year to Date	Previous Year to Date
201601	Garden	32	50	40	30	32	30
201602	Garden	38	32	50	35	70	65
201603	Garden	33	38	32	30	103	95

Figure 5-11. *Time transformation example*

This approach of time transformation is launched by the reporting tool, so you don't have to have the calculations precomputed in the database, which is the previous month sales or the year to date sales for a given customer, and it is something that is calculated when you launch your report that executes the SQL query. You ensure in this way that the calculus is refreshed at that time, but it is possible that you suffer some performance issues as far as you are multiplying the accessed rows, for example, in the case of Year to Date for December you will access to 12 rows to get a single one. You can also think in a model that allows you to do time transformations in a more effective way by precalculating the information in the fact table, so you could add columns to your fact table and add there the calculation for the previous month, previous year, etc. In order to do that, you cannot use a high-detailed base fact table, having a look on our base model, if you have the invoice identifier in the table it makes no sense to calculate the previous year invoice amount, because this invoice code is unique and won't have information for previous periods. You can precalculate these kind of analyses only at aggregated levels.

Fact Phases

When we are referring to fact phases we are talking about different views of the same fact. So, for a given fact as invoiced quantity you have the current value, the annual forecast from beginning of the year, the mid-year forecast review, target defined by sales responsible or comparisons among them, both as difference and ratios. You can also include time transformations inside this section as far as they could be considered as phases of the fact. In order to model them we have different alternatives:

- **New tables in the model**: You will need to use this strategy if the different phases of the fact are introduced at a different level, for example, in case of forecast facts where we usually don't have them at the highest level of detail. Also this new table could contain the existing fact in an aggregated way, in order to improve performance. We will talk about performance optimization in Chapter 7.

- **Add columns to existing tables**: Using this denormalized strategy you will get better performance but you can complicate the ETL process and you will require a BI tool schema modification.

- **Add rows to existing tables**: You can lose performance with this strategy but maintenance is easier.

At the end of the day, phase is such a new attribute, so you can use the same strategies than for the rest of attributes, normalized using it as attribute or denormalized model creating new facts for each attribute value.

Real vs. Potential

The base model defined allows us to have a potential view of data as far as it is following a snowflake structure. As far as we have a single version of a product, if we change the product category of a product the whole data history related to this product will change. But we can also have the possibility of saving the real view, which was the category of the product in the sales time, by saving the category in the fact table. It is important to save only relevant fields for this real analysis as far as it will use a lot of database space, we can have millions or billions of rows in our fact tables, so you will save having to repeat much information there if you add all your attributes to the fact table.

In order to choose which attributes are relevant for you it is required to evaluate different topics. You can save some attributes related to the sale, but maybe you are require to save also some hierarchical dependency to be able to do the analysis that you want. Let's retake here an example from a real case that we have found in some customers that we explained in Chapter 1, when we were talking about datawarehousing. We have been working in a project to analyze the sales force performance, how many customers were visited by each sales force member, how long they were a customer, activities performed for each customer, and so on. We had the requirement of having two views, real, analyzing who had visited the customer; and potential, analyzing who had the customer assigned. Within the same month, they were usually the same but over time it can change. The main problem that we found there is that we had available which was the employee that visited that customer to develop a given activity, but we hadn't saved which supervisor he had assigned; or to which area he was belonging, as far as we had only in the fact table the information about the employee, but not its hierarchy.

You need also to think about saving descriptions or not. Maybe you want to analyze data with the description that it had initially and not the one that it can have now, if it has changed.

■ **Note**   You need to be careful if you have the hierarchy information loaded into the fact table, because it won't be coherent, so you cannot define one-to-many parent-child relationships using this table, because at some point of time the product will belong to a category and after some time it can change, so the relationship won't be one to many. Also you need to be careful about descriptions; if a single attribute identifier has different descriptions it is possible that you suffer some inconsistence in your BI tool in spite of that the SQL generated can be correct.

There are also some risks by adding multiple options for your users, one of them is that your users have not the correct training to use it and they can doubt regarding the real meaning of both views, so some advanced options should be clearly defined in your documentation and in the training courses that you should impart to your users.

Historical Hierarchies

Related to the previous section you can also define some historical hierarchy tables that allow you to do have an intermediate solution between real and potential, by doing an approximation to real attributes without saving hundreds of fields in the fact tables. Inside a historical hierarchy you will save hierarchy status with the defined periodicity, daily, weekly, monthly, quarterly, or yearly; and by defining a relationship of key attribute and time attribute together with the fact table you will be able to see how the hierarchy was in a given period. The standard procedure is to have a base relationship table and then copy it each period saving an entire "photo" of the table each time.

In order to define which period you will use for the hierarchy you will need to balance accuracy of the hierarchy assignment with volume that you want to keep and resulting performance. It will depend on the size of the base relationship related table but most of the time having a monthly basis is enough accuracy for your users.

You can define also a strategy of date_from/date_to information where you have validity of each row based on the date when it started to be defined until the date when it ended. The main problem in this case required joins are not direct and you need to evaluate the proper hierarchy with between clauses, which can affect the performance and can complicate the BI model, especially if you define it inside the BI tool. In this case we would recommend using views to define the between relationship among tables.

Multiple Views Using a Table, Entity Isolation

There are times when you have similar attributes that can have the same possible values, so they could use the same table as lookup but they are different concepts. We can think of geographical areas, such as regions, countries, and cities, which are the same for customers and employees but they have different meanings. It won't be the same analyzing sales per customer region than sales per employee region.

In our base model we can find an example with fields that contain date information. We have defined multiple date fields, creation date, when the sales document is created; order date, when the sales document is ordered and invoice date, when the sales document is invoiced. They are currently defined as information fields, as far as there is no link with the time dimension through these fields, so you cannot aggregate with the current definition to upper time levels such as month, quarter, or year. In order to do that you will need to define time entities for all these date concepts; you cannot reuse the same time, month, quarter and year entities to define multiple attributes because concepts are different, and your BI system could perform undesired joins.

On the other hand it makes no sense to have different tables because they will contain the same information related, so in this case we recommend you to use the view layer to define different views over the same tables as shown in Figure 5-12. There you can see the example for the main relationship table but it should be done for all tables related with time in the model, t_l_month, t_l_quarter and t_l_year.

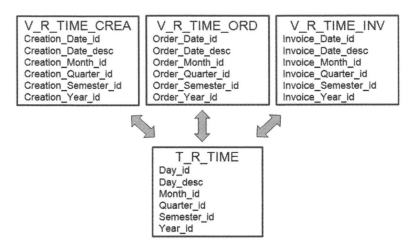

Figure 5-12. *Multiple time entities view layer*

Modifying Existing Structures, Adding More Sales Document Status

We have defined four statuses in the sales document workflow: Quoted, Ordered, Delivered, and Invoiced. We could complicate a little bit this model by adding two more statuses to this workflow, Paid and Returned, as far as Odoo allows us to save the payment and, in case of need, the refund of the money to the customer. This will add some complexity to the model and new columns to save payment date, quantity paid, amount paid, return date, quantity returned, and amount refunded. But also this change will allow us to analyze new KPIs such as the average payment days KPI or the return ratio.

We won't include this either in the base model so we don't complicate the example we are going to implement in this book, nor in the ETL part in order to not extend too much that chapter, but it is also interesting to understand that the model that we define is not static, it can be modified any time that you need. Just take into account that if you are going to modify a table that is loaded incrementally, you will need to reload the whole table if you want to have complete information for the new columns and the correct status informed for old invoices.

Currency Transformation

We have defined a small entity for currency just to have the description of each currency. But let's suppose that we are based in a multinational environment and we have transactions in different currencies. In order to be able to do an aggregated analysis of total amount of sales for the whole company, you will need to have all the information in the same currency.

From a data modeling perspective the easiest way to have this in place would be to have all the information in a single currency by transforming it in the ETL process, but maybe in some moment of time you are interested in seeing it in a different currency, so we are going to propose that you have some modeling options to cover this requirement depending on the particularities that you need to provide.

Potential currency: The simplest way to do the transformation is to define a table that contains source currency, target currency, and exchange rate. This exchange rate can be the latest available or an average of a defined period such as current month, previous month, current year, last year, last 12 months, or the whole history of the datawarehouse. The structure to be able to perform currency transformation with a potential view is shown in Figure 5-13.

t_f_sales				t_f_currency exchange		
Day_id	Currency_id	Invoiced_amount		Currency_id	Target_Currency_id	Exchange Rate
20160701	EUR	30		EUR	EUR	1,00
20160702	EUR	35		EUR	USD	1,09
20160703	EUR	30		EUR	GBP	0,89
20160704	EUR	40		EUR	CNY	7,36
20160705	EUR	50		EUR	JPY	112,97

Amount	Currency Transformation				
Day	EUR	USD	GBP	CNY	JPY
20160701	30,00 €	$32,70	£26,70	¥220,80	¥3.389,10
20160702	35,00 €	$38,15	£31,15	¥257,60	¥3.953,95
20160703	30,00 €	$32,70	£26,70	¥220,80	¥3.389,10
20160704	40,00 €	$43,60	£35,60	¥294,40	¥4.518,80
20160705	50,00 €	$54,50	£44,50	¥368,00	¥5.648,50

Figure 5-13. *Potential currency exchange*

Real currency: The real currency approach is similar to the potential one but adding the day in the currency table. In this way the join with currency table is done including the date, so you have the exchange rate that was official when the bill was invoiced.

Currency analysis has usually many variables in place. A potential approach using a single exchange rate is useful when you are analyzing business trends and you don't want to have affectation in your analysis of the changes on currency exchange rates. During 2014, the Dollar/Euro rate went from 0.72 to 0.95, so you could think that your U.S. companies were performing better during this year if you analyze data in Euros or your European companies are performing worse if you do it and then analyze it in Dollars. So if you use a static exchange rate for the analysis you will avoid this effect in your sales analysis. You need to take into account where your bank account is located and which is the currency used. If your bank accounts are located in the source country with country currency, whenever you want to consolidate money into a single account, you will apply the exchange rate that you have in that moment, so it makes sense to analyze data with the current exchange rate.

On the other hand, if you are getting money in the target currency as soon as you are paid because your bank account is using target currency, it will make sense to have a real currency transformation analysis. Another variable to take into account is the date field that you use to do the exchange. In order to have an accurate analysis you should use the transaction date, when you customer pays you the bill… but in this case we don't have it available in the original model, so the best option should be using the Day (remember, that combination among Creation, Order, and Invoice date depending on the status), at least if we don't add the payment date included in the previous section. Having a potential analysis also allows you to take into account the overall situation, evaluating if the exchange rate affects to your business. Something especially interesting if you have a business with multiple international transactions.

Regarding data contained in the real table we need to ensure that we have information for all the days; otherwise the join with this table could cause us to miss some data. We need to define also where are we getting this information because it quite possible that it is not available on your ERP system so one proposal could be a web parsing process that explores a public web to gather currency exchange data. Or even better, pull the data from some web service directly inside our ETL processes.

Geographical Information

Geospatial reporting and mapping capabilities are interesting features of almost all BI tools in the market, more or less integrated with the software, and most of the tools provide some way to show information graphically in a map. But in order to be able to show information in a map you will have some requirements regarding data that you should meet, a way to allow your tool to understand where it must locate figures or draw areas or bullet points.

You will have mainly two possibilities: your tool is able to geolocate the information based on a description or you must provide longitude and latitude to your BI tool, in other words, your model must allow saving this information into your database. In order to use the first option you must ensure that the descriptions that you are going to use to locate information into the map match with the expected ones by your tool; otherwise it won't be able to locate it. Regarding the second possibility we will have two main options: get the information from the source ERP or CRM systems if our sales force has some mobility tool to gather geospatial information or use some geocoding tool to get this information based on written addresses. You have multiple options in the market for geocoding, Google Maps geocoding API, ArcGIS, Mapbox – and it is quite sure that in the near future you have multiple other possibilities for that.

Data Modeling Tools

There are multiple modeling tools that can help you in your modeling process. In this section we are going to talk about two of them, Erwin DataModeler and MySQL Workbench that we are familiarized with and that we consider interesting also in their free edition. If you have the budget to go for a commercial edition of them you will have support and multiple added features that will facilitate still more your tasks, but with the free edition of them you will have also enough functionalities to go ahead with the current modeling scenario.

As usual during this book we need to warn you that this won't be an exhaustive exploration of data modeling tools capabilities, just an overview about basic functions that we think you will need to define your model.

Erwin DataModeler

Let's start with Erwin DataModeler as far as it has been the one used to develop our base model. This software has different versions, community edition for free and some commercial ones. As you can guess for this project we are going to use the community edition. In order to install it you can find it at `http://erwin.com/products/data-modeler/community-edition/`, where you have the Download Now link on the right part of the page. Download the proper version, 32 or 64 bit for your laptop and launch the installation of the executable file. Installation can be done by clicking next to all screens, as far as default components that are installed are the ones that you need to develop the logical and physical model.

Once installed you will be able to open the tool, available if you have kept the default path into Start button ➤ CA ➤ Erwin ➤ CA ERwin Data Modeler r9.64 (64-bit), at least if you have a 64-bit computer. Once opened, model creation starts by clicking on the New button, the first one in the top left. Then you will be prompted with a menu as shown in Figure 5-14.

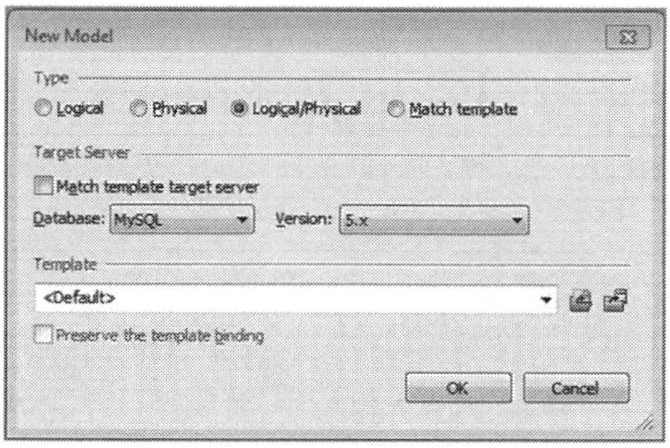

Figure 5-14. New Erwin model options

In this step you will be able to choose the model type that you want to implement (in this case we will choose Logical/Physical to create both types); the database that you will use, as far as once defined the model you will have the possibility of getting the creation code for all the required tables (we will choose MySQL, as far as MariaDB uses the same code, it is a continuation of MySQL); and some templates options as far as Erwin allows you to save templates to facilitate the creation of new models.

In Figure 5-15 you can see the layout of Erwin, having the menu bar in the top of the screen, two button bars just below, and then a model explorer where you can see model objects, and the design panel where you will define your model. In the bottom part you can see three windows: Action log where you can review the activity done in the model, Advisories with warnings and messages; and the overview section where you can see an overview of the whole model and center the window in the section that you are required to modify.

Note that by default you are designing in Logical model, but you can change it in the Logical/Physical selector that you have in the button bar, as you can see in Figure 5-15.

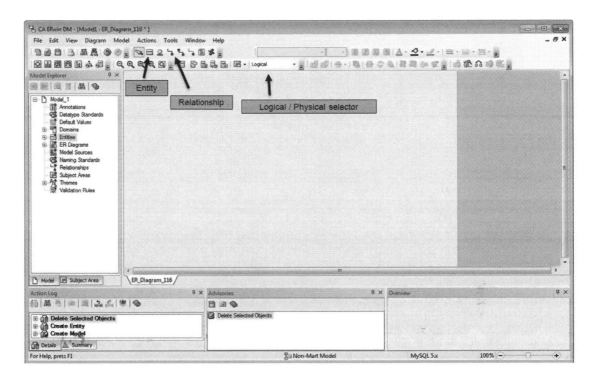

Figure 5-15. *Erwin modeling tool layout*

To draw the model we will use mainly the buttons Entity and Relationship. We will choose Entity (see Figure 5-15) and then by clicking in the design area we will add a new entity. It will set a name as default, E/1, E/2, E/3... You will be able to rename it with single click over the name in order to set the proper name, or with a double-click you will open the entity editor tab that will allow you to rename all the entities in a single screen, and also you will have multiple tabs for each entity to change some properties, some that will affect the table creation script like Volumetrics; others will affect the format in the design area, such as Style or Icon, and also you can have information about changes done in the entity with the History tab. Inside the Entity editor you will be able to add entities just by clicking on the new entity button, as you can see in Figure 5-16. Once we have all the entities defined in our model we will add the relationships that will link our tables. To do that you need to click on relationship button (one to many or many to many depending on the relationship type) and click from the smallest cardinality table to the bigger one (from the "One" to the "Many"). As example, in our model from Quarter entity to Time or from Time to Sales, etc...

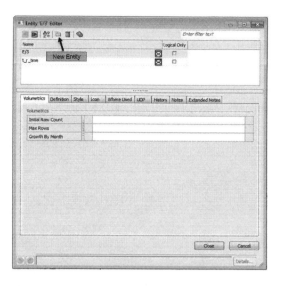

Figure 5-16. *Entity editor*

Once you have defined entities, you can open the entity to define the different attributes that will be related to columns in the physical table. In fact, when you add a new attribute you will need to choose which field type will be used.

In order to open the Attribute editor to define the attributes right-click on the entity and select Attribute properties. It will open the attribute editor for the selected entity where we will define the attributes that will contain our entity including the physical type that will be used in the creation script for the database. In Figure 5-17 you will have the attribute definition for t_l_customer table where we can see the different attributes and which of them will be key attributes, marked as Primary Key.

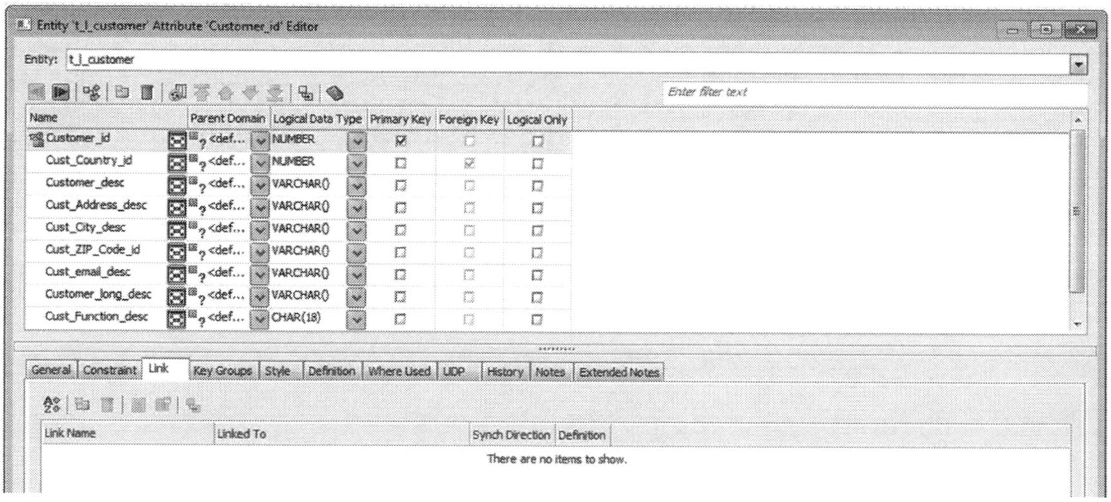

Figure 5-17. *Attribute editor for t_l_customer table*

When you define an attribute as a primary key, it automatically is defined as a foreign key in the rest of tables that contains this attribute and it is also marked as a primary key of related tables (in our example it marks customer identifier as a primary key in the Sales table). In case that this primary key definition is not correct in the related table, you can remove it from the attribute editor, in this case, t_f_sales table.

After defining all entities and relationships you will arrive to the model shown in Figure 5-6 inside previous section "Defining our Model."

If you want to change some physical characteristic of columns or tables, you can switch to Physical view, remember the Logical/Physical selector of Figure 5-15, and then you will have the option to open table and column properties instead of entity and attribute. It is especially interesting that the column properties tab, MySQL subtab, where you can define the exact column type, if the column is nullable or not and some other options that are MySQL specific.

Once your model is finished based on your requirements, you can easily generate a script that creates the model into your database, or directly link the Erwin model to a database and then create the required schema directly. To do that you need to be in the Physical view and go menu Actions ➤ Forward Engineer ➤ Schema, as shown in Figure 5-18.

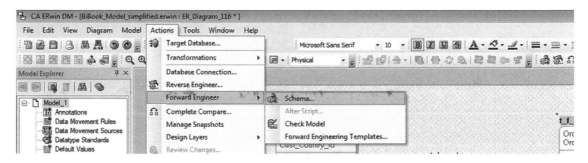

Figure 5-18. *Generate physical model*

With this option selected, a new menu opens where you can check which objects you want to generate and you can choose then if you want to preview the script, which will allow you to execute it wherever you need, or if you want to generate it directly, option that will require that you have connectivity to the database in this moment. In Figure 5-19 you can see this menu with multiple options to select object types, script options for each type, summary of actions that will be done, and multiple buttons to filter the objects, report what is going to be done, etc.

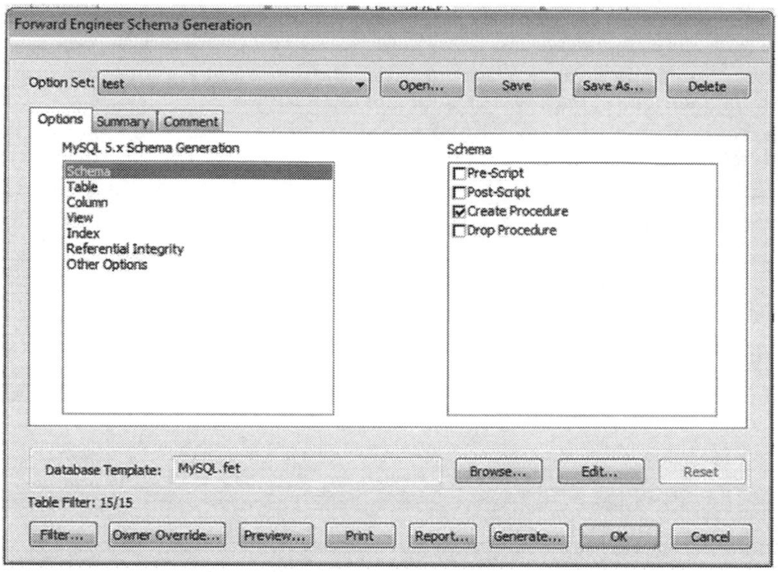

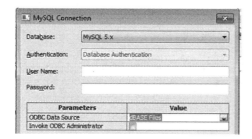

Figure 5-19. *Forward Engineer Schema Generation options*

If you select the Generate option a new menu will open asking you the connection parameters to the database, including database version, username and password and ODBC used, as you can see in Figure 5-20. Once you click on the Connect button, it will start database objects creation.

Figure 5-20. *Database connection*

MySQL Workbench

We will quickly review the MySQL Workbench tool that will help you to create diagrams, tables, and views in a friendly way, but it is dedicated only to MySQL databases (also to MariaDB as far as it is an evolution of MySQL), so we won't spend too much time on it. Almost all database vendors have its own editor, so it would be impossible to cover all options that you can find in the market for data modeling. But as far as MariaDB database allows us to use MySQL Workbench, we think that it is an interesting tool to know about. So let's try to briefly explain how to install it and some quick steps about how to develop models with MySQL Workbench. MySQL Workbench is available for free, but in order to get support if you require it, you would need to go for MySQL Enterprise support, so in this case it can be out of your budget, of course, depending on your budget size.

MySQL products have usually the Community edition that you can download and use it for free, and then if you want to have support from Oracle Company (that recently bought MySQL), you will need to pay a support fee. In a production environment we hardly recommend you to purchase support for you database, but in case of workbench this support is not so critical.

If you choose MySQL Workbench as your modeling tool option, the first step should be download it. To do this you can access directly to the download page, `http://dev.mysql.com/downloads/`and ensure that you are under Community downloads, in order to nit use a commercial edition that you should pay for. Once downloaded and installed you will be able to access to the tool, by default located in the Start button ➤ MySQL ➤ MySQL Workbench 6.3 (or whatever version that you download). If you click on the menu File ➤ New Model you will be able to create a new model. Once it is done you can add tables to the model until you add all of them, as you can see in Figure 5-21.

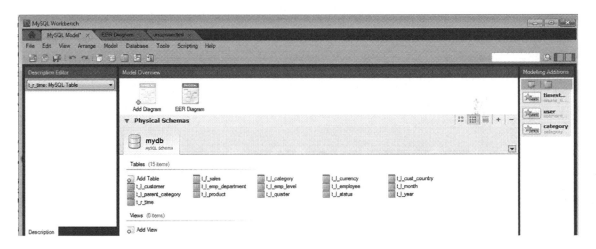

Figure 5-21. *MySQL Workbench model screen*

Also in this case, as far as the script was already created with Erwin, we have imported it. You can do it with the menu File ➤ Import ➤ Reverse Engineer MySQL Create Script. Once you have all tables in your model, you can create the Diagram that allows you to have a view of your model, as done in Erwin, defining also which tables are related by linking them with 1-N relationships, using the 1-N relationship button. You can see the result in Figure 5-22, as you can suppose is quite similar to the one defined in Erwin.

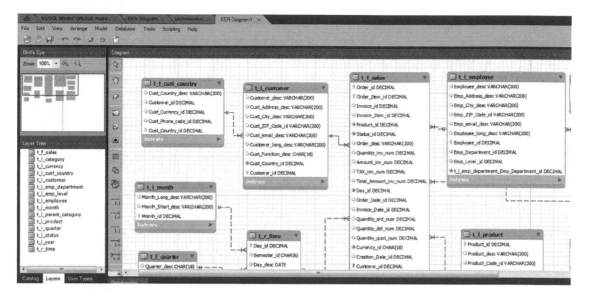

Figure 5-22. MySQL Workbench diagram editor

Preparing the ETL

If you have followed step-by-step instructions in this book, at this time you will have a database structure defined, and if you have used some of the modeling tools that we have just evaluated, you will have easily created this structure in the database. The next step should be fill them up, but in order to do that we need to know which source we have available for that. We will analyze the process for filling them up inside the next chapter, but before jumping there we are going to investigate the location of the information.

In our humble opinion, the best approach to do that is a top-down strategy, identifying source systems, source tables, which are the join among them if we need more than one source table to load our target table; and finally which source fields we will use and if there is any formula or calculation to achieve the target field. Also it is important in this step to provide a summary about load strategy for each table or any other condition of filter that we should take into account to fill up the table properly, such as which primary key we should use or which integrity checks we should perform either with a foreign key or with check processes.

Source Systems

In our case this task is quite easy as far as all the information that we will use in our basic model will come from our Odoo ERP system, so this task has been quickly done. But usually it is not so easy. We can have different modules of the same ERP system, Sales, Financial, Operations, Logistics, Human Resources, etc. Also we can have multiple sources as commented in the alternative models that we have explained in previous sections, such as web parsing to get the daily currency exchange rate, using a geocoding platform to get the latitude and longitude information, and we could think of many other systems such as CRM tools, external information providers, official statistics, flat files, csv or Excel files, or any other possibility that we could think about. It is interesting also to have available connectivity parameters, such as server name, database, port, username, password, file name, file location, file format, update frequency and also contact information for troubleshooting or doubts that could appear to define the ETL process.

Source Tables

We have been explaining how to load most of the tables and this information is also useful to include it in a summary table that will allow you to check easily where to go to get the information and how to implement the ETL process. But apart from this information let's go see which tables within our source system contain the information that we will require. So we will define for each dimension which tables are included; which source system we will need to go search; how the load is defined the for this table, incremental or full; which tables we will use as source and which conditions – they can be a filter or required joins among tables if we have more than one source table for a given target. Test way to gather all this information is to use a tabular format that contains all these fields, like the one shown in Figure 5-23.

Dimension	Table	Source System	Load Type	Source Table/s	Conditions
Sales	t_f_sales	Odoo ERP	Incremental	sale_order sale_order_line sale_order_line_invoice_rel account_invoice account_invoice_line account_invoice_line_tax account_tax	sale_order.id=sale_order_line.order_id sale_order_line.id=sale_order_line_invoice_rel.order_line_id sale_order_line_invoice_rel.invoice_line_id=account_invoice_line.id account_invoice.id=account_invoice_id.invoice_id account_invoice_line=account_invoice_line_tax (outer) account_invoice_line_tax.tax_id=account_tax.id
	t_l_status	Manual	Static	N/A	
Currency	t_l_currency	Manual	Static	N/A	
Customer	t_l_customer	Odoo ERP	Incremental	res_partner	
	t_l_cust_country	Odoo ERP	Full	res_country	
Employee	t_l_employee	Odoo ERP	Incremental	hr_employee	
	t_l_emp_department	Odoo ERP	Full	hr_department	
	t_l_emp_level	Odoo ERP	Full	hr_job	
Product	t_l_product	Odoo ERP	Incremental	product_product product_template	product_product.template_id=product_template.id
	t_l_category	Odoo ERP	Full	product_category	
	t_l_parent_category	Odoo ERP	Full	product_category	
Time	t_r_time	Autogenerated	Full		
	t_l_month	Autogenerated	Full		
	t_l_quarter	Autogenerated	Full		
	t_l_year	Autogenerated	Full		

Figure 5-23. *Source table list*

Source Fields

Finally we will need to go ahead in each table to investigate which source field we need to take to fill up each target field, which business rules and formulas we need to apply, and any condition that we think relevant to achieve the correct calculation. In Figure 5-24 we can find the formula for t_f_sales table that is the most complex in our model. We deliver this one just as an example, but all tables on the model should have a source field analysis like the one shown.

Table	t_f_sales	
Field Name	**Source Table**	**Source Field**
Order_id	sale_order	id
Order_Item_id	sale_order_line	id
Invoice_id	account_invoice	id
Invoice_Item_id	account_invoice_line	id
Customer_id	sale_order	partner_id
Employee_id	sale_order	user_id
Product_id	sale_order_line	product_id
Status_id	Calculated	if invoice_id is not null then invoiced else if qty_to_invoice <>0 then Ordered else Quoted
Order_desc	sale_order	name
Quantity_inv_num	account_invoice_line	quantity
Amount_inv_num	account_invoice_line	price_sub_total
TAX_inv_num	account_invoice_line account_tax	price_sub_total*amount
Total_Amount_inv_num	Calculated	amount_inv_num + TAX_inv_num
Day_id	Calculated	if invoice_id is not null then account_invoice.date else if qty_to_invoice <>0 then sale_order.date_order else sale_order.creation_date
Order_Date_id	sale_order	date_order
Invoice_Date_id	account_invoice	date
Quantity_ord_num	sale_orde_line	qty_delivered
Quantity_del_num	sale_orde_line	qty_delivered
Quantity_quot_num	sale_orde_line	product_uom_qty
Currency_id	sale_orde_line	currency_id
Creation_Date_id	sale_orde_line	creation_date

Figure 5-24. *Source field analysis for t_f_sales table*

Conclusion

Data model is the basis for your BI analysis. A correct model defined in this moment of the project will save you many headaches in the future, so we heartedly recommend that you invest time in this point to ensure that the model that you are proposing will easily solve the questions that your users will ask of your BI system. Following up with the proposal steps will ensure you a project done with quality enough and documentation to understand the reasons of the model structure; otherwise if you go ahead just by creating the tables that you think relevant we can ensure you that in the near future you will lose the source of many of the decisions made in the model. For sure you can need some Proof of Concept projects and tests without going through Business, Logical, Dimensional, and Physical models, but once you are clear enough that a model strategy will serve to meet your requirements, you should follow all steps in order to clearly organize the model definition.

Now that we have our model defined, let's go to see how to fill it up with some ETL tools and how to get profit from it with a BI platform.

■ ■ ■

## CHAPTER 6

ETL Basics

In the previous chapter, we developed our datawarehouse model. This was the first chapter that we get our hands on our solution. It is very important to have a correct understanding of that chapter and be familiar with the model in order to consolidate this knowledge that we will acquire for the next set of chapters, as we will have to build upon it. In Business Intelligence, usually we talk about three areas: Database and Datawarehouse design, ETL, and Reporting. We covered the first one, and are now we are just ahead of starting the second one. In this ETL chapter we will see how to Extract, Transform, and Load the data into our datawarehouse. Let's start!

Why Do We Need an ETL Process?

One may wonder, if we already have the data in our databases, why not point our reporting tools to these databases and start working in the reports and dashboards, right?

Well, life usually is not that easy. First of all, connecting directly to an operational data system is never recommended. The queries generated by reporting tools, or even by users, can be very time consuming while harvesting resources needed for the operational systems. However, there are more reasons. Sometimes the data can't be used in the way it is stored in the operational system. It needs some sort of transformation that is not possible in the reporting tool, or with pure SQL. And sometimes, and trust us, this is usually the case, especially when the company has a reasonable size, we need to integrate data from external sources: providers, point of sales, different databases, Excel or plain text files, or even data from the Internet.

In this given scenario, we need a way to amalgamate data, consolidate it, and store it in a format that we can query. Also, most of the time, we will need to go through a stage of data cleansing, as data from source systems is not always complete, correct, or suitable to be stored.

This approach has many benefits. Obviously, it complicates things as it requires us to develop these processes, but in exchange, we avoid hitting the operational systems; also, we can modify our data to suit our needs and we can apply different retention policies for the data if we want to, so it gives us an unprecedented degree of flexibility.

Details of the Solution

Before entering into too much detail, we want to present you with our idea of the final flow of the data. We have our transactional system, based in Odoo, where we will gather some tables and download them on the staging area. From that staging area we will perform all the cleaning aggregation, filtering, and manipulation needed to fill our final tables. This is usually the most common approach when it comes to data extraction and loading. This approach is usually chosen over others because it does not put pressure on the operational system. If we query it again, Odoo, the transactional system, may incur a performance impact. With this solution, all operations will be performed on the staging and dwh databases that are not part of the operational system, so the load will appear on these databases.

© Albert Nogués and Juan Valladares 2017
A. Nogués and J. Valladares, *Business Intelligence Tools for Small Companies*,
DOI 10.1007/978-1-4842-2568-4_6

Usually, the extraction will be performed at night or during off-peak hours, so the data transfer process that takes place between the transactional and the staging area does not impact the operational system on working hours, which are supposed to be the critical hours; and the hours where the system is under more pressure. Hence, the final schema of the solution will be something similar to what can be seen in Figure 6-1.

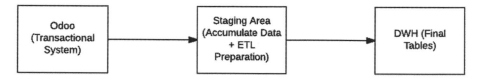

Figure 6-1. *Diagram of the solution. The data flow will start in the Odoo and finish in the datawarehouse*

Open Source ETL Suites

We have good open source databases, but probably we have even better open source ETL suites. In this book, we will explore two tools, and while we will concentrate more on Pentaho Data Integrator (also known as Kettle), we will see also some examples using Talend Open Studio.

We have chosen these two because they are mature enough, keep evolving all the time, and they are widely used so they have great support from the community. There are some commercial ETL tools, but believe us, unless you have a very specific requirement that cannot be matched by one of the free tools, there is no reason at all to pay for any. Usually they are easy to manage, powerful, and have a never-ending set of plug-ins to connect to any data source you can imagine, all for free. Also it is possible that if you are using some commercial database in your company, you have included licensing to use their own ETL tool, as happens with Microsoft SQL Server and the SS Integration Services tool.

Downloading and Installing Pentaho Data Integration

To start we first need to download and install the Pentaho Data Integration tool (PDI or Kettle from now on). We can get the PDI zip at their webpage:

`http://community.pentaho.com/projects/data-integration/`

Click on the Download Section, and this will transport you to the bottom of the webpage, where the latest version is shown, which is 6.1 at the time of writing. Click on the version name, and this will open a new window, which will trigger the download. The file is quite big, nearly 700 MB, so make sure you have a decent Internet connection before starting the download.

When the download is finished, we will have a zip file with several programs, in fact, java libraries and classes inside. For this chapter's purposes, since we will be designing jobs and transformations (later), we will be using a windows computer, with a graphical interface. We can use also any Windows or Linux system without problems, as long as we have the java JRE installed in our system (Version 8 will do the trick). Go to the Oracle webpage and download it, if you already don't have it installed in your computer. Later on, to schedule the jobs that we design, we will install the PDI in our Linux server, but since we have not installed any X Window component there, it would be impossible to use the graphical editor to design our jobs and transformations. As a summary, the server is ok to run the jobs, but to design them, we need a computer with a graphical interface.

Once we unzip the file PDI-ce-6.1.0.1-196.zip (or something similar, depending the current version) in any place on our computer, we can trigger the GUI editor by double-clicking the Spoon file. Make sure you're selecting the right one for your environment. If you're in Linux, you should choose spoon.sh, and before running it, make it executable with a chmod command. If you are in Windows, simply run Spoon.bat. If the script can locate the java in your system, the program will be launched. If you have trouble getting it executed in Windows, check the following advice from the webpage:

Edit the Spoon.bat file and:

Replace in the last line "start javaw" with only "java".

Add a "pause" in the next line.

Save and try it again

If everything is going well, we should see a screen similar to the one in Figure 6-2. If you see the repository window, asking you to create one, click cancel for the time being. We will create one later.

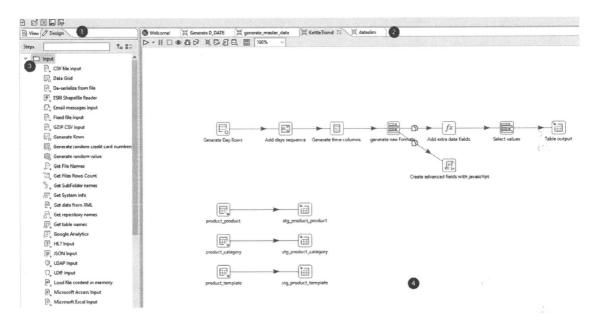

Figure 6-2. *PDI Main screen*

It is important that at this point we understand how the main PDI window is distributed. The main four areas we need to be aware of are these:

1. The Design Tab or the view tab. The first is to add new operations to the canvas, and the second is to see the already added components, the connections, and so on

2. The tab of transformations and jobs lets us switch from one transformation or job being developed to another.

3. The third, is where we can choose any new operation and drag and drop it to the canvas

4. The fourth part is the canvas. All our operations and flows need to be defined in the canvas. When we open PDI for the first time, it should have a blank canvas, but next time it will open the ones we were working on previously.

Understanding ETL Concepts

To understand an overall ETL design, first we need to understand a few concepts and objects used by ETL tools; these are presented in this subchapter, along the way to create or use them in PDI. Usually all ETLs have sources and targets of data, which is the place where the data resides and where we have to connect to read or write it; perform some actions and manipulations on that data, which are grouped into components usually referred as a transformation and an upper-level component, which is the coordinator of the flow between all transformations that is usually called job or workflow. It is now the right time to see these components in depth.

Repositories and Connections

As in most ETL tools, the repository is the metadata where all our PDI objects will be stored. The repository stores Jobs, Transformations, Connections, and much more. Until a few years ago, there were two types of repositories in PDI, Kettle Database Repository and Kettle File Repository, but recently another one, Pentaho Repository has been added. This latest one is the new recommended one by Pentaho, as far as it is the best one when working on a large deployment. Since this is not our case, we have evaluated two options, using Kettle Database Repository or Kettle File Repository.

In a business environment, it makes sense to have the repository stored in a relational database. This usually simplifies things, as we do not need to copy jobs and transformations into folders, but lets Kettle to manage them automatically. We need to remember to back up this database schema all the time, as if it is lost, we will lose all Kettle deployments.

Since we are still learning about PDI, and we do not want to invest time in new database schema creation, we will use the File Repository, which is easier to manage. Basically, we will define a folder in our system to drop the jobs and transformations. Note, however, that it is not necessary to create any repository for PDI to work. We can work directly with files, and this usually the easiest way in a non-production environment.

To create a repository simply follow these steps: From the Tools menu, select Repository and then Connect, or press CRL-R; this will make the repository as in Figure 6-3 appear, now still empty.

Figure 6-3. *The Repository list in PDI*

There is a plus sign in the upper-right corner of the dialog; let's click it, and a new dialog will appear, asking us which type of repository we want to create. You can see this in Figure 6-4.

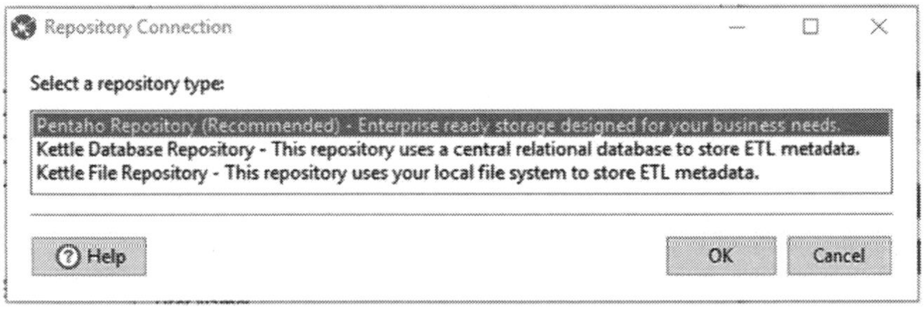

Figure 6-4. *The Repository list in PDI*

151

For the database repository option, we will select the second option. To work with a file repository, select the third. In our project, for the time being, we will work with files, but it is highly recommended to use the database option in a production environment.

■ **Note** **If you have opted for the database repository,** remember to create your database schema, and have your username and password for this schema ready as you will need these details to configure the Kettle Database Repository.

We now need to define our connection to the database. And here the first problem appears. We still haven't defined any connection. This is not a problem as we can create one, by clicking in the New button from the screen. If you look at the Connection Type you won't find MariaDB, but this is not a problem as we explained in the previous chapter: MySQL drivers are 100% compatible. So, let's select MySQL and then we are ready to create the connection. I will use a database named repo, with user repo and password repo. Check Figure 6-5 for the fields that need to be filled.

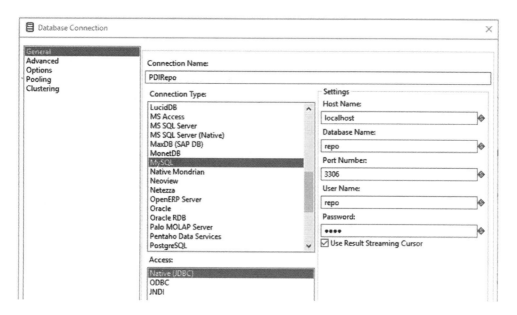

Figure 6-5. *Filling the connection details to connect to our MariaDB database*

Um, we haven't finished yet! Now is where the "bad" news come. By default, PDI comes without drivers to connect to too many databases, including MariaDB and MySQL. In order to be able to establish the connection, we will need to download the Java driver for it. If we do not download it and try to connect or test the connection, we will see a message like the one in Figure 6-6 complaining that the driver can't be found.

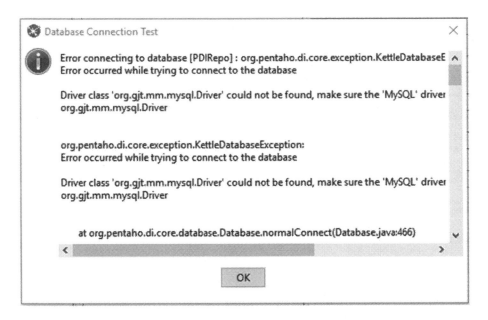

Figure 6-6. MariaDB/MySQL driver not found in Kettle installation

To get the driver, we need to go to the MySQL webpage that hosts the Java driver, simply navigate to the following page: http://dev.mysql.com/downloads/connector/j/ in your browser and download the zip file with the driver. If you are using Linux, it may well be possible that this driver is already installed in your machine; if not, you can try to run any package manager to download and install it, or follow this same procedure to copy it inside the program classpath.

Once downloaded, we are mainly interested in the jar file called something like this: ***mysql-connector-java-5.1.39-bin.jar***. Then copy this file inside the ***lib*** path of your PDI installation. Then click Cancel in the database connection dialog in PDI, and close PDI, as otherwise it won't work. We launch the Spoon executable again, and this time if we copied the file in the right place, we can test the connection, and we should see a similar message to this:

```
Connection to database [PDIRepo] is OK.
Hostname       : localhost
Port           : 3306
Database name  : repo
```

We have our first connection ready. Note that connections work always in the same manner. It does not matter if it is a repository or a data connection. The procedure to set them up is always the same. Make sure you have the driver in the lib folder, choose the appropriate database flavor, and fill in the details, naming the database connection appropriately!

If we opt for a File Repository, we only need to select the parent folder where we will create our repository, which can be any folder on your computer; give it a name and then a description. Look at Figure 6-7 for details:

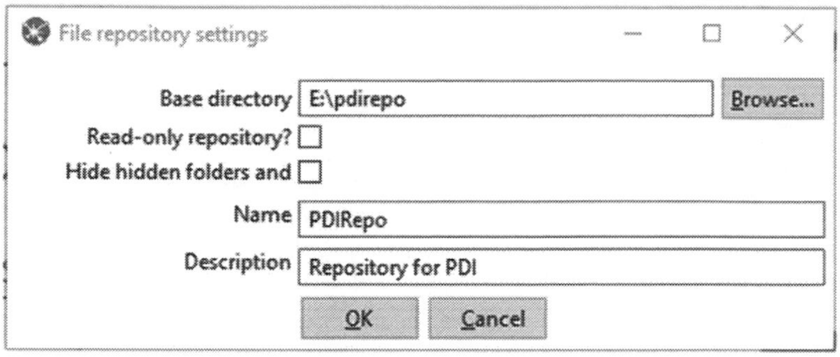

Figure 6-7. *File Repository settings in PDI*

After that, we hit OK and the repository will be created and we will be redirected back to the welcome screen.

Transformations, the Core of Kettle

The transformations, along with jobs that we will see later, are the objects that we use to define the operations with the data. Usually a job will contain one or more transformations, so it is better to understand them first.

A transformation is a chain of steps. A step is an action that will be performed usually over data. But it is not only restricted to that. We have steps that interact with variables (internal from PDI or system variables), others that interact with files, and others simply manipulate the data we are dealing with. There are dozens of different steps in Kettle. Some of them run directly in the PDI engine, whereas others run in the file system or the operative system or even with external or remote tools: interact with databases, flat files, Excel files, access files, xml, web servers …or even system variables. We won't be commenting all of the PDI steps, as this will require at least one entire book, we'll only see part of them. We will concentrate basically on the steps we need.

The steps are grouped by category, depending on the function they do. We have input, output, transform, utility flow, scripting, lookup, join, and even Big Data steps. We will basically work with input and output steps, to read and write data from the database and external files, and we will also use some Transform steps, which will let us to modify our data, some flow steps (especially the filter rows step) and likely we will use some of the utility steps that basically let us operate with files: zip them, send emails, write log files, and so on. There is another set of steps, called scripting, that is extremely powerful. It lets us specify code directly inside a step. This code can be written in SQL, JavaScript, or Java. But this will require a more advanced knowledge and it is not intended to be seen in this book.

■ **Note**   Mastering PDI can be an art. There are a couple of interesting books out there, especially two by María Carina Roldán that are of interest. One is a more general introduction and the other goes a bit further. These are called *Pentaho Data Integration Beginner's Guide* and *Pentaho Data Integration Beginner's Guide*, both in their second edition and published by Packt. Whereas they do not cover the latest version available, 6.x, this is not a problem, as minor changes have been introduced only recently.

Transformations need to be placed in a specific order in the canvas. And then they need to be linked so the output of the predecessor is the input of the posterior transformation. Sometimes it is possible to specify more than one output. In this case PDI will ask us if we want to duplicate the output between the posterior tasks or we want to copy the output to every following step.

It is also possible to have conditional outputs. Some steps divide the input between two outputs, depending on a Boolean condition specified. If the condition is met, the input that satisfies the condition will be sent to one step, whereas the input that does not satisfy the condition, will be redirected to a different step. We will see this later with an example of a transformation.

Jobs or How to Organize a Set of Transformations in a Workflow

The Jobs in Kettle are containers of transformations and linked actions that are also specified through dragging and dropping steps into the job canvas. However, the type of steps we can find in the job view are a bit different from the ones we can find in the transformation view. In the job view, we have some steps grouped by its category. General, which are the main step components of jobs, Mail, File Management, Conditions (For flow control), Scripting, Bulk Loading, BigData, Modeling, XML, Utility, File Transfer, and File Encryption – the most important ones.

Each job needs to have a starting point. A good practice is to use the success step at the end. And for debugging, we can use the Dummy Step. We'll see more later. We can find the start step in the General group, so to start designing a job, we need to drag and drop the start step into the canvas. Once this is done, then we can add Transformation steps, which will be the ones doing most of the work, and link them by moving the mouse over the step, waiting for a second and clicking on the "exit" sign in the step. Once clicked, do not click again, and drag and drop the mouse to the following step. Once done this will create a link between the first step to the second one. We can see how to link them in Figures 6-8 and 6-9.

Figure 6-8. *Moving the mouse over the step to make the menu appear. We will hold the mouse and click the exit icon (the last one)*

Figure 6-9. *Linking steps. Once we clicked the "exit" button, we drag and drop the arrow to the next step*

It is important to note that there are alternative ways of linking steps. One may be selecting the two steps, right-click on the first one (the source one), and clicking "New Hop." This will create exactly the same hop.

Once a hop has been created, we can edit, delete, or disable it. To do so, simply right-click on the arrow of the hop, and a contextual menu with these actions will appear. As we explained previously there are steps that can create conditional hops that are executed by evaluating an expression, and depending on the result, execute one or another step. In general cases, most of the steps will have the unconditional mark, so a lock icon will appear over the arrow. This means that the target step will be executed regardless of the result of the previous step. The process to link transformation steps are exactly the same. We will be creating hops all over our Jobs and Transformations.

Create and Share a Connection

To start developing with PDI, if we want to connect to databases as in our case, we need to create the connections first.

To create a connection in PDI, we can follow the same steps as we did previously, when we explained how to set up a connection to a new repository metadata. During this book we will use mainly two connections. One to the PostgreSQL where Odoo resides and another to the MySQL database that will host the datawarehouse. Now, is a good moment to create these two connections and let them set up for the remainder of the book. Once we create the connection, we will share it, so all our jobs and transformations will be able to use it, without the need of defining a new one in every single job or transformation. Let's see how to do it.

Before starting, we already have installed the MySQL java connector if we followed the instructions in the connecting to the repository section. But for the Odoo metadata connection, we need the PostgreSQL java library. By default, this one is shipped with the program, at least in the latest versions, but if you have trouble with the connection, please make sure the driver is in the correct folder.

In the file menu, click on New, and then on the new menu, click on Database Connection. A similar menu to the one previously seen will appear. In the General tab, we will make sure we choose PostgreSQL connection, and we will fill in the details we already know, like the username, database name, password, and the IP address of the server that holds the database. Then we will give the connection a name, for example, Odoo, and then we can click the Test button to make sure we can connect without any issues. If all is going well we should see a successful message indicating that the connection can be established. We click ok and our connection will be created.

From this point there is only one step remaining, which is basically sharing the connection. This is a straightforward task. In the left view, make sure you have the tab View selected, instead of Design, and then look for the folder called Jobs or Transformations, depending if you have opened a job or transformation in your canvas. Inside you will find your current job or transformation, and again inside you will see a folder called Database Connections. If you browse the folder, you should see your new connection. You can see the database connection icon in Figure 6-10.

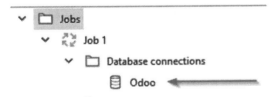

Figure 6-10. *The database connection*

At this point, all we need to do is hit the right-click on the connection and click on share. If everything is going well, the only thing we will notice is now that the connection name appears in bold. This means that our connection is now shared, and despite this if we did not see anything much different, we will be able to use it across all our jobs and transformations. If you don't see them in other jobs or transformations, close them and reopen it. You should now be able to see the connection.

■ **Note** By default, if you work in a repository based environment, the connections are automatically shared, so there is no need to perform these steps. If you work with isolated files, then this is necessary.

The Global Picture

We want to perform an analysis based on what products we are selling. The idea is that this analysis, once done, will be able to answer some of these questions: which products are the most sold, which employee is the highest performing one, and what is the evolution of our sales year after year.

To be able to do this analysis, we need to define an ETL process that gets some data from our operational system, manipulates it, and stores it in our datawarehouse. Later on, we will be able to plug some fancy graphical tools against our small datawarehouse database, and perform analysis that will help us in getting answers to our questions.

In this chapter, we will focus on the first part, extracting the data from Odoo, and loading it into our datawarehouse database. The steps will be the following: load all related product tables that we need for our analysis; load the customer- and employee-related tables, so we can track the products sold by employee or purchased by customers, or even altogether; the sales data, which will form our base fact table; and then we will need some time series to give all this some meaning. As we explained, we want to see the evolution of the sales as well as other time trends, for example, seasonal sales and other aspects, so we need to build a time table, because extracting only the dates in a plain format is not suitable, and won't let us do any aggregation by week, month, quarter, or year. Let's start by the product-related tables.

The Product, Product Category, and Product Parent Category Tables

There are three tables in the model that are related to products and its categories. The t_l_product table, t_l_category and t_l_parent_category. These are not exactly the same tables we can find in Odoo metadata but they are similar. In order to generate them, we only need three tables from Odoo, as we seen in the last section of Chapter 5. These tables are called product_category, product_product, and product_template. We will directly download these three tables to the staging area and then we will later build the queries to generate them.

We can include this in the previous transformation, or we can create a new one. Since we are only working with the staging area, we can group all these in the same transformation.

Drag and drop a Table Input step to the canvas and do not connect it anywhere. This will be the first branch of our transformation. Since we are not enforcing primary keys in the staging area, we can load them in parallel without any errors.

Double-click on the table input step, select the Odoo connection to the metadata we created previously, and click on the Get SQL Statement button. On the new dialog, display the Tables folder and look for the first table, called product_product. Select the table and click OK. PDI will now ask us if we want to include the field names in the SQL query. If we choose not to include them, they will be replaced by an asterisk, but this is not usually a good practice, so we will answer YES. A query similar to this one will be generated:

```
SELECT
  id
, create_date
, weight
, default_code
, name_template
, create_uid
, message_last_post
, product_tmpl_id
, barcode
, volume
, write_date
, active
, write_uid
FROM product_product
```

Now it is time to click on the Preview button to make sure we can retrieve the data from Odoo. I everything is going well, a grid with the data will be shown. Click on Close after verifying it. Before closing the step, rename the step name with the product_product name.

At this time, we are only downloading the tables to our staging area, so we are not performing any transformation yet. For this, we only need now to drop a table output step in the canvas. Connect both steps by dragging and dropping an output arrow from the table input to the table output step.

Time to configure some known things. For this purpose, we will name the output and the target table name as stg_product_product. We will choose the appropriate connection and will check the Truncate Target Table checkbox.

We now have almost configured our destination but again, we still need to create the table structure. This time, unfortunately, PDI makes a mistake. If we hit the SQL button and try to create the table with that create statement, we will see that the three date columns have been defined as UNKNOWN types as seen in the following code snippet:

```
CREATE TABLE stg_product_product
(
  id INT
, create_date  UNKNOWN
, weight DOUBLE
, default_code LONGTEXT
, name_template LONGTEXT
, create_uid INT
, message_last_post  UNKNOWN
, product_tmpl_id INT
, barcode LONGTEXT
, volume DOUBLE
, write_date  UNKNOWN
, active BOOLEAN
, write_uid INT
)
;
```

At this point we have a couple of possible options:

- Fix the create table statement by choosing the correct datatypes, and run it manually on our target database with the help of any client.

- Remove these fields as we don't need them for our model, and there is no need to get them from the metadata.

Any of the two specified options should work. But since the purpose of this book is to explain things in an appropriate manner and explain how to deal with unexpected things, we will opt for the first possible solution. An updated create statement for this is shown next:

```
CREATE TABLE stg_product_product
(
  id INT
, create_date  DATETIME
, weight DOUBLE
, default_code LONGTEXT
, name_template LONGTEXT
, create_uid INT
, message_last_post  DATETIME
```

```
, product_tmpl_id INT
, barcode LONGTEXT
, volume DOUBLE
, write_date  DATETIME
, active BOOLEAN
, write_uid INT
)
```

Run this statement in your staging database, and rerun the transformation. If everything is going well, you should see it finishing successfully. Go back to the staging database and run the following query to make sure the records have been appended, and the date formats are preserved:

```
select * from stg_product_product;
```

Repeat the same procedure for the product_template and product_category tables.

■ **Note**   To make things easier, you can select with the control key pressed at both steps (the table input and output) and copy and paste it in the canvas. Then you only need to edit a few things, changing the table names in both the input and output step, and you're ready to go.

You will probably face the same problem with the UKNOWN data type as we faced before. Simply change them for the DATETIME type in your create table scripts and you are ready to go.

The final result of your transformation should be as seen in Figure 6-11:

Figure 6-11.   *The generate rows step in our transformation canvas*

The Customer and Customer Country Tables

The way we will deal with these two tables is almost the same as we did with the products. The first step will be downloading them to the staging area from our Odoo metadata and then we will do a different transformation to get the required data from the staging table and accumulate it in our datawarehouse. Again, there are many strategies for this, being truncate + inserts and upserts (mix between update and insert, existing keys are updated, new ones are inserted) the most common ones.

There is not enough space in the book to detail every single transformation, but the steps to do are very similar to the ones in the previous set of tables. In this case, initially we will look for one table in the Odoo metadata: res_partner, which contains information about employees, our own company, providers, customers, and the like. There are fields on this table that help us identify which kind of entity is every entry in the table, like the company_type field, which specifies if the entity is a person or a company, or the redundant is_company flag. The supplier, customer, and employee columns can also be used to tailor the data we download from the metadata and to avoid downloading data we don't need.

CREATE THE CUSTOMER TRANSFORMATION

Once we know the procedure, it is the same steps all the time. As seen, we clearly need three tables to be able to fill the two we have in our datawarehouse. These are hr_employee, res_country, and res_currency.

To help you, we will give you some clues. If you started in the res_partner you are right on track, but reading this table data is not enough. Here are some clues on how to follow:

- In order to get the remaining data for the t_l_cust_country, in the res_partner there is a column named country_id that is a foreign key of the res_country table, where we can get the country of the employee.

- In the same res_country table, we have another column, currency_id, which is a foreign key to the res_currency where we can get the currency name of the customer invoices.

We leave up to you how to create the transformation to download these three tables in the staging area, but as always the code is available in the github repository so you can always check your result against the solution we provide, and if you aren't able to get it to work, you have a workable copy there to use.

The Employee and Employee Category and Employee Department Tables

If we have successfully completed the customer transformation, the employee one will have no secrets for us. We need to replicate exactly the same steps we did previously but instead of using the flags in the res_partner table to gather the customers, we now need to use the hr_employee to grab our employee details. Make note that it may be necessary to install the "Employee Directory" Application in Odoo, not to have the sample data, but to check them in the interface, just in case we want to play with. It will be helpful for this exercise.

There is an extra step though: to gather the employee function description, aka the role of the employee, we need to join the hr_employee table with the hr_jobs using the job_id column to gather the name field from the latter table, which is the role hold by the employee in the company.

After this, we need to gather some data from the hr_department, to gather the employee department. But at this point the only thing we need to do is to identify where to get these fields, so we can directly download all three tables into our staging area, and on the final transformation we will perform the joins.

The Fact Table: How to Create the Transformation for the Sales

Until here, most of the transformations have been straightforward. Things will now become a bit more complicated, but don't worry, we will gradually increase the complexity. We have now all the dimensions but the time one. Let's focus in now getting our more meaningful data, and the data we need to analyze our company insights.

To understand this transformation first, we need to understand how we will approach the sales in our company. We already explained this in Chapter 5, but just to refresh it, we will create orders for our customers and when the order is ready, we will deliver the goods and invoice our customers. This invoice is obvious and will have a number of products conforming product lines, and the total invoiced per item line, including taxes. Then we have all the relationships with our dimension tables, like products, employees, customers, and the manual status table. This is required to link an order and an invoice to our dimensions so we can query the database to know who sold what, to whom, and which products are the most sold ones.

To construct this table, we need to get data from a few tables in the Odoo metadata. These are:

- For the order data, we need to get data from the sale_order and sale_order_line.

- For the invoices, the account_invoice and account_invoice_line.

- For relations between the two tables, the sale_order_line_invoice_rel.

- For tax reasons we want also to download the account_tax and account_invoice_line_tax.

Before downloading these tables, here we need to think about it first. If tables are not very big, we can download fully into our staging. However, this is sometimes not the case, and our tables may be very big. In that case, instead of downloading the full tables we can put a condition applied on some timestamped field, like the write_date field, and only download the recent data.

For accomplishing this, we follow the standard procedure: we drag and drop the table input step into our canvas, select the table as usual, and then we answer yes to the question if we want all columns being specified in the select clause. After that we only need to add an extra sentence at the end of the query, to download only recent data. If the load processes run fine every day, something we will work on the scheduling chapter, and we can opt for downloading the data of the previous day. As a safety measure, if we work fine with the update insert processes that load the data from the staging area to the datawarehouse, we can download two days of data, just in case we have a non-business day and for whatever reason the process fails. Feel free to play a little with the time part of the equation. You know better than us your requirements, and also take into account the volume of data you are extracting.

The statement should be something similar to the following:

```
SELECT
  id
, origin
, create_date
, write_uid
, team_id
, client_order_ref
, date_order
, partner_id
, create_uid
, procurement_group_id
, amount_untaxed
, message_last_post
, company_id
, note
, "state"
, pricelist_id
, project_id
, amount_tax
, validity_date
, payment_term_id
, write_date
, partner_invoice_id
, user_id
, fiscal_position_id
, amount_total
```

```
, invoice_status
, "name"
, partner_shipping_id
FROM sale_order
where write_date > current_date - INTERVAL '2' DAY;
```

With that in mind, we can do the same for the rest of the set of tables, so we only download the really needed data to our staging. Just to note that you need to make sure you select the truncate target table option in the table output not to keep adding data in the staging table.

For the remaining four tables mentioned at the beginning of this point, we can proceed with exactly the same pattern. Now it is time to move on to more complex things: creating the time dimension.

Creating the Time Dimension

It is time to move to more complicated matters. What you will learn in this chapter is something that almost always you will need in any ETL process. If you recall from the previous chapter, where the ER model for the datawarehouse was seen, there are four time tables. Among these tables we are first interested in filling the t_h_time table, which contains five fields. These are day, month, quarter, semester, and year. After having this one filled, we will think about filling the remaining three, as these will be completed directly from the main one. Before starting to write the transformation to generate that data, we need to define a few things that are important. These are the following:

- Define what will be our first day in the table and our last one. We can define this dynamically or statically. If we opt for the first option, we will need to use a merge step, as the data will be different every day. To simplify things, we will opt for a fixed range of dates, as we won't be reloading this table every day. We will use the strategy to load once, for the entire life of the project.

- Define the formats of the fields, and the primary key of the table. To select the primary key of the table, we do not have too many choices. It has to be a unique value, and the only one of the five is the code of the day. The format chosen will be YYYYMMDD, which is simple to read, and we avoid storing dates on our system, especially as a part of a primary key, which is usually not a good option, both for performance reasons and for comprehension and usability.

Having these things clear in our mind will help us and simplify the ETL process for the time dimension table. Let's start!

The first step is to open the PDI if we have it closed, then we should go to File, then on the submenu click on New and then on New Transformation. A blank transformation canvas will open.

Generate as Many Days as We Need to Store

The first step is create an entry for each day we want to store. We decided in the previous introduction to go against updating the time dimension table. So, we need to account for a large period of time in this step. Considering we opened our shop on January 1, of 2016, we will hope that the shop will remain open for at least 10 years. So, we need to create 365 days (approx.) * 10 years = 3650 entries in our table. One for each day of this 10-year period.

PDI has a convenient step to generate rows based on a range introduced by the developer. As you may be thinking, the name of the step is called Generate Rows, and can be found inside the Input group step. We can use the browser, or we can look for it manually on the Input folder. In any case, we need to drag it from the Design tab, to the canvas.

Generate Rows

Figure 6-12. *The generate rows step in our transformation canvas*

Once we have the step in our canvas, it is time to edit it, in order to configure it for our needs. Just double-click it and a new screen will open where we can fill in multiple details and tune it up. We will fill it up as shown in Figure 6-13.

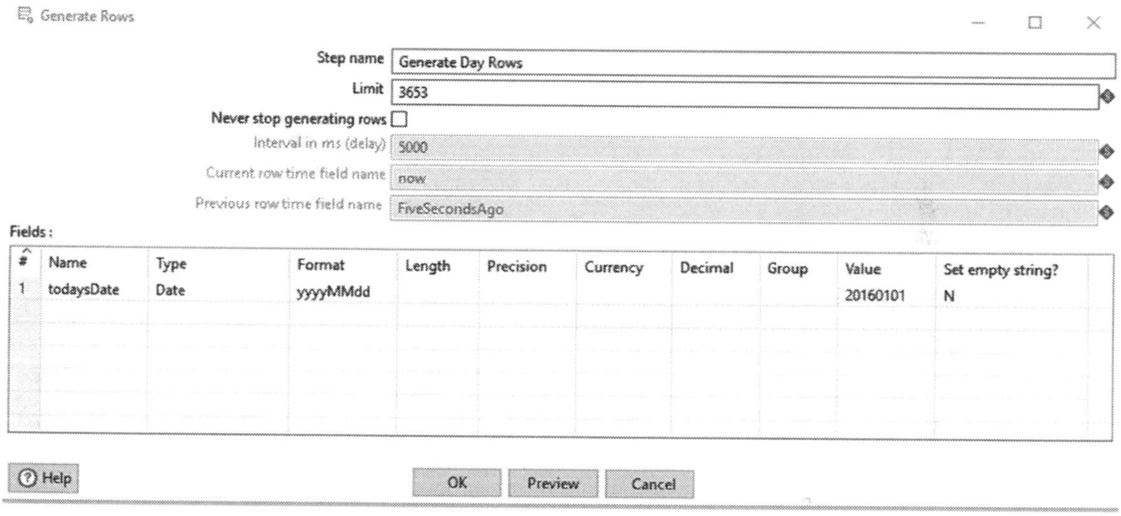

Generate Rows		— ☐ ✕

Step name	Generate Day Rows
Limit	3653
Never stop generating rows	☐
Interval in ms (delay)	5000
Current row time field name	now
Previous row time field name	FiveSecondsAgo

Fields :

#	Name	Type	Format	Length	Precision	Currency	Decimal	Group	Value	Set empty string?
1	todaysDate	Date	yyyyMMdd						20160101	N

⑦ Help	OK Preview Cancel

Figure 6-13. *The generate rows step, conf to create 10 years of data*

The first required field is *Step Name* and we should name it appropriately. While we can leave the default value, this is usually not a good practice, as it is likely that at some point we will need to modify or look again at this transformation and it will be difficult to understand it. We have chosen to name it: Generate Day Rows. You can use any name you think of.

The second field is very important too. This will be the number of rows generated. As we want to generate 10 years of data, let's put 3653, days which is exactly the number of days between January 1, 2016, and December 31, 2025. Leave the other fields untouched.

A grid then is displayed in the bottom part of the dialog. Here we create the columns of data we want to generate. Since we want to generate a set of rows for each day in the calendar, we basically only need one column right now, the day. We will care later about the month, semester, quarter, and year. Just focus on getting first the day ready, and then we can manipulate the data to generate the other columns we need.

So, the only thing we need is an entry on the grid. In the name column, we will enter todaysDate, in the Type we will enter Date and in the format, we will enter yyyyMMdd, which will store our date in a convenient format. Do not worry about the datetype, we can change it later, and in fact we need to, as we still need to operate with the date to get the other fields. The last but very important column is the Value column. Input here the starting value, with the correct yyyyMMdd format. In our case this will be 20160101, which translates to the January 1, 2016.

Before moving to the following steps, and before clicking OK, make sure that the dialog is equal to the one shown in Figure 6-13.

It is now time to make sure all data exiting this step is correct and it is according to what we defined. To explore the data without running it, we can click on the Preview button, and a new dialog with the data will appear on screen. By default, only the first 1.000 rows are previewed. Make sure these match the format we selected as shown in Figure 6-14. Do not worry that we have exactly the same row repeated all the time as this is what is expected. Then, in the following step we will operate on each row to generate the appropriate data.

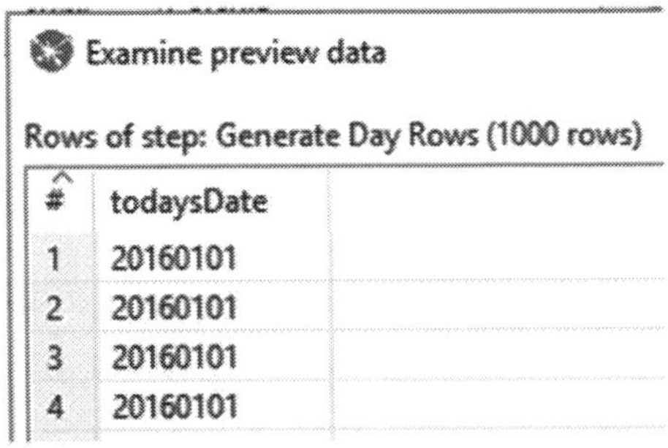

⚙ Examine preview data	
Rows of step: Generate Day Rows (1000 rows)	

#	todaysDate
1	20160101
2	20160101
3	20160101
4	20160101

Figure 6-14. The Preview option of the generate data step

Ok, we have now as many rows as days we need to add into our final table, but still now, the information is pretty useless. We have the January 1, 2016, repeated all over and this is not what we want. No worries. PDI provides us another Transformation step called Add Sequence. The utility of this step is crucial as it will let us create a sequence, starting in the number we want, in our case zero, and operate with the incoming data and this sequence. You see it, right? Yes! we will add the date plus the sequence number, and store them as our new day output. So, we will start with the 1st of January 2016, and will keep adding one day more to each entry in the step data. Let's see how to apply this logic in the PDI step.

In the Design tab, browse for the Transform folder, and inside, we find the Add Sequence step. Then we need to drag and drop it to the canvas. Before editing the step, it is always good to link it with the previous step we have, as some controls of the step, depend on reading data (usually metadata about the data it comes) from the previous steps, and without that there is no way PDI can use them. To link both steps, we click on the first one, and on the new menu that will appear, we hit on the exit button, and drag the arrow over the second step. Once done we drop the mouse and the link will be created as we saw before.

We can now double-click the second step, and configure it. Again, we need to name it, so this time we will use "Add days sequence" as a descriptive name. The name of the value we chosen is valuename, but can be anything. We will leave the other fields unchanged except the start at value, which instead of 1, we will change it to a 0, as otherwise the first day of our calendar table will be the January 2, 2016, and this is not the desired result. The result should be the same as that in Figure 6-15:

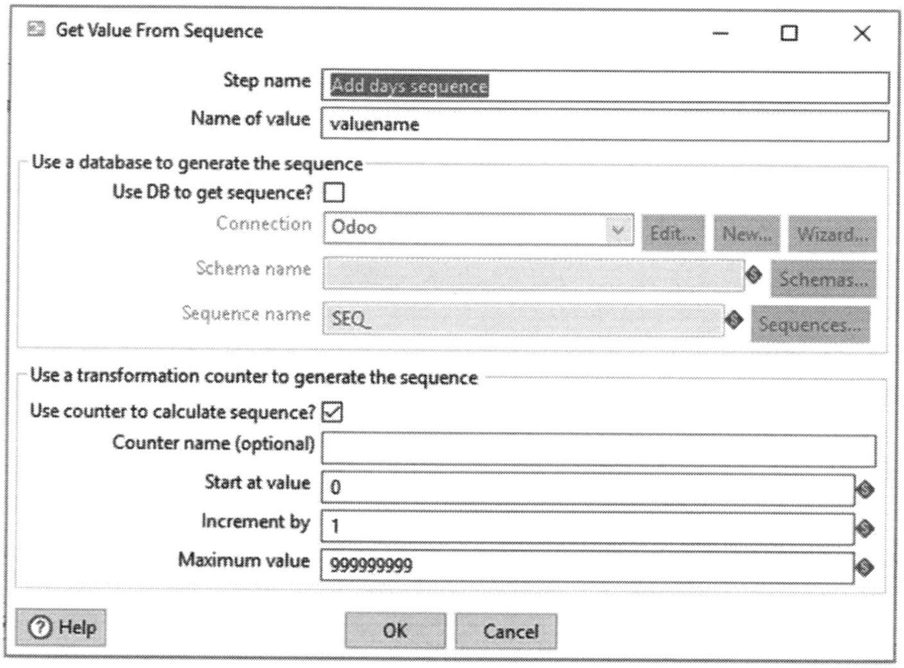

Figure 6-15. *The Add Sequence step, configured to add one day to each row*

Unfortunately, this step does not have a Preview button, but this should not stop us from previewing. Fortunately, there is another way of previewing data in every step, though it requires running the transformation. Right-click on this second step and look for the preview option. Click it and a dialog will appear. Then click on Quick Launch. This will run the transformation up to this point and will show us the results on screen. Make sure the output is the same as that in Figure 6-16.

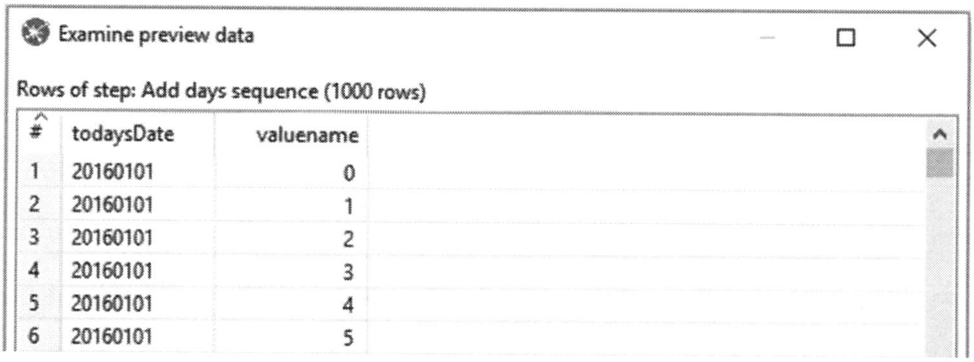

Figure 6-16. *Previewing a step that does not have the Preview button*

As we see, a new column has been added that contains the generated sequence, plus the previous data. You may imagine what we are going to do in the following step, right? The things are starting to be clear now. Time to hit the Stop button, to stop our transformation execution.

■ **Note** Never hesitate to use the options to preview rows as many times as you want. Having to modify several steps because we are carrying incorrect data from previous steps may be very annoying. Make sure the output of each step is what you're trying to achieve by using the preview option frequently.

The next step, as you may be wondering, is obviously adding the two columns into a new column. This is the trickiest step, as it is the one that sums up the previous ones and where we will need to make sure that all is tuned appropriately to have what we need. If instead of a date, we already had done the conversion into a number, adding 31 to 20160101 would generate an incorrect date as the 32nd day of January, which does not exist, so this is why we operate first and convert after that.

Using the Calculator to Compute Some Other Fields

Let's move to the interesting step. We are ready to use one of the most powerful built-in steps in PDI. The calculator step. Look for it under the Transform folder and drag and drop it to the canvas. As a step name we have chosen Generate time columns but you can use any name you want. As we did in the previous step, make sure that a link is created between the second and third steps. At this point we should have a chain of three steps, linked one after the other.

The calculator step is similar to what we have seen until now. It has a grid where we can define fields. These fields defined where there will be our output steps. To generate the output steps, we can use some built-in functions and/or, fields from the previous step, or variables, which we will see later. Let's start creating a new output field. The final appearance of the calculator should be like the one shown in Figure 6-17. But let's see the explanation of every field.

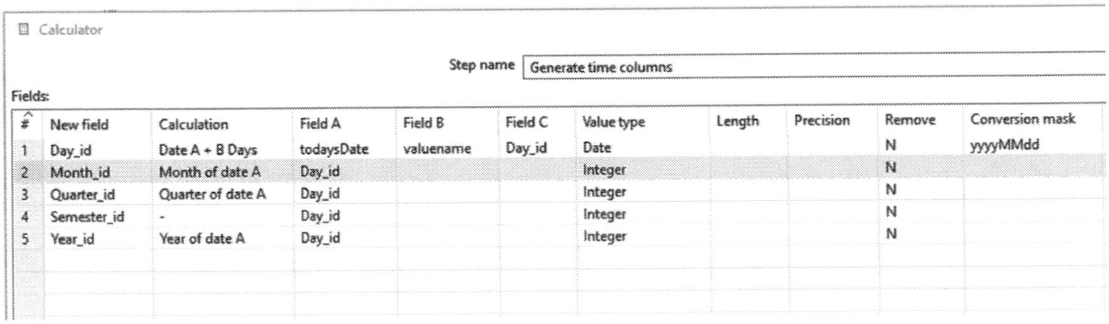

#	New field	Calculation	Field A	Field B	Field C	Value type	Length	Precision	Remove	Conversion mask
1	Day_id	Date A + B Days	todaysDate	valuename	Day_id	Date			N	yyyyMMdd
2	Month_id	Month of date A	Day_id			Integer			N	
3	Quarter_id	Quarter of date A	Day_id			Integer			N	
4	Semester_id	-	Day_id			Integer			N	
5	Year_id	Year of date A	Day_id			Integer			N	

Figure 6-17. *Calculations done in the calculator step of PDI*

Under the new field column, write Day_id. This will be the name of our first column in the output of this step. Under the calculation column, hit the drop-down box and look for the operation called Date A + B days that will let us add a number to the date we already have. In order to make it work, we have now three columns, called Field A, Field B, and Field C. Field A will always be the output field or the result of the operation. Field A will be the first parameter for the operation, and field B the second. Note that there are some operators that only work with one parameter. In this case, we only need to use Field B column and let Field C column remain empty. Since our operation is C= A+B, we need to fill all three columns.

In Field A column, we will choose todaysDate from the drop-down list. This is how we named the generated Date in the first step. In field B we will select valuename, which is the way we named the sequence in our second step. In field C, we will select the same name we gave to our new column, Day_id, as we want to store the result of the calculation in this column. In value Type, we will select Date, as we still need the date format for the other column operations. Make sure the conversion mask column is set to yyyyMMdd as this is our desired format.

With that, we have our first field of the table (the first column) almost ready. But, if you remember, the target table had five fields, so we still have four to go. Fortunately for us, PDI provides again a set of built-in functions that will allow us to easy calculate the remaining fields.

Month_id, Quarter_id, Semester_id, and Year_id are the remaining fields that we need to add so far. No problem with these as we only need to specify the calculation and the result is direct. For month_id, we are interested in the number of the month for the year, so January will be 1, February will be 2, and so on. In the calculation column, we need to select Month of Date A, and the fieldA will be Day_id, as we are computing it over our day sequence. The value type will be an Integer as that's obvious.

For the Quarter_id we will use the same logic. There is a calculation called Quarter of Date A, and we will specify Day_id as well in the fieldA column. Again, the datatype is an Integer.

For the Semester_id we have a problem, as Kettle unfortunately does not have a built-in function to calculate the semester. So, we need to find a way to specify the semester. Again, it is a shame that we cannot apply a MOD function in our Semester column, as with an easy operation we would be able to compute it. So, this time, we cannot use the calculator for calculating this field so we will use the dash, which means that we will leave the field empty. We will solve it later.

The Year_id again has an easy solution, as there is a there is a calculation called Year of Date A, and again, its result is an Integer.

Calculating More Complex Fields and Dropping Unneeded Ones

Ok, so far we have almost all calculated except the Semester. But if you recall from our ERD model, we had four time tables. It is time to look at the columns we need to fill in the other tables. Previously we explained that these three other tables will be filled from the big one we are building, so we need to have all our needed data for these tables as well, included in the macro time table we are building. A quick look to the diagram, will reveal us that we need the Month description: January, February ..., not just a number, the quarter description that will be Q1, Q2 ... and the year description. For the year, usually it does not make sense to have the year spelled in letters, so for this we will use exactly the same value as the year number: 2016, 2017, and so on. These conversions can't be done on a calculator step, so we need something else.

Here we introduce another very important step in PDI. This step is called select values, but usually it does much more than that. It lets us to transform our data without having to use any piece of code. You will find it under the Transform group. Let's see how this works.

The Select values step has three tabs. The first one, where we can select which fields of the previous steps we want to propagate to the exit of this step, and if we want to change the name of the field, the length or the precision. There is a useful button here called Get Fields to Select, which will load all previous fields into the grid. Then we can remove or add whatever we want, but usually hitting this button is a good starting point. Make sure you have already connected the previous step to this one, as otherwise the button won't work. We can again use the preview button to make sure all still works.

It is the right time to see how to drop a field and do not propagate it to the exit of the step. If we clicked the button, you will see that the first field that we used to generate our dates, our cloned field that always contains 20160101, which we called todaysDate is still on our transformation. Since this field was only used as a base to add the sequence number, and we already have the day formed, we don't need it anymore. So, the first thing to do in this step is to remove the todaysDate field from the grid. Right-click on it and select Delete Selected Lines, or press the Del button on your keyboard. Exactly the same applies for the valuename field, which contains our sequence number. We can delete it from the grid, too, as it is not needed anymore.

▓ **Note**   Probably you will be wondering why we are not using the second tab of this step called Remove. It's a good question and depends what you prefer. If you don't select a field, that field won't be propagated, so it is the same as removing it. Which option to use is up to you.

Fine, let's recap. We still have missing the semester, the quarter description, and the month description. Let's start by the latest one.

Another function of the Select values step is found in the third tab. The Metadata tab lets us change the metadata of a field. The metadata of a field is all that data that explains how a field is. In this case, we can change the type of the field as well, from a number to a text and so on. Here we will change our day in format 20160101, to only use the month, and change the month from a number to the actual name of a month in plain English. To accomplish this, first we need to go back to the first tab and duplicate the Day_id field, as we will operate over this field and we still need the 20160101 representation, as we need it for another column in the database. So, let's add a sixth fieldname in the grid, which will use the Day_id as a base fieldname, but renamed to monthName, as this will be the result of the transformations. We leave the length and precision columns empty as we aren't changing anything there. Check Figure 6-18 to make sure it matches what you're developing.

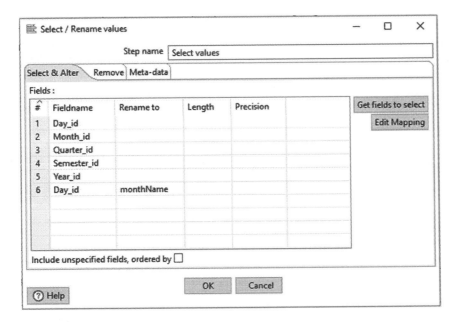

Figure 6-18. *Creating a new field to compute the month name*

Then we go to the third tab, the Metadata tab, and we use the newly created field monthName as a fieldname. The type of this field will be a String as we plan to store text, and here it comes the magic. The Format column lets us specify the format of the outputted text. If we use the drop-down button, we won't see it but there is a special format to convert the month part of a date to a Text. This format is MMMM (four m's in uppercase), so let's use this format. There is also another column that we can use, the Date Locale column, which will use the locale specified here to output the name. Since we want to use English month names, we can use any English locale: en_us will do, but also en_GB and much more. The result should be the same as could be seen in Figure 6-19. Then launch a preview to make sure you have the expected results.

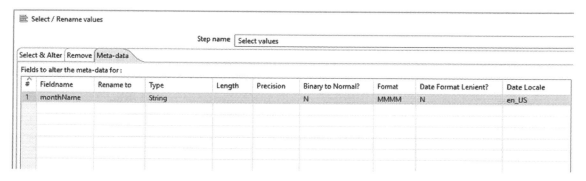

Figure 6-19. *Transforming the metadata of a field. From a date to a String to store the month name*

There is another field in the diagram that we still haven't talked as it is almost the same as the one we have. It is the abbreviated name for the month. Fortunately, we have the format MMM for this, so we need to repeat exactly the same steps to get the month name, but with the difference that this time our Format column will only contain three M's. we will name this field monthShort. We leave this as an exercise for you.

Doing Advanced Manipulations

With that we are only two fields to go. The semester and the Quarter name. As in life, there are many ways to accomplish this task. Either way requires us to present a new step. Since both steps are deemed important and you should know about them, I will present the two possible ways of doing it. The first one entitles using the Formula step that can be found inside the Scripting step and the second one uses probably one of the most powerful steps in Kettle, but also the most difficult to master, which is the Modified JavaScript Value, which can also be found in the Scripting group. Let's focus first on the Formula step, as it is the easier one to understand.

The formula step

It is time to drag and drop the formula step into the canvas and create a link between the last step in the chain, which was the select values to the new dropped step. In this step, we have two tasks. One is to create the quarter description that will be something like Q1, Q2, and so on, and another one which is calculating the semester. Let's focus on the first.

Double-click on the newly added formula step. First name the step. We have used "Add extra date fields" as a name. Since we are creating the Quarter Description, in the new field we will put QuarterDesc as a name. In the formula step, we can double-click, and a popup will appear. In the upper part of the screen we can write our formula, whereas in the left we have the list of formulas and transformations available to us. If we click on one, in the bottom part of the dialog, a help page will be shown with the syntax and examples for each formula. Admittedly this step is not very straightforward to use, so with some practice you can write the formula directly in the upper box. We will use two functions: one is the concat formula that is done by using the ampersand & keyword; and then another to convert an integer, which is the quarter ordinal to a text, as the concat function operates with two text fields.

It is important to note that the fields that come from previous steps cannot be referenced by their names directly, so we need to add them between brackets []. Also, any text we want to concatenate needs to be defined between double quotes. So, the resulting formula will be the following: "Q" & TEXT([Quarter_id]). If we preview the step, we should see that now we have our quarter description ready.

Time to go to the remaining field, which is the Semester. There are many ways to calculate a semester. One is a conditional function, like IF or a CASE statement. If the value of the month is between 1 and 6, then the semester is 1, otherwise it is 2. Another one is trying to do an integer division of the month number between 7 and add one to the result. Since this is not the easiest and most straightforward approach, let's use a conditional statement to calculate it.

We will name the new field Semester. We will use simple logic to calculate it. If the month is less or equal than 6, then the semester is 1, otherwise it is 2. The formula should be this one: IF ([Month_id] <=6; 1; 2). Check if that matches Figure 6-20.

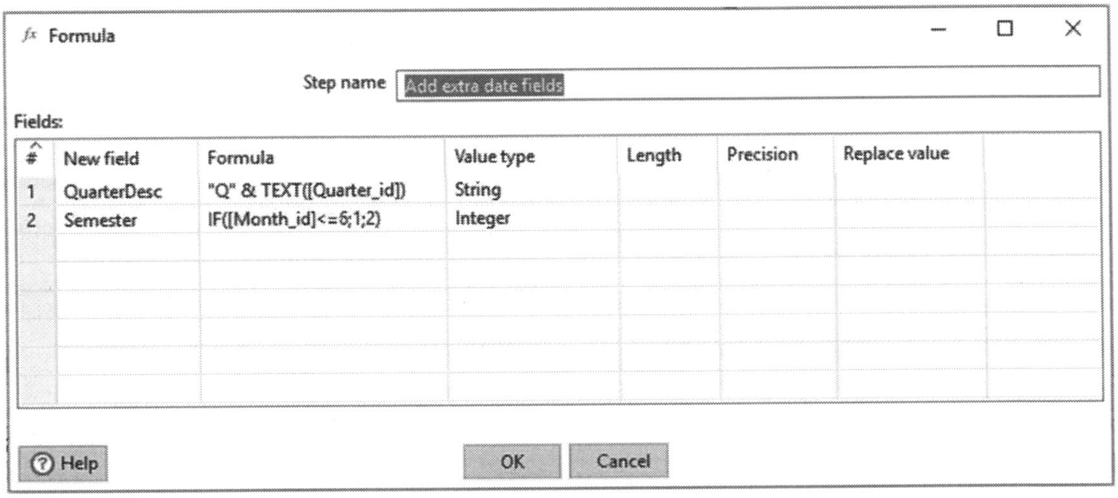

Figure 6-20. *Formulas used to calculate quarter and semester*

The Modified JavaScript Value step

This is one of the most powerful steps available in PDI. The modified JavaScript value step lets us use JavaScript code to perform any operation we need, create new fields, and combine fields. It obviously requires knowledge of JavaScript as we need to use this language to use this step. Since it is not the purpose of this book to teach JavaScript to you, we will only present the calculations we did for this step to accomplish the same results as in the Formula step. If you drag and drop the step and link it in your canvas, you are ready to start. Double-click on it, and you will see a text entry in the step. Simply copy this code, and then click on the button in the bottom called Get Variables, which will create the output steps for the two fields that you have defined.

```
var QuarterDesc = 'Q' + Quarter_id;
var Semester = 0;
if (Month_id <=6){
Semester = 1;
}
else{
Semester = 2;
}
```

If all is well, you should see the same shown in Figure 6-21, and the preview should yield the same results as in the calculator step.

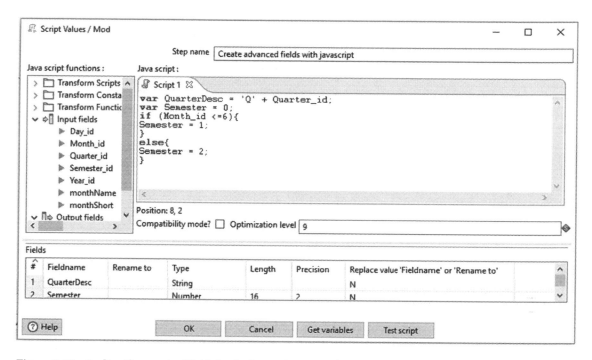

Figure 6-21. *Coding the required fields in the Javascript step. Check that the fields are propagated to the next step in the bottom grid*

Fixing a few things

If we preview our transformation, we will note that we have repeated values for our months and quarters. How we can differentiate Q1 from 2016 and Q1 from 2017? As it is right now, it is not possible without looking at other fields in the same row. Same happens with the month. We will face trouble inserting this information, as our dimension tables need to contain unique values for each record. Usually we will enforce this by the use of a primary key, but sometimes in a datawarehouse this is not used, and it is the ETL process that is the one that contains the logic to avoid these situations. Also, we want to include in our month descriptions the year as well, to make it more clear.

There are many places to fix this. We may fix it when we build the ETL that reads the data from the staging area and writes it to the datawarehouse by using some SQL expression to manipulate the fields; we may go back to previous steps and solve it there; or we may add an extra step and solve it right now. Depending on the case, one solution may be better than other, but sometimes it does not matter.

In order to not make the things more complicated, lets fix the previous steps.

1. Go back to our select values step, and change the two month formats MMMM to MMMM yyyy and MMM to MMM yyyy. Note the space between the M and y.

2. For appending the year to the month, either we use again a new formula or JavaScript step at the end or we do a metadata transformation using the Select Values Step. We will opt by the latter. In our select values step, we go to the tab Select & Alter and create a new row with fieldname Day_id and rename to newMonth_id. Then we go to the metadata tab and add a new row with fieldname newMonth_id of type string and format yyyyMM. This will create a new month format of the type YYYYMM, which is what we need for the datawarehouse, as we will see later.

3. Fixing the quarter is a bit more complex as there is no date pattern for it. But we can do it either in our JavaScript or formula step.

    a. For the formula step, first modify the Quarter description by appending the year at the beginning. The formula should now read: TEXT([Year_id]) & "Q" & TEXT([Quarter_id]). With that we would fixed the description but not yet the id. Create a new row called newQuarter_id with the following formula: TEXT([Year_id]) & TEXT([Quarter_id]). Check Figure 6-22.

    b. For the JavaScript step, we need to do a similar fix. The JavaScript code will be changed to the following and a new entry should be added in the bottom grid to propagate the newQuarter_id as in Figure 6-23.

```
var QuarterDesc = Year_id + 'Q' + Quarter_id;
var newQuarter_id = str2num(num2str(Year_id) + num2str(Quarter_id))
var Semester = 0;
if (Month_id <=6){
Semester =  1;
}
else{
Semester = 2;
}
```

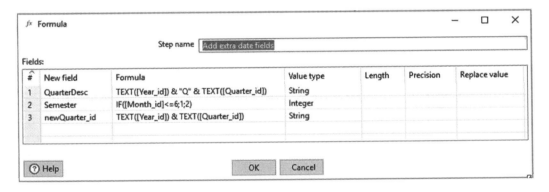

Figure 6-22. *The final Formula step*

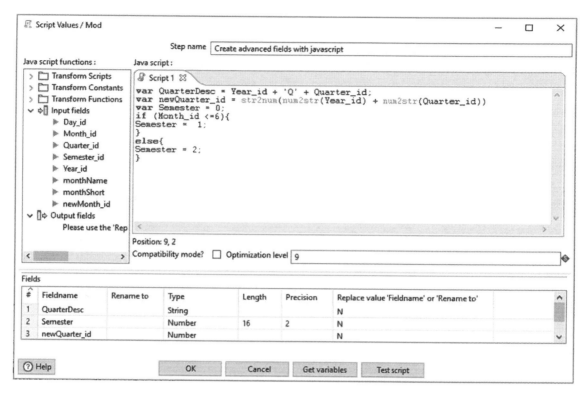

Figure 6-23. *The final Javascript step*

The semester does not need any fix as we will store a 1 or 0, as we don't have a dimensional table to it. If you want to include the year, too, this is left as an exercise.

The last step

Now that we have all what we need, it is time to generate the last step. Well in fact, the last two steps, as the latter will be a database output step to store the result of this in our staging schema.

First we need to drag and drop another *select values* step in our canvas. We will link it with the current chain of steps, and we will use it for a dual purpose. First, to select all the fields we finally want to store and discard the ones that are of no interest. Secondly, we can use the metadata tab as well, to change the datatype of a column if we need so. We can also let the last step, or even a database implicit conversion, change the type of the data, but it is a good practice to do it directly on the ETL, as the others may have unexpected results.

At this step, we need not to propagate a couple of fields. These are Month_id, Quarter_id, and Semester_id, as they are not in the form we want, or do not contain data at all. So either we don't include them in the list or we add them onto the Remove tab. Then we need to rename three columns to replace the three that we have not propagated to the output. So, apply the following renames:

- The fieldname newMonth_id will be renamed to Month_id

- The fieldname newQuarter_id will be renamed to Quarter_id

- The fieldname Semester will be renamed to Semester_id

Apart from this, we will also be reordering the fields. This is not really mandatory but we prefer to do it for clarity. The final state of this step should be as seen in Figure 6-24.

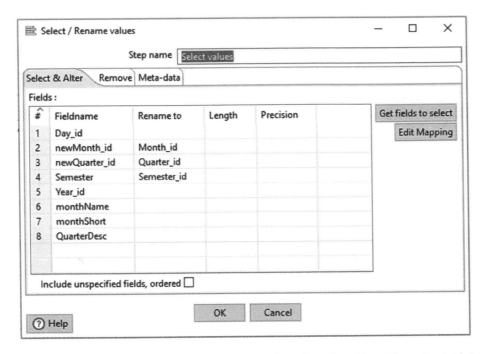

Figure 6-24. *The last but only one step. Selecting the values of our Time Dimension table in the stagging area*

The only remaining thing is to change now two fields, so we store one as a String, though it should be an integer, but we will do the conversion later, which is the Day_id, and another one, the new Month_id which will be an INT instead of a string. These can be seen in Figure 6-25.

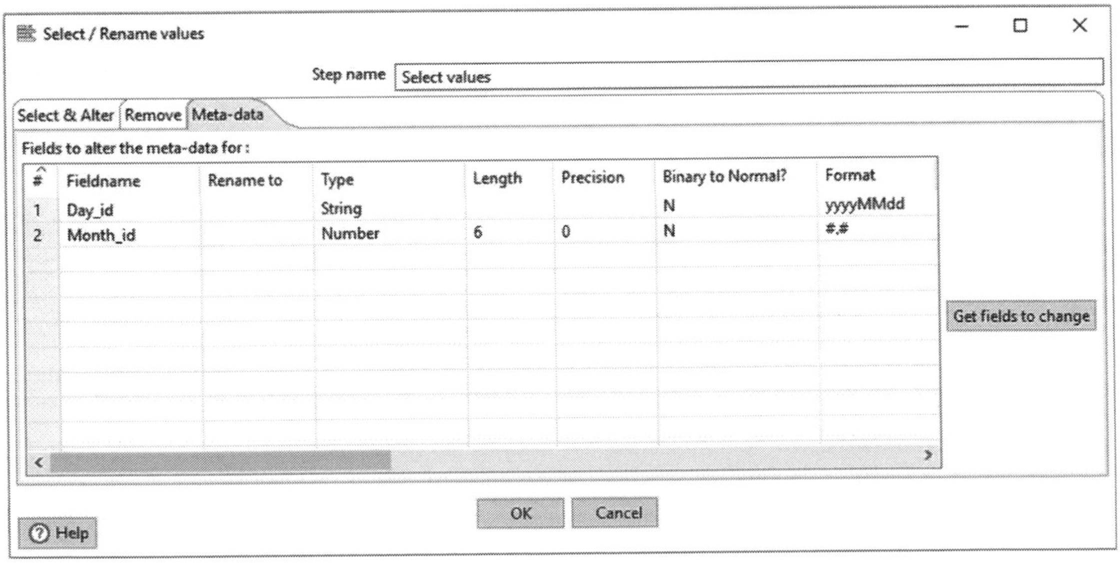

#	Fieldname	Rename to	Type	Length	Precision	Binary to Normal?	Format
1	Day_id		String			N	yyyyMMdd
2	Month_id		Number	6	0	N	#,#

Figure 6-25. *The last metadata changes*

Now it is finally the moment to create our last step. For this, we go to the Output folder in the design view and drag and drop the table output step in the canvas and then we link it.

In this step, we need to configure a few things: naming the step appropriately; selecting the connection to the staging area we defined previously; selecting the schema, which in case of MySQL/MariaDB is the database we created, so our staging database, as a target table name set stg_time, and check the truncate table, as we said before we will be loading this table only once. If for whatever reason we want to load it later, checking this box we will make sure that all previous data is discarded before adding the new one. We can leave the remaining options as they are.

If you look at the bottom part of the dialog, you will see a button called SQL. If you hit this button a suggested create table statement will be shown, and we can decide to run it against our staging database. If we look at the fields we will see that PDI has tried to decide which of the datatypes are defined for each field. This is sometimes not correct or suboptimal, as it is the case this time again. But since in the staging area we are not much concerned about it, we accept them, and hit the execute button. The table will be created as it is, but the data is still not loaded. The result should be something like this:

```
SQL executed: CREATE TABLE test.stg_time
(
  Day_id TINYTEXT
, Month_id INT
, Quarter_id TINYTEXT
, Semester_id INT
, Year_id INT
, monthName TINYTEXT
, monthShort TINYTEXT
, QuarterDesc TINYTEXT
)


1 SQL statements executed
```

We can now close the dialog, as all is ok. Then going back to our transformation, it is time to run it and have the data loaded in the staging area. Go to the action menu and click on run, or press F9. The transformation will start to run, and the Run dialog will appear. Leave all as default and click on the run button.

After a few seconds, the transformation should end successfully. In the bottom screen of the PDI look for the Step Metrics tab and make sure we have 3653 rows generated. The output should be very similar to Figure 6-26.

#	Stepname	Copynr	Read	Written	Input	Output	Updated	Rejected	Errors	Active	Time	Speed (r/s)	input/output
1	Generate Day Rows	0	0	3653	0	0	0	0	0	Finished	0.0s	135.296	-
2	Add days sequence	0	3653	3653	0	0	0	0	0	Finished	0.1s	32.910	-
3	Generate time columns	0	3653	3653	0	0	0	0	0	Finished	0.1s	27.885	-
4	generate new Formats	0	3653	7306	0	0	0	0	0	Finished	0.1s	53.720	-
5	Add extra date fields	0	3653	3653	0	0	0	0	0	Finished	0.2s	22.006	-
6	Select values	0	3653	3653	0	0	0	0	0	Finished	0.2s	17.733	-
7	Create advanced fields with javascript	0	3653	3653	0	0	0	0	0	Finished	0.2s	17.733	-
8	Table output	0	3653	3653	0	3653	0	0	0	Finished	2mn 43s	22	-

Figure 6-26. *The steps of the transformation and rows read and written*

We can now go to our staging database and query the new table to make sure the data is already there:

```
SELECT *
FROM stg_time
```

And the result is the 3653 records in the table. So, all ok. Checking the final output, all looks ok, as can be seen in Figure 6-27:

Examine preview data

Rows of step: Select values (1000 rows)

#	Day_id	Month_id	Quarter_id	Semester_id	Year_id	monthName	monthShort	QuarterDesc
1	20160101	201601	20161	1	2016	January 2016	Jan 2016	2016Q1
2	20160102	201601	20161	1	2016	January 2016	Jan 2016	2016Q1
3	20160103	201601	20161	1	2016	January 2016	Jan 2016	2016Q1
4	20160104	201601	20161	1	2016	January 2016	Jan 2016	2016Q1
5	20160105	201601	20161	1	2016	January 2016	Jan 2016	2016Q1

Figure 6-27. *The result of our time transformation*

Connecting All of It Up

The last step in our ETL process is the most complex one. Not just because we need to create a complex transformation but because we have to select the right data and perform the diverse joins between tables to get exactly the data we need, in the required format, and place it in the correct tables of our datawarehouse. For this, we will create a new transformation. In PID, go to File ➤ New ➤ Transformation. A blank transformation will appear on our canvas.

The Time Tables

Time to start our final transformation. We will start filling the time tables. We had only one table in the staging area to fill the four tables we have in our datawarehouse to time related concepts; we need to derive data from this staging table. We will start filling the t_l_year table.

The t_l_year table only contains a year and its description. If you recall when we discussed the time dimension, we said that it did not make too much sense to store a description for a year unless you want to store the year in text mode. This is not something we are interested at this point, so we will use the year number in both fields, at this time. Yes, it may look a bit silly but with that, we just keep the same schema for all tables.

To start, just drag and drop a table input step to the canvas. Choose the staging database connection, and write down this query:

```
select distinct
Year_id Year_id, Year_id Year_desc
from staging.stg_time;
```

We are selecting all different years from the table and using the year number in both fields. With the distinct clause, we are making sure we do not violate any primary key in the year_id field, nor introducing duplicates that can cause Cartesian products in the reporting stage later on.

Everything is going well; if we preview the output, we should see the 10 rows with the 10 years 2016-2025 on screen, duplicated in two columns with different names. So, that's it! We are done for the first table, easy, right? We know we only need to connect this to a table output and choose our dwh database as a connection, and chose our t_l_year table as a destination like as in Figure 6-28:

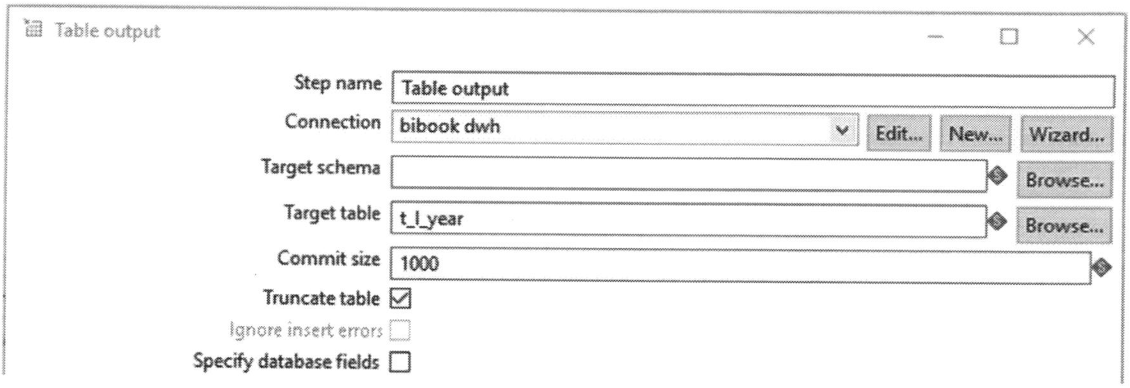

Figure 6-28. Linking our table input from the staging to our final table by using a table output step

---

■ **Note**   If we specified a constraint in our data modeling tool for the tables, in this case we cannot use the output table with a truncate option checked, as the database engine will complain, as some other tables use the t_l_year table as a master. We will receive a error message similar to this one: "Cannot truncate a table referenced in a foreign key constraint (`dwh`.`t_r_time`, CONSTRAINT `R_11` FOREIGN KEY (`Year_id`) REFERENCES `dwh`.`t_l_year` (`Year_id`))". In this case, we need to use another step, called Insert/Update or also known as Upsert or Merge, which we are going to discuss for some other parts of the transformation.

Note that sometimes if we do not specify the field names in the input table statement, or we use some that do not match the final column names in our datawarehouse, we have the option to check the specify database fields in our table output. This will enable us to use a tab, called Database Fields in the same step, where we can map the origin sources with the target column names. Using the Get Fields button may be helpful and then it is only a matter of creating the relationship between the incoming and outcoming fields or column names.

The next table can be the quarter. For this one we use the same idea but using the fields QuarterDesc and QuarterId from our stg_time table. Then we do the same with our t_l_month_table.

The t_r_time one is a bit trickier. If you remember, we said that we still do not have the Day in English format in our staging table. So, it is time to generate it. You can also generate values from the data in the staging by applying built-in functions in the database or by applying SQL operations to the data we already have to generate new data. In this case, we are applying a built-in function to extract the day name, in English format from a date stored in date format. Using the Day_id field, we can generate our full day name. Just use the following snippet code as a table input for our t_r_time table:

```
select distinct
Day_id, Semester_id, DATE_FORMAT(Day_id,'%W, %D of %M %Y') Day_Desc, Month_Id, Quarter_id,
Year_id
from staging.stg_time;
```

Make sure this time you select the specify database fields in the output table step and map the columns between the input and the output accordingly. And we have now a perfect table, with all the columns filled.

The Product Tables

The next set of tables we can fill are the product tables. These consist of the t_l_product, t_l_category and t_l_parent_category. The t_l_category is a direct select with the stg_product_category from the staging. We only need to select the needed columns: id and name. The same applies for the t_l_parent category table. One may wonder that for the t_l_product table we can just do the same. Just wait! That's not all. Think about what could happen if we stop selling a product, and delete them from our transactional. If we choose a truncate + insert strategy, this product, as it does not exist anymore in our transactional, will be lost. Ok, for future sales this is not a problem, as we are not selling it anymore, but what will happen with already loaded sales information? Probably you will already know the answer: we will miss these sales, or at least the lines that refer to this specific product, when we cross the table with the product table. This is something we have to avoid.

First focus on the input table. We need to join three tables to retrieve all the data we need. The join statement we need to define in our input table is the following:

```
SELECT
  a.id,
  a.name_template,
  a.default_code,
  IFNULL(b.description_sale, b.name),
  b.categ_id,
  b.type,
  b.list_price,
  b.active,
  c.parent_id
FROM staging.stg_product_product a, staging.stg_product_template b,
staging.stg_product_category c
where a.product_tmpl_id=b.id
and b.categ_id=c.id;
```

Then it is only a matter of connecting it to a destination step. If you read the previous note, you will be aware of a possible solution. Instead of truncating and inserting all registers from scratch, we can use another step called Update/Insert aka Merge or Upsert for the ones that have more relational database experience.

Configuring this step is a bit more complex, but nothing impossible. The first fields are similar to an Output Table step. Just give the step a name, select the dwh connection, write the name of the table to use as a target: t_l_product, and then pay attention to the first tab named the key(s) to look up the value(s). Let us explain what this means by first understanding this step. What does this step is, for every input row, it tries to find a matching row in the destination table. Which columns are compared are the ones defined in this first table. Since we mainly work by id's, what we should compare here is the id of the product from the staging table against the id of the product in the dwh table. If they match, it means that the product does exist, and so, we have two options: doing nothing, as we already have the product and we don't need to add it; or update any value: in our case, we are only storing the product description. It may well be that the name of the product has changed, or there was a typo, so in these cases we may want to update it. The fields to update need to be specified in the second table of the step, along which will be the input column that will be used for updating it.

If the comparison fails, because the row does not work, then the step, will insert the missing record in our final table. Since we are not deleting anything nor truncating, there is no risk of losing information that has already been cleaned from our operational system, and we have a system in place that will keep adding any new products that appear.

If this sounds a bit confusing, let's check Figure 6-29 for an example of how the product category table upsert should be implemented.

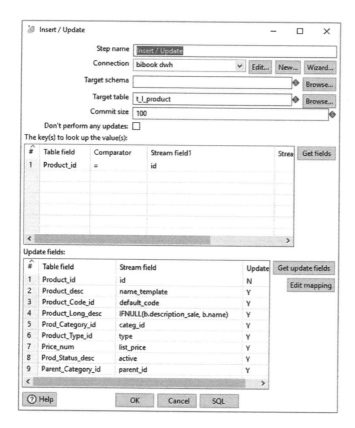

Figure 6-29. *Defining the lookup key and the update fields in an Insert/Update step*

Just as performance advice, in these situations, where multiple lookups are involved, it may pay to have an Index defined on the id key, on the datawarehouse table, so the lookups are performed much faster. If we already defined that column as a primary key, then we don't need to worry, as the database already took action, to ensure that an index enforces the primary key.

▤ **Note**   There are other ways to accomplish the same, but they are difficult to understand for novice people. In the Datawarehouse group of actions, you will find a step called Dimension Lookup Update, which is more powerful than the step we used, and the same can be accomplished in a similar manner.

The Employee and Customer Tables

Once finished with the product we still have two set of tables to fill: these are the employee- and the customer-related tables. They are somewhat similar, and follow the same strategy as the ones exposed previously. We encourage you to try to fill them by defining the steps as an exercise. Here are some pointers.

FILL THE EMPLOYEE TABLES

Related to employees we have three tables to fill: t_L_employee that will hold data from the employees; t_l_employee_department, which contains data about the departments in the company; and the t_l_emp_level, which contains data about the possible job description of each employee. The latter two can be full loads by means of a truncate/insert. You can use an Insert Update step as well, if you think that some departments or job titles may disappear from time to time. These are likely small tables so there should be no problem, and they are a direct query over the stg_hr_departments and stg_hr_jobs. Make sure you do a select distinct.

For the first one, we need to perform an incremental load. What we need to do is very similar to what we did for the product table. The table is a direct select from the staging one; and we do not need to do any transformation or compute anything new, just select the required fields.

FILL THE CUSTOMER TABLES

About the customer tables, the process is the same than in the employees table. We need the two tables only from the staging. These are: stg_res_partner and stg_res_country. The procedure is also the same as we discussed for the previous ones, with the t_l_cus_country being loaded in full, as we do not expect the list of countries to be modified (though it may well be!) and the customers an incremental load with an upsert.

The Sales Table: t_f_sales

The last table to load is the fact table. This table will contain all sales of the company, at a product line level. This means that we need to gather items, as we saw previously from seven different tables. While we can load the seven tables from the staging one by one and perform the joins in PDI, it does not make much sense if we can take advantage of the SQL skills we have. Again, we will drag and drop and Input Table step to the canvas, and introduce the following query, which joins the seven tables and collects the required data for filling the fact table:

```
select a.id Order_id, b.id Order_Item_id, d.id Invoice_id, d.invoice_line_id Invoice_Item_
id, IFNULL(d.product_id,b.product_id) Product_id,
  IF(d.id is NULL,IF(b.qty_to_invoice<>0,'Ordered','Quoted'),'Invoiced') Status_id, a.name
Order_desc, d.quantity Quantity_inv_num, d.price_subtotal Amount_inv_num,
  d.line_taxes TAX_inv_num, d.price_subtotal+IFNULL(d.line_taxes,0) Total_amount_inv_num,
DATE(IF(d.id is NULL,IF(b.qty_to_invoice<>0,a.date_order,a.create_date),d.date)) Day_id,
  DATE(a.date_order) Order_Date_id, DATE(d.date) Invoice_date_id, b.qty_delivered Quantity_
ord_num, b.qty_delivered Quantity_del_num, b.product_uom_qty Quantity_quot_num,
  b.currency_id Currency_id, DATE(a.create_date) Creation_Date_id, a.partner_id Customer_id,
a.user_id Employee_id
  from stg_sale_order a
join
stg_sale_order_line b
```

```
on (a.id=b.order_id)
left outer join
(select a.id, a.date, b.id invoice_line_id, price_unit, price_subtotal, discount, product_
id, quantity , c.order_line_id, e.amount*b.price_subtotal/100 line_taxes
from
stg_account_invoice a
  join stg_account_invoice_line b on (a.id=b.invoice_id)
  join stg_sale_order_line_invoice_rel c on (c.invoice_line_id=b.id)
  left join stg_account_invoice_line_tax d on (b.id=d.invoice_line_id)
  left join stg_account_tax e on (d.tax_id=e.id)
)d
on (b.id=d.order_line_id )
```

Then we add an Insert / Update step as a target step and we fill in the key fields as shown in Figure 6-30, and the remaining fields as specified in the same picture, too.

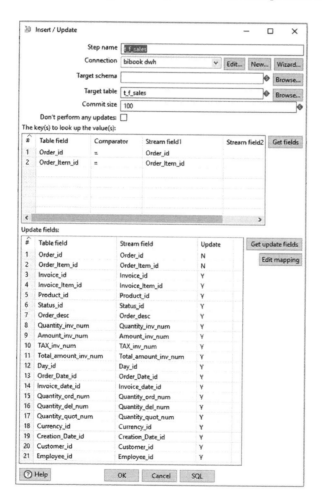

Figure 6-30. *Filling up the fact table t_f_sales*

This may look a bit intimidating at first but it is no more than applying what we have learned so far, with some basic transformations that we saw in the SQL chapter. We are now finished with all our data loading transformations.

Designing the Job

Once we have the two transformations ready, we only need to define a job. We will instruct Kettle to run the first transformation, and in case of an error it will send an email to a specific mailbox, which can be ours. If the execution of the first transformation has finished OK, then it will move to the second one. This one is the latest we designed, and will copy data from staging to the datawarehouse tables. Again, we need a step after the transformation execution just in case something went wrong. If everything has finished OK, then we can also send an email to alert people that the load has finished and the data is already available to be analyzed.

It is true that designing a workflow (or a job) is a part of an ETL process. However, we decided to add another chapter in this book titled "BI Process Scheduling: How to Orchestrate and Update Running Processes." In this chapter, we will see how to create the Job, schedule it, and create all control mechanisms to send the emails and notify people. We hope that this gives you a respite, and you recharge batteries for seeing things at a higher level, the job level.

Open Source Alternatives to PDI

At the beginning of the chapter, we were discussing Open Source ETL programs. For the development of this chapter we used PDI (or Kettle) from Pentaho. We think it is possibly the best Free/Open Source ETL software on the market. However, in some cases, it may be interesting to explore alternatives. A very good one is called Talend and his flagship product Talend Open Studio for data integration can be downloaded from: `https://www.talend.com/products/talend-open-studio`

The idea of Open Studio is pretty much similar to PDI. With the graphical designer, we can write all the transformations we need and then compile it. The output is a bit different from PDI in the fact that it generates Java Code, but all the concepts and methodologies learned in this chapter apply in the same manner.

The reason we choose PDI over Talend for this chapter is because we believe that for new users, PDI may be easier to grasp, at least at first. Where both feature a drag-and-drop functionality, we believe that the PDI interface is more clear and minimalistic. Each step has a similar set of options and most of the steps work out of the box. You only need to connect an input with an output and you're almost done. Talend has many options to configure so in some cases it may be powerful, but it should be foolish to underestimate PDI as it is a very powerful tool also. And you can enhance or add new functionality with the plug-ins that are shared in the community website or found across the Internet.

All in all, what tool to choose is sometimes a personal decision. If some knowledge of either tool is already present in your team, stick to this particular solution. If no prior knowledge can be found within your organization in any of these tools, our recommendation is to start with PDI. In any case, before starting, make sure to check on the Internet or help forums about information about the transformations you need to develop before starting them and finding they are difficult or almost impossible to develop with a specific tool.

In Figure 6-31 you can see the main Talend window and the main data connectors available in the tool in the right part of the screen.

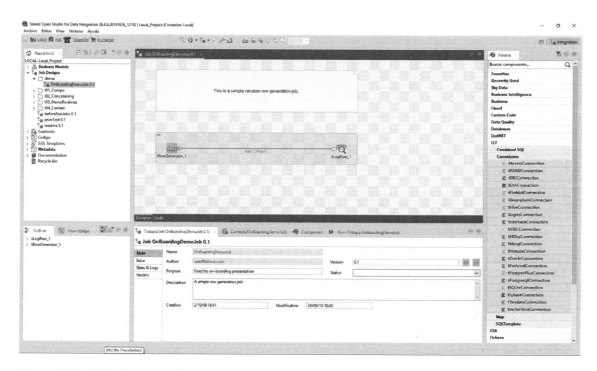

Figure 6-31. *Talend main window*

Conclusion

In this chapter, we had a look at what an ETL process is. The previous chapter served as an introduction to our data model, and this laid the fundamentals for this chapter. We know that there is quite a lot to process as writing an ETL is no simple task, but hopefully all the transformations designed in this chapter as well as all the steps and methodologies used are now clear. Basically, the concept is moving data and at the same time applying the required transformations. To accomplish this, in order not to impact production systems that may be processing important orders or invoicing, we download the data during off-peak hours to our staging database. Once there we opt basically for two possible strategies: either perform incremental loads by using an Insert/Update step that is suited to large tables, especially fact tables or to dimension tables that records can disappear from the operational, and let the Truncate/Insert strategy for the dimension table and tables that do not tend to change very much, and where there is no risk of losing values.

If you survived this chapter, you are ready to discover probably the most beautiful part of a BI project. The reporting part is the one that presents the users results based on the data we are working with. In the following chapters we will develop reports and dashboards where the user interactively will be able to play with the data we accumulated in this chapter. Also, we will be creating the necessary processes and scheduling them to make sure our data is up to date and all works like a clock.

CHAPTER 7

■ ■ ■

Performance Improvements

In a mid-sized company, it is highly possible that the volume of data we own is not that big. However, if we are a retail company, it is likely that our transactional can have a good amount of transactions, especially if we have several years of data. Whether that is the case of your business or not, it is important to have some exposure to improve performance when working with databases. This is the reason this chapter was added to the book: to provide you with some insights about how to speed up your processes. Whereas this is not a very extensive compilation, it should be enough to start.

In this chapter, we will look at several performance optimizations both at the database level and at the ETL chain. Let's start where the data resides (Database) and then move on to the calculations.

Database Optimizations

There are many operations that can be done on a database to improve performance. Also, there is another set of actions that can be taken that while by themselves won't improve performance directly, they can help indirectly, like enabling data compression. Let's first present a list of recommendations and explain why they are important when working with databases.

Avoid Using Dates for Join Keys (and Primary Keys Too)

This recommendation is especially true when the data contain also the time, or it is a timestamp field or even worse if it contains a time zone portion as part of the data. These keys can represent problems when matching two tables, especially if they are different data formats in different databases. Each database has its own method of storing dates, and the comparison of these fields can be tricky.

Usually, if we want to use a date, it is sometimes useful to add a new column, or to simply store it as a number. Obviously, we have to respect the original value of the key, so if it contains a time portion, and this part is representative, we need to take it into account.

Using strings is sometimes not as good as a choice that to use plain integers, especially if they are so long, or are of a variable length. The use of the so-called surrogate keys (autogenerated keys) when there is no obvious primary key in the table is also usually a good idea in these cases.

Also, take into account that to enable partitioning, there are sometimes restrictions on the datatype of a field. We will discuss partition later in this chapter, but this can also be a good reason for the decision of a specific column datatype.

© Albert Nogués and Juan Valladares 2017
A. Nogués and J. Valladares, *Business Intelligence Tools for Small Companies*,
DOI 10.1007/978-1-4842-2568-4_7

Analyze the Tables in Your Database

All database engines need statistics updated on the data distribution, shape, and other things of each table of the database. These statistics are used by the database optimizer to choose the order of the joins, determine whether to use an index or not (you can read more about indexes in a while), and perform a range set of internal optimizations to resolve a query.

If these statistics are not present, or still worse, they are stale, the optimizer may come up with a suboptimal plan. This means that a query will potentially perform much worse than it should. To ensure that decisions made by the optimizer are as accurate as possible, we want to keep the statistics of our table fresh. If the total amount of data to be retrieved by the query is small this is not so important, but when tables start growing, this can be a very important factor. A bad plan can be several orders of magnitude worse than the optimal plan. This can lead to very poor performance or even query hanging if we deal with lots of data.

Every single database has its own toolkit to collect statistics. In MySQL/MariaDB we can use, as in some other databases, the ANALYZE TABLE table_name; command. In Oracle, we need to call the dbms_stats package and so on for other databases.

The complete syntax for MariaDB is as follows:

```
ANALYZE TABLE table_name [,table_name ...]
 [PERSISTENT FOR [ALL|COLUMNS ([col_name [,col_name ...]])] [INDEXES ([index_name [,index_
name ...]])]]
```

But in most cases, we will use the basic analyze: for the entire table and all dependent indexes there is the following statement:

```
ANALYZE TABLE table_name PERSISTENT FOR ALL;
```

And if using an old version of MariaDB/MySQL then use:

```
ANALYZE TABLE table_name;
```

Sometimes, it may be needed to do it outside the MariaDB CLI or gui editor. In this case, MariaDB and MySQL have a bundled tool that permits you to analyze tables from the outside. This tool is called mysqlcheck, and apart of analyzing a table can check, optimize, and repair one as well, depending on the switch used on the program invocation. For analyzing a table the call should be the following:

```
./client/mysqlcheck [OPTIONS] database [table]
```

So, for analyzing a table, the syntax should be as follows:

```
./client/mysqlcheck -h host_where_the_db_resides -u username -ppassword -a database_name
table_name
```

And a real example will look like the following:

```
bibook@bibook:~$ mysqlcheck -h localhost -u bibook -pxxxxxx -a dwh t_f_sales
dwh.t_f_sales                                    OK
```

And please, note again that there is no space between the -p and the password, like in the client connection.

One last thing you can do, apart from analyzing tables, is defragmenting them. This command is very important when the data of the table has been deleted and inserted back many times. Similarly in a hard drive, data may not be placed contiguously in the storage engine, making the fetch of that data more difficult and slower than it should. This command is also interesting to reclaim wasted unused space of the table. Again, each database vendor has a specific procedure for this, but for MariaDB tables, you can use the following syntax:

```
OPTIMIZE TABLE  table_name;
```

Or the equivalent command line call

```
./client/mysqlcheck -h host_where_the_db_resides -u username -ppassword -o database_name
table_name
```

An example is shown below:

```
bibook@bibook:~$ mysqlcheck -h localhost -u bibook -pxxxxxx -o dwh t_f_sales
dwh.t_f_sales
note     : Table does not support optimize, doing recreate + analyze instead
status   : OK
```

■ **Note** If the table is big, these commands may take some time, especially the optimize command if it tries to re-create and analyze later. Use them during maintenance windows or when you know you are not querying the tables.

Indexing, or How to Speed Up Queries

Some columns on a table can be part of an index. An index is a data structure that allows fast retrieval of values. Think of an index of a book. If you want to find a specific topic or word, you go to the index and find the exact page quickly instead of having to go through all the pages. In databases, the concept is very similar. We would like to create indexes on fields that are regularly accessed, so the query optimizer does not need to scan all the data but look in the index to know where the data is and directly access the location. Even better, depending on the case, if all the needed information is already on the index, in some databases it is not even necessary to access the blocks of data.

There are several different types of indexes, depending on how they are implemented and certain ones are better than others. Again, this depends on your database vendor, so please check the manual to confirm these types of indexes are available before trying to use them. Here, in this chapter we will basically explore the two most common: the standard ones or B+ tree indexes, and the bitmap indexes.

Standard or B+ TREE Indexes

B+ TREE indexes are the most common indexes. The elements of the index are stored in sorted order. The idea is having a structure like a tree, with a root at the top, and then pointers to leaves, which at its own turn are pointers to other leaves, until the very end level, which is only a leaf with no pointers, and are the leaves that contain the last values. If the data is ordered, it is very easy to traverse the index and reach the leaf with the value we want.

We are not going to get into too much detail about how physically these indexes are stored and used, because it is not the purpose of the book and they depend on each implementation in every single database, so check a specific manual from your vendor if you are interested in it. For the purpose of this chapter it is enough to understand that the data is ordered, and this is a great example of speeding up a query; and when all data requested is in the index, there is no need to check the data blocks. Just a picture to illustrate, so you can now understand a bit better how it works. Figure 7-1 shows the index over a column that has number values, between 0 and 31, but not all of them are present. The row will contain a partition of this list, so when querying the index, the first thing to do is check the number we are looking for against the number in the root. If the number we are looking for is lower than the root we will follow the left path, otherwise we will go the right path.

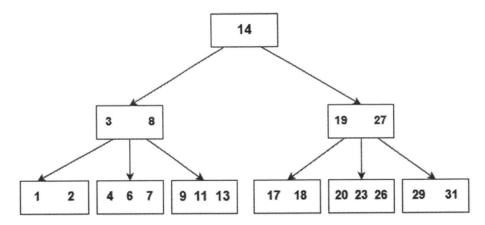

Figure 7-1. *A BTREE index. The search starts in the root node and depending on the number comparison, the left or right path is followed*

Let's see now a practical example. Imagine a large table, usually the fact table we designed in previous chapters, where you have a detail of the customer and product detail at invoice level. As we said previously, this may well be a very large table, especially if you run a retail business, with loads of transactions every day. At some point our data warehouse table will start growing out of control, and while we may decide against indexing a few columns because we usually use most of the data for our analysis, for some people maybe, their analysis is mainly focused in a few columns, for example, to implement a counter of products purchased by each customer, and they want to have quick access to these. Or it may well be possible to create derived table facts, for example, tables with accumulated data at weekly or monthly level, and we want to index these dates to quick retrieval over a period of time or for a specific customer.

■ **Note**   For time analysis, partitioning (which we will see later in this chapter) may work much better that indexing, especially in a datawarehouse environment, as there is a limited set of options. If we have very few customers that generate a lot of data, partitioning may also be a choice, but it makes more sense at first sight to index the customer column.

Let's come back to our sample table. Right now, this table has nearly 1.5 Gigabytes of data, and approximately 6 million rows. It is increasing at a pace of more than 3,000 records a day. So, yes, some pretty decent company we are running, right?

For this example, I've indexed the code of a product. This probably won't be the brightest idea, as you will probably refer to a product by its name, but if you recall from our star-snowflake datawarehouse design, we had joins by id's. So, querying one table directly by retrieving the id of a product looks absurd but when joining tables, if the ids are indexed, we speed up the queries a lot. But don't go further yet; just start with a simple table, and check the power of an index.

```
select * from fact_table where productid = 56894
/* Affected rows: 0  Found rows: 2  Warnings: 0  Duration for 1 query: 0,047 sec. */
```

Let's try not to use the index and see what happens:

```
select * from fact_table IGNORE INDEX (id_idx) where productid = 56894
/* Affected rows: 0  Found rows: 2  Warnings: 0  Duration for 1 query: 5,235 sec. */
```

As you can see, there is a massive difference, right? Ok, 5 seconds may not look too much, but imagine larger tables, joins, and subqueries with other queries or tables and this can start growing exponentially!

On the other side, having an index has some sort of performance impact. If an index is present in a table, DML operations like deletes, updates, and inserts will have some penalty. Usually this is not a problem, especially when these operations are done during the batch loading stages, and only affect the datawarehouse, but on an operational system maybe adding an index that causes severe overload inserting transactions is not acceptable. But it is as always in life: a tradeoff between several aspects, in this case, speed during querying against speed during batch loading. On a datawarehouse, it may well be worth it, because usually you will insert once, but query multiple times. A good option for improving data load (that could be explained also inside the next section on ETL optimizations) is just to drop the indexes during the load and re-create them once finished.

■ **Note**   We have used the IGNORE_INDEX hint in MySQL/MariaDB. This instructs the optimizer not to use the index to resolve a query. If you do not need an index anymore, drop it, even if not used; the index is maintained when there are inserts, updates, or deletes on that table, affecting performance.

Bitmap Indexes

Bitmap indexes are other types of indexes widely used in datawarehouses. These indexes do not resemble a tree like BTREE indexes, but they store the different values of a column, encoded in a bit sequence. Let's imagine we need a real-time analysis in our fact table t_f_sales that is becoming to be very big only to detect which orders are in Quoted status but not yet invoiced.

The possible values for the Status_id column, are Quoted, Ordered, or Invoiced. This represents a small possible value of outcomes so in this case a bitmap index may be better suited than a btree, useful for high cardinality columns. We want our query to scan all entries in the fact table that are in Quoted status, so we can really jump on the order detail and take a glance all the order details so our employees or the finance guy can check why these are not yet invoiced, and proceed to invoice if necessary. For the status case, imagine we only have four possible status values that are unknown, quoted, ordered, and invoiced. We need 2 bits to represent the status as 2 bits * 2 possible values per bit (1 or 0) = 4 combinations. The index will be created using a structure similar to the one in Table 7-1.

Table 7-1. *Bitmap index for the status_id*

Invoice Num	Status_id	Bitmap
1	unknown	00
2	quoted	01
3	quoted	01
4	ordered	10
5	invoiced	11
6	ordered	10
7	quoted	01

With a structure like this we can have a small index with all possible statuses. These indexes tend to work well in a star schema design, or a snowflake one like ours. Usually, they are placed on the foreign keys of the fact table, as these refer to the dimension tables, which usually are much smaller tables. For example, if we only have a few number of customers, but these generate a large order of invoices, it is possibly clever to have a bitmap index on the foreign column that references the customer id on the t_f_sales table.

Unfortunately, MariaDB nor MySQL yet support bitmap indexes at this point. It is an already requested feature but not yet clear when it will be implemented. You probably need a relational database like Oracle with the enterprise version to support it, which is probably not in your scope, due to licensing costs, but it is something to be aware of.

■ **Note**   There is a good comparison between both types of indexes in an oracle webpage article that may enlighten you. This article is available here: `http://www.oracle.com/technetwork/articles/sharma-indexes-093638.html`

FULLTEXT Indexes

MariaDB and MySQL do support FULLTEXT indexes. These indexes come in handy when you need to index text information columns: columns that contain CHAR, VARCHAR, or any of the TEXT variants. These indexes are really powerful when you want to do searches on text columns but you don't have the exact information contained in columns.

A FULLTEXT index can be placed on one of these products and speeds the query a lot when used with a syntax like the following: MATCH (product_description) AGAINST (expr [search_modifier]). Imagine our dimension table for products, the t_l_product. At some point, we launched a new product but with different packaging and a different number of units. This will generate a lot of SKU codes, one for each specific version. Yes, all products are the same, but in our product table, we will need to have an entry for each specific packaging. The name of the product will likely be similar, but the packaging name will be added in the product description. Product A, 0.5 ml; Product A 1l; Product A 4x4 package; Product A promotional 1.2l; and the like. If we want to speed up the search over the product description column, the FULLTEXT index is the choice. This index, instead of storing or using the entire value of the description field in the product table, will store partial words of the description.

Unfortunately, they do not work with LIKE clauses (the index is not used) so it is of limited usage in a datawarehouse environment, especially in the reporting stage, unless you use a tool that is really wise enough to make use of. On an ETL, it may well have some use, especially if you code the query and add the MATCH and AGAINST clauses, instead of using the LIKE clause, which is the preferred option used by most of the reporting tools available out there.

For an example, and the limitations of the index, please check the MariaDB page where the index is described here: `https://mariadb.com/kb/en/mariadb/fulltext-index-overview/`

Partitioning

When tables start to grow, you usually need to define indexes. However, sometimes indexes are not the best option, especially when you want to retrieve a large set of data. Indexes are indicated usually when the values retrieved are less than 5% of the total, and are also good for random access.

When you want to retrieve a considerable number of records from a table, and you identify the pattern of data access is by some specific columns, and sometimes even a range of values for these columns, you can increase performance a lot by using partitioning.

Partitioning consists in selecting one or more columns and slicing the original table into smaller pieces depending of the value of that column. Partitioning is widely used for time attributes, Day, Month, or Year usually. But having a partition per day doesn't mean that you need to have a partition for each day: you can use ranges to define that all days in the month are inside the same partition. Your question could be, why do we use daily partitioning to store months? Why don't we create partitions by month? The main benefit of partitioning is when you use the same field for filtering as for partitioning. So you need to analyze your processes and your reports to check which field is the most used for filtering. If your reports and processes are filtering by day your partitioning system will work better if it is partitioned by day, and then you can use ranges that don't require a partition for each day. Let's see an example about partitioning.

Imagine we have our fact table, the t_f_sales. As you are aware at this point, we have several columns that define an invoice line. One of which, is the invoice data. Imagine now that our finance department is very interested in the orders that have been invoiced during the current month. At this point, we are interested in all product lines that have an invoice date matching the current month. If we partition the t_f_sales table by the invoice date, and we specify a partition for each month, we will be able to retrieve the current month records without the need of an index, and even better, in a way that the database will easily locate the registers belonging to this month, without caring at all registers that do not match the condition (i.e., from past months). Imagine we want to access the partition of December 2016, then we will be accessing the one specifically for this month. An example can be seen in Figure 7-2.

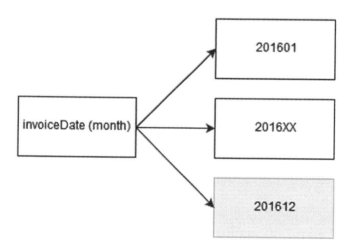

Figure 7-2. *Partitioning the sales table by invoice month*

We will see right now how partitioning is implemented but hope this sounds good. The only drawback is that sometimes the query does not have a standard access pattern. Sometimes you will be looking by the invoice date, another by the customer, and another by an employee or product. As always, there is not a magical way that suits all, so it is a matter to decide which is the best combination possible, and if needed, perform tests.

As clearly stated, partitioning works well with big tables, especially when we need to retrieve a considerable amount of data, but it can also have other applications, like ease of maintenance, when we want to deal with historical data, as we can store each different partition in a different file.

Performing maintenance, data cleanup, and automatic purges are easier to perform when we have partitioned tables. Besides performance, these are other reasons why partitioning is widely used in data warehousing.

To create a partitioned table, we only need to use the same create table statement but at the very end, we need to add a new clause, to specify partitioning, its type, and the partition expression. Something like this needs to be added at the end of the create table statement:

```
PARTITION BY partition_type (partitioning_expression)
(
        PARTITION partition_name VALUES values_condition,
        ...
)
```

Partitioning Types

Now that we understand how partitioning works, it is time to present the most common types of partitioning available in most database engines. These are:

- Range partitioning;

- List partitioning;

- Hash partitioning, which won't be explained here, as it uses a mathematical hash function to distribute the data evenly in partitions.

Range Partitioning

As the name implies, range partitioning consists of creating range of contiguous values that are non-overlapping and group them together in one partition. These values will be defined by the partition expression.

Let's come back to our t_f_sales example. Imagine we want now to partition the table by the day_id date. Our financial guys want to see the type of products sold every month, to check if there are some stationary products, which experience more orders or sales at some point of the year or during a specific season. We want to be able to access a specific month and retrieve all products and the count of the number of them sold for that specific month. As commented previously, getting data from a specific month will work better in a daily partitioning if we filter using a between statement from the first day of the month until the last day of the month.

As it looks clear from the business requirement, this query can benefit by using partitioning, when we partition by the day_id date. Again, choosing the day_id date as the partitioning key may be suitable for this query but not for others. So please take into account all business cases and try to know as much as possible in advance which will be the table access method, as a bad design in this stage can severely impact the performance of the system later on. For this exercise purposes, we have been told this will be the main access to the table, so let's see how to implement the range partitioning on our fact table.

We will be creating a specific partition (based on day_id) for each month. In our case, we will be creating partitions for all 2016 months. For this purpose, note two things. First, to create a range, we will convert a specific date to a day code. To do so, there are many ways, but MariaDB/MySQL have a function called TO_DAYS that returns the day number since the first day of the first month of the year 0. With that we will be able to specify the partitioning function and our upper bounds based on the dates we input. Since we want to have the specific data for each month together, we specify as the upper bound the 1st day of this month. With that, each partition will hold the records that do not belong to the previous partition, up to a value less than the first of the following month.

The second aspect we have to think about is how to name our partitions. Here there is no rule, so we can name them as we want, but again, common sense should prevail, so name the partitions in a meaningful manner. The DDL code to create a partitioned by range t_f_sales table is the following:

```
CREATE TABLE dwh.t_f_sales_byrange (
  Order_id INT(11) DEFAULT NULL,
  Order_Item_id INT(11) DEFAULT NULL,
  Invoice_id INT(11) DEFAULT NULL,
  Invoice_Item_id INT(11) DEFAULT NULL,
  Product_id INT(11) DEFAULT NULL,
  Status_id VARCHAR(8) DEFAULT NULL,
  Order_desc VARCHAR(255) DEFAULT NULL,
  Quantity_inv_num DOUBLE DEFAULT NULL,
  Amount_inv_num DOUBLE DEFAULT NULL,
  TAX_inv_num DOUBLE DEFAULT NULL,
  Total_amount_inv_num DOUBLE DEFAULT NULL,
  Day_id DATETIME DEFAULT NULL,
  Order_Date_id DATETIME DEFAULT NULL,
  Invoice_date_id DATETIME DEFAULT NULL,
  Quantity_ord_num DOUBLE DEFAULT NULL,
  Quantity_del_num DOUBLE DEFAULT NULL,
  Quantity_quot_num DOUBLE DEFAULT NULL,
  Currency_id INT(11) DEFAULT NULL,
  Creation_Date_id DATETIME DEFAULT NULL,
  Customer_id INT(11) DEFAULT NULL,
  Employee_id INT(11) DEFAULT NULL
)
ENGINE = INNODB
AVG_ROW_LENGTH = 455
CHARACTER SET utf8mb4
COLLATE utf8mb4_general_ci
PARTITION BY RANGE (TO_DAYS(day_id))
(
        PARTITION t_f_sales_201601 VALUES LESS THAN (TO_DAYS('2016-02-01')),
        PARTITION t_f_sales_201602 VALUES LESS THAN (TO_DAYS('2016-03-01')),
        PARTITION t_f_sales_201603 VALUES LESS THAN (TO_DAYS('2016-04-01')),
        PARTITION t_f_sales_201604 VALUES LESS THAN (TO_DAYS('2016-05-01')),
        PARTITION t_f_sales_201605 VALUES LESS THAN (TO_DAYS('2016-06-01')),
        PARTITION t_f_sales_201606 VALUES LESS THAN (TO_DAYS('2016-07-01')),
        PARTITION t_f_sales_201607 VALUES LESS THAN (TO_DAYS('2016-08-01')),
        PARTITION t_f_sales_201608 VALUES LESS THAN (TO_DAYS('2016-09-01')),
        PARTITION t_f_sales_201609 VALUES LESS THAN (TO_DAYS('2016-10-01')),
        PARTITION t_f_sales_201610 VALUES LESS THAN (TO_DAYS('2016-11-01')),
```

```
        PARTITION t_f_sales_201611 VALUES LESS THAN (TO_DAYS('2016-12-01')),
        PARTITION t_f_sales_201612 VALUES LESS THAN (TO_DAYS('2017-01-01'))
);
```

>> DML succeeded [2,384s]

Let's now insert our data from the t_f_sales into the new range partitioned table. To do so we will use a simple insert as select statement:

```
insert into dwh.t_f_sales_byrange select * from t_f_sales;
```

>> 36 rows inserted [0,268s]

And now, let's check with a select that our data is ok:

```
select count(*) from dwh.t_f_sales_byrange;
```

>> 36 rows selected

Everything is working fine. It is, however, a good practice to include a catch-all partition at the very end, with the highest value it can contain, thinking of the future. Otherwise if we try to enter a record and there is no partition to store that record the system will complain. You can see the error here when inserting a record for January 2017, as there is no matching partition created. Let's insert a fake order 99 for an item 999, ordered and invoiced the 1st of January 2017:

```
INSERT INTO dwh.t_f_sales_byrange
(Order_id, Order_Item_id, Invoice_id, Invoice_Item_id, Product_id, Status_id, Order_desc,
Quantity_inv_num, Amount_inv_num, TAX_inv_num, Total_amount_inv_num, Day_id, Order_Date_
id, Invoice_date_id, Quantity_ord_num, Quantity_del_num, Quantity_quot_num, Currency_id,
Creation_Date_id, Customer_id, Employee_id)
VALUES
(99, 999, 15, 24, 69, 'Invoiced', 'SO999', 1, 320, 48, 368, '2017-01-01 00:00:00', '2017-01-
01 00:00:00', '2017-01-01 00:00:00', 0, 0, 0, 3, '2017-01-01 00:00:00', 7, 1);
```

And the error thrown is the following:

```
Error (50,1): Table has no partition for value 736695
```

This basically means that the system has been unable to find a partition to store that value. As we said, partitioning requires maintenance. So, we should create in advance all the required partitions. If we did not create them, we can always add them later. Let's just create one for January 2017, and try to run the insert statement again:

```
ALTER TABLE dwh.t_f_sales_byrange ADD PARTITION (PARTITION t_f_sales_201701 VALUES LESS THAN
(TO_DAYS('2017-02-01')));
```

>> DML succeeded [0,732s]

Now we try to insert again the record and this time should work:

```
INSERT INTO dwh.t_f_sales_byrange(Order_id, Order_Item_id, Invoice_id, Invoice_Item_id,
Product_id, Status_id, Order_desc, Quantity_inv_num, Amount_inv_num, TAX_inv_num, Total_
amount_inv_num, Day_id, Order_Date_id, Invoice_date_id, Quantity_ord_num, Quantity_del_num,
Quantity_quot_num, Currency_id, Creation_Date_id, Customer_id, Employee_id) VALUES
(99, 999, 15, 24, 69, 'Invoiced', 'SO999', 1, 320, 48, 368, '2017-01-01 00:00:00', '2017-01-
01 00:00:00', '2017-01-01 00:00:00', 0, 0, 0, 3, '2017-01-01 00:00:00', 7, 1);
```

```
>> 1 row inserted [0,119s]
```

List Partitioning

We have seen so far range partitioning, which is a good candidate for working with sequences of numbers or dates. But sometimes you don't want to work with ranges but with specific values or a list of values. In these cases, the list partitioning comes to rescue.

The syntax is very similar to the range partitioning, but we will be specifying values instead of a range. At the end of the CREATE TABLE statement we will add the following clause:

```
PARTITION BY LIST (partitioning_expression)
(
        PARTITION partition_name VALUES IN (value_list),
        ...
        [ PARTITION partition_name DEFAULT ]
)
```

Let's imagine now a new requirement. We want to quickly get all orders by the currency type invoiced. We can have euros, dollars, and so on, and each guy in the company manages a different region, so the invoices that he or she will track are different from the others. At this point we have two possible currencies: 1 and 3. Value 1 is Dollars and 3 Euros. But we might have different values. We need to get all of them as we will need to create a partition for each. Also, take note that the value in the expression needs to be a number. So, for example, creating a partitioning for the Status_id of the order: Quoted, Ordered, Invoiced, will not work unless we translate these statuses to numbers (id's). To implement the requirement, we decided to partition the t_f_sales table with List partitioning, creating two partitions (one for each currency_id). Here, in a list partitioning, we do not have a catch-all partition, so we need to make sure that the value always will be in one of the partitions; otherwise we will be shown an error. The code to create the partition is as follows:

```
CREATE TABLE dwh.t_f_sales_bylist (
  Order_id INT(11) DEFAULT NULL,
  Order_Item_id INT(11) DEFAULT NULL,
  Invoice_id INT(11) DEFAULT NULL,
  Invoice_Item_id INT(11) DEFAULT NULL,
  Product_id INT(11) DEFAULT NULL,
  Status_id VARCHAR(8) DEFAULT NULL,
  Order_desc VARCHAR(255) DEFAULT NULL,
  Quantity_inv_num DOUBLE DEFAULT NULL,
  Amount_inv_num DOUBLE DEFAULT NULL,
  TAX_inv_num DOUBLE DEFAULT NULL,
  Total_amount_inv_num DOUBLE DEFAULT NULL,
  Day_id DATETIME DEFAULT NULL,
  Order_Date_id DATETIME DEFAULT NULL,
```

```
  Invoice_date_id DATETIME DEFAULT NULL,
  Quantity_ord_num DOUBLE DEFAULT NULL,
  Quantity_del_num DOUBLE DEFAULT NULL,
  Quantity_quot_num DOUBLE DEFAULT NULL,
  Currency_id INT(11) DEFAULT NULL,
  Creation_Date_id DATETIME DEFAULT NULL,
  Customer_id INT(11) DEFAULT NULL,
  Employee_id INT(11) DEFAULT NULL
)
ENGINE = INNODB
AVG_ROW_LENGTH = 455
CHARACTER SET utf8mb4
COLLATE utf8mb4_general_ci
PARTITION BY LIST (Currency_id)
(
        PARTITION t_f_sales_1 VALUES IN (1),
        PARTITION t_f_sales_3 VALUES IN (3)
);
```

■ **Note**   We can have different possible values mapped to each partition. If the guy in charge of the invoicing in euros, UK pounds, and other European currencies is the same, we may put all these currency_ids into the same partition by putting all possible id's inside the in clause separated by commas: VALUES IN (3,4,5,6) for example.

Then now, we can insert the values. Everything is going well, so we should get no error message:

```
insert into dwh.t_f_sales_bylist select * from t_f_sales;
```

```
>> 36 rows inserted [0,149s]
```

Partitioning Considerations

As a good practice, it is always useful to add a catch-all partition at the end. This will ensure that errors like this do not happen:

```
ALTER TABLE dwh.t_f_sales_byrange ADD PARTITION (PARTITION t_f_sales_max VALUES LESS THAN
(MAXVALUE));
```

```
>> DML succeeded [0,543s]
```

So now, even trying to add a record for 2018 will not fail:

```
INSERT INTO dwh.t_f_sales_byrange(Order_id, Order_Item_id, Invoice_id, Invoice_Item_id,
Product_id, Status_id, Order_desc, Quantity_inv_num, Amount_inv_num, TAX_inv_num, Total_
amount_inv_num, Day_id, Order_Date_id, Invoice_date_id, Quantity_ord_num, Quantity_del_num,
Quantity_quot_num, Currency_id, Creation_Date_id, Customer_id, Employee_id) VALUES
(99, 999, 15, 24, 69, 'Invoiced', 'SO999', 1, 320, 48, 368, '2018-01-01 00:00:00', '2018-01-
01 00:00:00', '2018-01-01 00:00:00', 0, 0, 0, 3, '2018-01-01 00:00:00', 7, 1);
```

```
>> 1 row inserted [0,068s]
```

Now imagine that we just want to historify the newly created partition. You have exported the data for the old partitions and you want to delete them from the database to make space for new records. Admittedly this does not make too much sense, as you will always want to delete the oldest data, not the newest, but it is just a sample. Let's run the delete partition statement:

```
select count(*) from dwh.t_f_sales_byrange where order_id = 99;
>> 1

ALTER TABLE dwh.t_f_sales_byrange DROP PARTITION t_f_sales_201701;

>> DML succeeded [0,432s]

select count(*) from dwh.t_f_sales_byrange where order_id = 99;

>> 0
```

As you can see the data attached to that partition is now gone. Be careful when implementing purging as there is no easy way to rewind that.

Using the EXPLAIN Sentence

When we write a query we tell the database which tables to access, joins to perform, but we do not specify any order or instruct the database to use any index nor any different structure. The database has some freedom to choose what it thinks is the best way to resolve the query we present to it. As we saw in this chapter, in order to have good execution plans, we need to have statistics, may need to have some columns indexed, and also, we can use partitioning. There are further optimizations that we will see in this chapter but at this point we have enough material to present the EXPLAIN sentence. This sentence, preceding any query, instead of executing it, will tell us exactly which access plan is chosen by the database engine. With that we can easily see if an index has been used, partitioning, or we are querying all the data in a table in bulk. Also, we can see the order of access to the table and some more interesting inputs.

Let's first try with accessing the first partitioned table by a specific Day_id:

```
SELECT  *
FROM dwh.t_f_sales_byrange
WHERE day_id BETWEEN STR_TO_DATE('01/09/2016', '%d/%m/%Y') and STR_TO_DATE('30/09/2016',
'%d/%m/%Y');
```

This will return us all the records we have in September 2016. To make sure partitioning is working we can call an extended explain version, called explain partitions, which will show us where the information is picked from (we removed some extra information):

```
explain partitions select * from dwh.t_f_sales_byrange where day_id between STR_TO_
DATE('01/09/2016', '%d/%m/%Y') and STR_TO_DATE('30/09/2016', '%d/%m/%Y');
```

Table 7-2. *Result of the explain partitions*

id	select_type	table	partitions	type	Pos_keys	key	rows	Extra
1	SIMPLE	t_f_sales_byrange	t_f_sales_201609	ALL	null	null	9	Using where

As you can see in Table 7-2, the partitions column indicates from which column the data has been gathered. The engine has been smart enough to determine which partition accessed from the table. Now let's try to access a partitioned table by one of the fields that is not part of the partition key. Let's imagine that we want to access the list partitioned table we created for currencies, but using dates. We expect the database not to be using partitioning, so all partitions will be accessed because with the information supplied the engine is not able to use partitioning. This behaviour can be seen in Table 7-3:

```
explain partitions select * from dwh.t_f_sales_bylist where day_id between STR_TO_
DATE('01/09/2016', '%d/%m/%Y') and STR_TO_DATE('30/09/2016', '%d/%m/%Y');
```

Table 7-3. *Result of the explain partitions on a table that is not prepared for that acces*

id	select_type	table	partitions	type	Pos_keys	rows	Extra
1	SIMPLE	t_f_sales_bylist	t_f_sales_1, t_f_sales_3	ALL	null	36	Using where

So, you can see it access both partitions. Let's try now to gather a specific currency_id from the t_f_sales_bylist partitioned table. We can check the expected behaviour in Table 7-4.

```
explain partitions select * from dwh.t_f_sales_bylist where currency_id = 3;
```

Table 7-4. *Accessing the t_f_sales_bylist table with a specific currency_id value*

id	select_type	table	partitions	type	Pos_keys	rows	Extra
1	SIMPLE	t_f_sales_bylist	t_f_sales_3	ALL	null	18	Using where

So, we clearly see again that partitioning is working. It is only accessing the partition with the value 3 in the currency_id column. The same can be seen with indexes and other aspects of a query. If an index is being used, it will also be shown. I created an index by order id in the table. If we ask the system to retrieve a specific order let's see what happens. Table 7-5 reveals this time an index is used to resolve the query:

```
explain select * from dwh.t_f_sales where order_id = 7;
```

Table 7-5. *Accessing the t_f_sales by a specific order id. This forces the engine to use an index for performance*

id	select_type	table	type	Possible_keys	Key_len	ref	rows	Extra
1	SIMPLE	t_f_sales	ref	idx_t_f_sales_lookup	5	const	4	Using where

The possible_keys column shows us which is the index used to resolve the query.

Now it is time to have a look at what happens when more than one table is involved. Have a look at the following query:

```
explain select * from dwh.t_f_sales s, t_l_customer c, t_l_employee e
where s.Customer_id = c.Customer_id and s.Employee_id=e.Employee_id
```

Table 7-6. *Accessing the t_f_sales by a specific order id. This forces the engine to use an index for performance.*

id	select_type	table	type	Possible_keys	rows	Extra
1	SIMPLE	c	ALL	Null	5	
2	SIMPLE	e	ALL	Null	22	Using join buffer (flat, BNL join)
3	SIMPLE	s	ALL	null	36	Using where; Using join buffer (incremental, BNL join)

And with that view, we can see the specific order of the join in the Table 7-6. The column table tells us the first one accessed and so on. With that, we can also see if indexes are being use and try to determine where the problem could be coming from.

■ **Note** Tuning a query is complex and requires practice. The order of the tables in a join, the use of partitioning and indexes and statistics, as well as many other factors are decisive. For more information on performance and how to interpret plans you can check the documentation of MariaDB or MySQL. The following link is a good starting point: https://dev.mysql.com/doc/refman/5.7/en/explain-output.html

---

Views and Materialized Views

Views are some sort of query calculations that are used to create something like a virtual table. In fact, the table does not exist, as the data is computed at query time based on the SQL code of the view. They are useful to hide complexity from queries to the user, and to be able to compute specific calculations without having to recode the same query over and over.

Materialized views are a step further and apart of holding the calculations and code and can contain actual data. This type of view is mainly used for performance reasons. If we have a query involving very large tables, the results will take likely too long to complete. This causes a problem, especially in a datawarehouse environment, when a business user is executing a report. It is not acceptable to ask the user to wait for half an hour, until the query finishes. To avoid such situations, a possibility is to create a materialized view, containing not just only the query logic, but also the result of these calculations.

A scenario like this helps solving complex or time-consuming calculations, as these calculations can be computed in advance. Imagine the report case we were describing. We know that our finance employee will run a specific report every day, in the morning and in the afternoon. We know, as he complained before, that the report takes too long to execute. We can precompute the result of these calculations and store them in a materialized view. Then we only need to schedule a refresh of these calculations (in fact, a refresh of the materialized view) at some point in the day, before he executes the report.

The drawback of a materialized view is that it contains only a snapshot of the data at the point in time when it was last refreshed. So, if we want fresh data, we need to refresh the materialized view first. Refreshing will take the same time (and probably a bit more) than the execution time of the query, so this clearly benefits queries that are executed many times a day, or where the data does not change too often, so we avoid refreshing the materialized views all the time.

Unfortunately, neither MySQL nor MariaDB support Materialized views at this time. But if you're using a different engine, this is clearly a reason to consider it. For MySQL and MariaDB, a possible option is to use stored procedures that emulate the code of a view, and storing the results in a regular table. Developing a full stored procedure is outside the scope of this book, but it can be quite useful. You have a good starting point at this website: http://www.mysqltutorial.org/mysql-stored-procedure-tutorial.aspx

HINTS

We saw an introduction to indexes when we were looking at the Indexes section. At this point of the chapter we used the NO_INDEX hint to ask the engine not to use a specific index. This is a good solution – to instruct the optimizer to not use an index even when it is fully operational – because you know that it will perform better than using it. That's good but it is not the use case for where hints were built. There are many hints, and each group of hints has a specific duty, but the main idea you need to grasp is that hints are query modifiers and that you influence the behavior of the optimizer engine by using them. By influencing the optimizer, you're favoring specific plans.

There are mainly two categories of hints in MYSQL and MariaDB: Optimizer Hints and Index Hints. Unfortunately, the first group is only present on MySQL Version 5.7.7 or newer. Most of the readers already familiar with Oracle database will find these very interesting, as they are quite similar to Oracle ones. With these, we can affect optimizer operations, materialize part or an entire subquery, so it is not computed every time for every value and so on. Whereas these are not available at hint level in MariaDB (they are at database level, but will affect all the queries, so you may need to change things between queries), we still have a way to influence the optimizer in MariaDB, by specifying the join order of the tables, which is usually one of the first aspects to try when tuning a query, along with the join method. The keyword STRAIGHT_JOIN placed after the SELECT statement or more common after the FROM statement will tell the optimizer to join the tables in the order they appear in the query. With that hint, you have a good way to exactly instruct the optimizer in the order of the tables used. Ideally, you would start by the most restrictive table, and end up with the less restrictive one, avoiding fetching the bigger one at the very beginning and carrying on a lot of (probably) unneeded registers from the very beginning. But it depends on each case.

■ **Note**   Optimizer hints are not available in previous releases of MySQL either in MariaDB, but you can use the optimizer_switch parameter. By using this parameter, you can control almost everything, including the type of joins and so on. In the following link, you have the supported values for both databases, depending on their version: `https://mariadb.com/kb/en/mariadb/optimizer-switch/`

The second group, available in both databases are Index Hints. These, as the name implies, are hints to control the usage of indexes. We have three possibilities: USE; IGNORE, or FORCE the use of an index. By appending these keywords just after the table name, like we did in the examples belonging to indexes, you can control whether the optimizer is able to choose or not an index. These are powerful, but require you to be completely aware of what you're doing, and require specific knowledge of how indexing works. We recommend not using them unless necessary, as likely performance will be worse than letting the optimizer choose the right approach.

■ **Note**   For more information on index hints, you can check these two articles: `http://dev.mysql.com/doc/refman/5.7/en/index-hints.html` and `https://mariadb.com/kb/en/mariadb/index-hints-how-to-force-query-plans/`

Denormalization

Denormalization is a concept that we already saw in the book, but it is an important one to have in mind. Denormalizing is the process of increasing performance by having some values repeated in the database, hence avoiding extra joins between tables, and by having data duplication in a controlled manner.

By denormalizing a database, we are increasing the performance, benefiting by having less join operations and faster retrieval of data at the expense of using more space or having some data stored many times and by adding the possibility of having data inconsistencies. This strategy is widely used in datawarehouses as usually you will be priming select statements over insert, updates, and deletes. DML sentences, like the previous try, will have more data to process in a denormalized database, so it will perform worse in these scenarios. This is something acceptable, as a datawarehouse should have much more select activity than DML activity.

Disabling Triggers and Constraints

It is not a very good practice, but to gain some more performance, it may be possible to disable constraints and triggers in your tables. However, do this carefully as disabling a constraint like a primary key may cause no issue when you are adding duplicate values in a table. So it can have an undesired effect. Only do these if you are completely aware of the collateral issues that this solution may cause.

ETL Optimizations

Sometimes it is easier to solve performance problems at the source of the data, in our case either our database holding the operational system, something that is not recommended; or in the datawarehouse database. If the issue is related to a report or a query running slow, that should be the place to fix it. But sometimes we can also experience performance issues in our ETL processes. While this is not as bad as having them than in the reporting part of the BI implementation, it can cause massive problems like delays in having the previous day data loaded, or the inability at all to have this data in the datawarehouse. In this subchapter, we will give some advice on how to solve these problems.

Offloading Operations to the Database

The first idea when working with ETL is to offload all possible work as closest to the data is. In our case, we want the database, in this case, the staging area, to perform as many works as possible. Usually this improves efficiency, because you have fastest access to the data.

If this does not improve the performance, then probably the bottleneck is in the database. Have a look at the first part of this chapter and try to identify if any if the performance improvements we explained can be put in place in the database.

Check the Network Links

If it is not possible to move work to the database, then we will need to look at the ETL. If you used PDI (Kettle), there are some tips that can improve the performance. First of all, make sure the network connection is good enough between your data sources (databases usually) and the datawarehouse and the ETL server. If we perform transformations and manipulations of the data in a fast manner, but the network links are poor, we may think it is a problem with the database or the ETL by mistake. For this it is a good reason to copy a large file between servers and compute the download and upload ratio. If we use Linux this can be easily achieved over ssh by using put and get. If it is a Windows server, we can try to copy a file from two different network drives with any sharing protocol, or over ftp and check the times.

Performance Tips with PDI

If the above tips do not work or improve the situation, let's move to a few tweaks we can do in PDI to try to improve our ETL performance.

Increase the Number of Rows in the Rowset

In every transformation, you can configure a parameter called number of rows in rowset. If you experience issues with slow performance, you might try to increase the parameter. By default, in new versions of PDI this parameter is set to 10.000 but older versions had a much lower value. As always in performance, it is a matter of testing. In theory, if you have transformations that process the same set of data over and over, or does repetitive work, increasing this number will increase the throughput. See Figure 7-3.

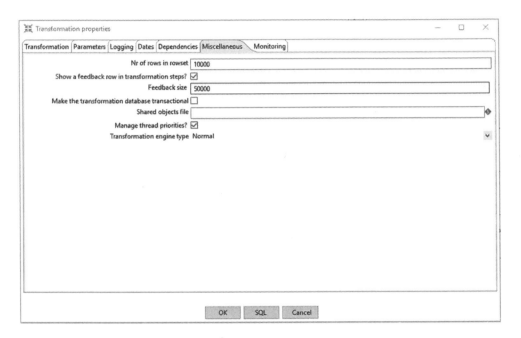

Figure 7-3. *Transformation miscellaneous properties*

Parallelize Transformations

When you are designing a job, instead of running all transformations one after another, check if it is possible to create branches with steps or transformations that do not have dependencies with any other steps or transformations that have not run yet. If that is the case, these steps or transformations are good candidates to parallelize.

■ **Note** Parallelizing things does not always improve performance. In theory, doing things in parallel can increase performance in most situations but it is not always the case. Imagine a first scenario, where at the same time we are acquiring data from a remote database, and manipulating data in another branch that was already acquired. The first will depend on the bandwidth between the source database and our etl server, where the second is all about our etl server horsepower. In that case, parallelizing the job can bring a considerable performance gain.

Bulk Loads and Batch Updates

Usually all database steps and or database connectors have an option to enable bulk mode. Bulk mode is intended to work with lots of data. Instead of inserting one register at a time, a bulk insert groups several inserts into a command, and the command is sent to the database to perform several small statements at a time. Most of the time, using a bulk loader is a very good option, as grouping small statements into bigger ones means only one transaction against the database instead of multiple ones, meaning that less overhead is generated, and, of course, increasing performance in a considerable manner. Figure 7-4 shows the available bulk loaders in the recent versions of PDI.

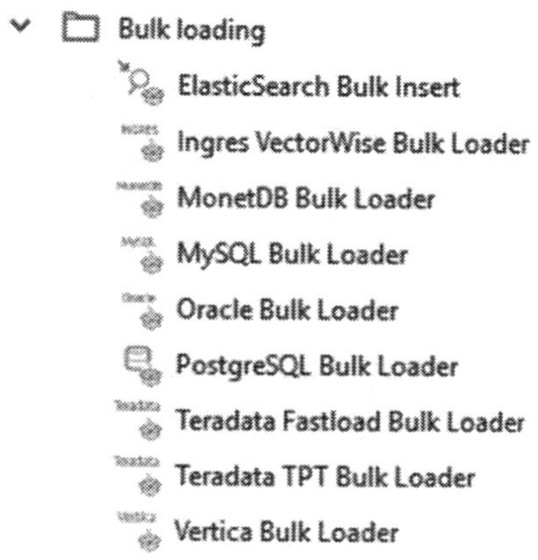

- ∨ ▢ Bulk loading
  - ⬡ ElasticSearch Bulk Insert
  - Ingres VectorWise Bulk Loader
  - MonetDB Bulk Loader
  - MySQL Bulk Loader
  - Oracle Bulk Loader
  - PostgreSQL Bulk Loader
  - Teradata Fastload Bulk Loader
  - Teradata TPT Bulk Loader
  - Vertica Bulk Loader

Figure 7-4. *Available bulk loaders in PDI*

Apart from using the bulk mode step, sometimes you can also specify several parameters in the database connection to improve performance. Each database has a different set of parameters so check the documentation of your database vendor to find the parameter and add it in the connection properties.

■ **Note**   There is a good resource on how to increase performance considerably in MySQL/ MariaDB databases by adding a few parameters in the connection. These parameters are mainly two: useServerPrepStmts and rewriteBatchedStatements. In this blog entry, you have an explanation and a guide to include the parameters and enable compression as well to your MariaDB/MySQL PDI connection: `https://anonymousbi.wordpress.com/2014/02/11/increase-mysql-output-to-80k-rowssecond-in-pentaho-data-integration/`

Apart from bulk loading, there are other parameters that can be changed or tweaked. A good option (enabled by default) on table output steps is to enable batch updates. There is a tick box in the main tab of the output table step to enable this feature. In the same dialog, playing with the commit size may help too. A large commit size should improve performance a little bit, but it can increase the time that committing a transaction takes. Also in some databases if there is not enough memory assigned to the database this can produce problems, as that data has not yet written to the data files and it is still on memory structures on the db that can run out of space. You can check both options in Figure 7-5.

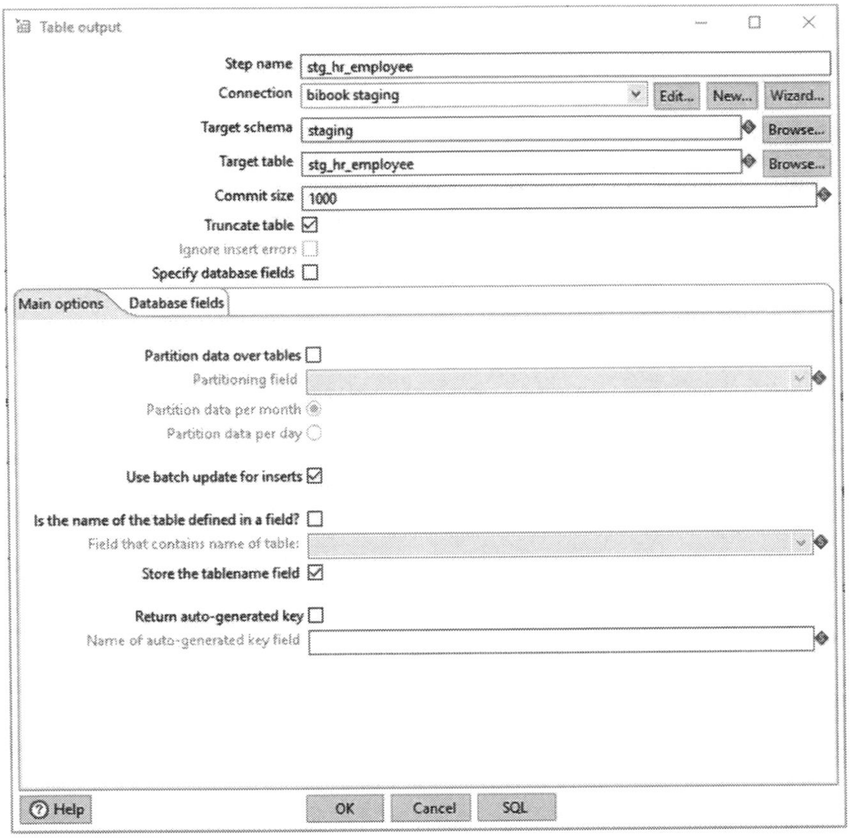

Figure 7-5. *Enabling the batch update for inserts and the commit size value in the table output step*

Ordering in Merge Operations

Some ETLs work much faster when there is a merge operation or an operation that consists of joining data from different sources if the data is already sorted. Some ETLs just sort the data when they perform some of these operations, but if they detect that the data is already sorted on a previous step, there are sometimes optimizations to avoid reordering the data. Having the data sorted in a previous step may improve the performance on a merge step.

■ **Note** In a post in the Pentaho support website, there is a performance checklist that is well worth checking. Make sure you read it carefully if you experience problems in your Jobs or Transformations. The post is available in the following link: `https://support.pentaho.com/hc/en-us/articles/205715046-Best-Practice-Pentaho-Data-Integration-Performance-Tuning-`

Conclusion

Until this point, we have seen how to deal with databases and ETL processes. We know it has been quite a journey to get there. This chapter has gone a bit beyond common knowledge to present several performance tuning aspects that can speed up ETL processes quite a lot. We did not go into too much level of detail but at least you will be able to work a first step of troubleshooting. The concepts presented here intend to be some basic guidelines, but they may not be enough to solve your problems. Our advice is to check on the Internet, forums, and other support places.

We still have a chapter pending that will act as glue: the scheduling and orchestration chapter. But before putting it together, we bet that you are impatient to see what we can do with the data we already collected. This can be seen in Chapter 10. In the following chapter, we will see how to work with reporting tools, and how to develop reports and dashboards that will answer business questions.

CHAPTER 8

∎ ∎ ∎

The BI Reporting Interface

We are advancing across the book and if you have been following the installation steps and instructions of previous chapters, you are arriving to the funny part of the BI platform implementation. By the end of this chapter we will have been able to analyze information across our BI platform in a graphical and intuitive way. The BI reporting interface is considered sometimes as the BI solution, but without all the previous work we have already done we could hardly have something to analyze inside our BI tool.

By the end of this chapter you will be ready to answer business questions such as which products perform better, which of the top five customers have better Net revenue vs. Gross sales ratio, which products are related among them in terms of sales, how much stock do we have in our warehouses, and how many months we cover with this stock based on previous year average sales. All these questions and any others can be answered with BI tools if we develop the required visualizations.

General functionalities and concepts have been evaluated in Chapter 1, so we will do a quick introduction and we will focus during this chapter on the design itself of the BI part. In order to do that we will use different tools that will allow you to choose which solution fits better to your needs. In order to have available a free or low-cost solution for the BI interface we have mainly two possibilities: use an open source tool or use some free edition of commercial tools. Open source solutions allow you to implement a free solution with the guarantee that it will remain free in the future, but with the risk of having lack of support in case your project grows up and you want to set a scalable solution for a big number of users. Free editions of commercial solutions will allow you to evolve from the free edition to the commercial one and get support from the company (of course, after paying licenses and support price).

We have some experience with open source tools such as Pentaho or Jaspersoft solutions, and as in main commercial tools, these two vendors have also a commercial version. In the open source solution you don't have ad hoc reporting possibilities; you can only write some queries to launch over the database and then use the results in a near-to-programming interface, which is not very intuitive for end users. So what we are going to explain in this chapter is how to use free editions of commercial tools such as Microstrategy Desktop, Power BI, and Qlik Sense as free editions of Microstrategy, Power BI PRO, and Qlikview platforms.

How to Choose the BI Tool

We were talking also during Chapter 1 about BI approaches, and we saw that you have available strategies of **Query & Reporting** to launch ad hoc reports, **Information Sharing** to deliver information to all the company, **Dashboarding** where you can play in a single screen with multiple selectors and panels, and **Data Import** and **Data Discovery** tools that allow end users to have a user-friendly interface to add its own data and investigate trends and patterns. This classification can be combined with the project strategy that you use for your BI project and after doing this analysis, you will be able to decide which BI tool fits better for you. In order to correctly decide, you will also need to take into account if you want to go for a free solution or if you have the budget to purchase a commercial edition. If in this point you think that you can go for a pilot project and then you are convinced that you will have the budget for licensing to go for the big project,

© Albert Nogués and Juan Valladares 2017
A. Nogués and J. Valladares, *Business Intelligence Tools for Small Companies*,
DOI 10.1007/978-1-4842-2568-4_8

then a commercial option with a free solution for the pilot would be a good strategy. It is not our aim to sell you any commercial tool but usually commercial tools are more robust, with support, with more investment for development, and innovation so it should be something that you should consider as a valid option if you are in this situation.

Your project strategy can be a classical approach with static reports based on a structured datawarehouse and focused on big developments at a database level that requires technical knowledge to develop and implement user requirements, or new data discovery approaches that are focused in visualizations and high dynamism for the end user, giving him autonomy by importing data himself and playing with different visualizations until he finds the required patterns.

Classical reporting projects require also a development of an infrastructure inside the tool that causes long term and big efforts in project development until the end user can see the results. By contrast, there are more robust solutions, especially for big environments with hundreds of users and thousands of report executions per month.

On the other hand, new BI platforms focused on user auto-service allow them to investigate the insights of the data, by adding investigation capabilities that facilitate the task of getting relevant information from our data, such as drilling, paging, filtering, selecting, and crossing information from multiple sources. Throughout this book we are analyzing tools for small/medium scenarios so we expect that you will be possibly thinking of a pilot project before going ahead for a bigger project. Due to that we will see tools focused on data discovery that will perform fine in these small/medium projects. Once you have done this pilot, you will be able to evaluate if the product tested fits your needs or if you prefer to go for licensing to evolve this pilot inside the chosen tool, or if you want to evaluate other platforms targeted for bigger environments.

Basics tool for classical analysis are reports and documents where users can see plain data filtered, ordered, with the ability to export to Excel, using BI as sourcing for Excel. The basic tool for data discovery analysis is the dashboard. While you can create simple reports that allow you to do static analysis with more or less detail, a dashboard contains a lot of information available but it only shows portions and aggregations of the information and then it allows filtering, segmenting, aggregating, or drilling across the information to achieve meaningful analysis. While reports and documents are used as an intermediate tool to extract and analyze data, the dashboard can be the final result in the sense that you can show it directly in a meeting and perform some analysis to show your findings inside the data. Based on this circumstance, visual appearance and design of the dashboard have at least same importance like the data inside. Let's analyze some best practices that will help you to develop useful dashboards.

Don't worry about choosing. There are solutions, especially commercial ones, which have all the approaches available so you can do a pilot project importing Excels to see how it looks like and then developing a project with all the related structure inside the BI tool, tables, attributes, dimensions, facts, hierarchies, metrics, filters, etc.

As this book is focused on free or cheap tools we will have limited possibilities, going just for free editions of commercial tools that are focused on allowing data discovery for end users using dashboarding capabilities. We won't want to enter in development using tools like Jaspersoft or Pentaho because it would require a bigger book. So let's see how to build nice and useful dashboards using data discovery tools in the next section.

Best Practices in Dashboarding

While reports are plain and there are few considerations for design as far as the result is usually a table or graph, when you are creating elaborated documents and dashboards, there is a set of recommendations that we usually apply during our developments.

Within the recommendations that we are going to analyze in this section you will see that some of them are quite logical and others that will seem arbitrary. Especially this second group of recommendations is based on our experience in designing dashboards and the user feedback from these dashboards. A dashboard must be intuitive, user friendly, which don't require any instruction manuals to know how to use it. If your users need help to analyze the data inside the dashboard, then it is because the dashboard is not clear enough. If you follow the set of recommendations below we are pretty sure that you will get successful results and positive feedback from your client, which will be the end user of the information.

Starting from Top Left

We read from the top left to the right and then to the next line. At least we expect that you read this book in this way; if not you won't understand anything about this. When information is graphical and you have a screen with multiple graphs and grids of data, we tend to do the same, so it will be interesting to locate the most relevant information or graph on the top left of the screen. In the same way the logic of the analysis should follow that order. If you define a graph that is used as selector of the rest of dashboard, it is recommended that it is located in the top left, as shown in Figure 8-1. In this example the area selected in the Heat Map graph will filter the Category distribution grid and the Evolution graph. Also the category selected in the Category distribution grid will affect only to Evolution graph. If you open this dashboard without any other instruction, it would be difficult for you understand that selecting a category you could change data shown in the Heat map. If you want that Category, the distribution grid acts as selector of the Heat map and the Evolution graph would be much more intuitive if you locate it on the top left.

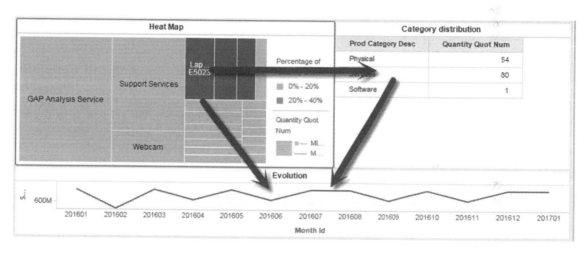

Figure 8-1. *Analysis workflow*

Joining Related Information

Inside a company, almost all information is related to a greater or lesser extent. The number of employees will possibly affect the amount of sales, profit, or operations. Also external data can be related with internal data, and the population of a region will affect the overall sales that our company is performing in this region. So trying to set all related information in a single dashboard is in fact impossible. Anyway, what we need to try is to have a complete analysis inside a single screen containing the majority of relevant information. Of course you cannot see all the information in a single screen so we will try to use selectors, filters, dependent visualizations, and drilling to be able to see as much information as we can in a single screen. As commented in Chapter 1, it is important to take into account which is our audience. We can develop an initial dashboard with a summary of most relevant KPIs in the company that will be shared with all employees and then develop other dashboards that have more concrete information and details of a single area, department or team, adding the required level of detail in each dashboard that will be targeted to its own audience.

Focus on Relevant Data

A single screen has limited space available so it is important to focus only in those KPIs that offer more added value to the decision-making process. It is important to follow up those KPIs that our management has decided as company goals. Based on the hypothetical target of increasing our sales 20% versus the previous year we will need to focus on sales measures. If the objective has been defined based on amount then we will need to focus our reporting on sales amount, if the objective has been defined in quantity we will need to focus on quantity, if the objective is in volume we will need to focus on volume. Crossed information can be useful but not relevant for the analysis so it can be avoided or at least moved from the initial screen of the dashboard. If our management has defined an objective of reducing overall costs in all our processes we need to focus on cost KPIs, segmented by cost centers, cost elements, cost nature, etc. If the goal of our company is to achieve a market share over our competitors we will need to analyze market share, new products in the market, new competitors that can appear, or external factors that could affect the market share such as regulatory affairs.

Added to this focus on relevant data it is important also the usage of meaningful KPIs. It can be important to know that we have 25% of market share for a given product category but it will depend on the main competitor share so we will need to see the comparison between both KPIs rather than the single market share. If we have a target of reducing our costs it will be relevant that we have decreased them below $100,000 our global costs, but it will be more relevant if we compare it with the target that we have fixed. If our target is increased 20%, our sales will be relevant to the real increment of sales compared with this 20% of target.

Formatting Recommendations

It is important which data we are going to show but it is also important how we show it. Over time it is more and more common to see the role of graphical designer inside BI development teams to help in dashboard design and user experience of end users, and make sure we are fully aligned with this strategy. It wouldn't be the first time that we see a really useful dashboard that we are showing to an end user and the first comment is, "ok, we can see all the required information together, we can play with the information and it will be very useful for our department but titles are in blue and our corporative color is green." So we would like to go ahead with some recommendations about formatting. You can think that most of the following recommendations are perfectly valid when you are creating slides in PowerPoint and you would be right. At the end of the day a dashboard can be used as a presentation document so the same rules apply for both cases.

Corporative Integration

Following a corporative style in dashboard design is the formatting element most important that we need to take into consideration if we want to obtain positive feedback from our internal customers. We will, as an organization, feel better identified with a document, screen, dashboard, or program if we see the corporative style present in it. So it is important to show company logo, that colors in the dashboard are corporative ones, that we use corporative font type, corporative images, etc. If your BI tool allows it, it is recommended also to have links to the corporative web page or Intranet inside the dashboard. Also we can locate some documentation about how to use the dashboard inside the Intranet and locate inside links to our BI tool.

Regarding one concrete topic, the chosen font type, also an added recommendation, it will make it easier to read the dashboard if we use a single font type in all texts that appear inside. It can be necessary to change font size but all of them should be with the same font type.

Related Data Alignment

As commented, there are some recommendations that seems to be obvious but we prefer to discuss them because it wouldn't be the first time that we see something similar. In Figure 8-2 you can see an example of a dashboard with two graphs located next to each other, both of them at category level. But as you can see, categories are not aligned between graphs. It is quite obvious that the best way to compare both graphs is aligning the categories, but in this case you have a graph sorted by Sales KPI and the second one is sorted alphabetically, so it is difficult to see the relation between sales and net revenue for a category; your view must move up and down between graphs.

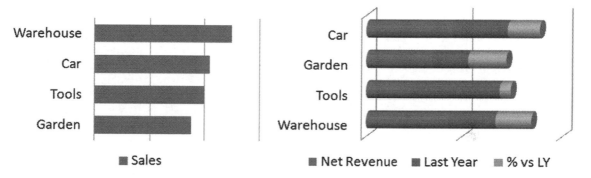

Figure 8-2. *Graph misalignment*

You should try to keep the alignment both horizontally and vertically if the detail is similar across graphs and grids. Of course you should try to keep alignment in size of the objects and position, and sometimes it is not easy because the relation between frame and internal figure is not the same. It is important for this alignment that the BI tool that you use allows you moving objects at the same time that you see data because you can see how the final result look. These kind of technologies are usually named WYSIWYG, acronym of What You See Is What You Get, because the final result is the same that you are seeing while editing.

Conditional Formatting

As commented in previous sections it is important to focus only on relevant data and the usage of relevant KPIs that allow us to compare real data with our budget. But, what about the possibility of using conditional formatting to remark that some data is more relevant than the rest? In most BI tools you have the possibility of using some way of conditional formatting to remark that values that are more relevant than the rest, for example, we could mark in green which cells have achieved the objective, in orange the ones that meet 80% of the objective, and in red the rest of them. Conditional formatting will facilitate the decision-making process by helping to identify where to act. You will find it with different names across different BI tools, thresholds, alerts, or conditional formatting but it is something that you can usually use. You have an example of usage in Figure 8-3.

Category distribution

Prod Category Desc	Quantity Quot Num
Physical	54
Services	
Software	

Figure 8-3. *Conditional formatting*

Intensity vs. Color

The previous example clearly shows a reverse example regarding what we want to comment on now. This color combination, yellow, light green, and red, is much too harsh, and it can also affect the real understanding of data below these colors. It is quite possible also that you find now another problem by using different colors; if the dashboard or document is printed in black and white, it can be difficult to differentiate which cells are the relevant ones if they don't have a part with different colors and different tones. So the recommendation is in general to use different tones of the same color, as you can see in Figure 8-1 in the Heat map graph where we are using different tones of blue (we prefer to specify because it is quite possible that you are reading this in black and white) instead of using different colors as you can see in Figure 8-3.

Visibility of Data

Another recommendation related to the previous one is to use contrast between graphs, grids, texts, and cells containing the information, frames, and background in order to have a proper visibility of data. Try not to use an excess of framework structure, rectangles, lines, figures, and images that can disturb the user and cause loss of attention on the important data. Also 3D graphs, shadows, or 3D effects could improve graphical quality of the report but they can affect the data analysis by hiding the data behind.

Graphic Usages

Inside best practices we would like to open a special section to talk about which graph type fits better in each analysis as far as you will see each set of data has some graph type that matches better than others. As general recommendations when using graphs, we would consider these:

- Limit the usage of circular graphs because they are difficult to analyze, only in some specific cases can they be useful.

- Try to use graphs that have an efficient usage of available space, limiting the usage of legend, axis, or data labels to those graphs that really require using them to understand the graph.

- In case of using data labels try to use them without decimals or keep them as minimal as possible, or if you are showing thousands or millions try to use K and M notations (10K instead of 10,000, 5M instead of 5,000,000).

- Use effective graphs. If due to data nature, the most effective graph for all the required information is Bar graph, use four Bar graphs. Don't try to add variety to the graph types shown just for adding variety.

Let's analyze also some use cases of graphs that can help you in the definition of your dashboard.

Pie Chart – Percentage Contribution for a Few Values

The usage of a pie graph, as commented just before, should be limited to some conditions. It must be targeted to analyze the contribution of each value to the overall amount (percentage analysis) for a small number of values, and it is especially useful if you want to show that one of the values is much higher than the rest. In Figure 8-4 you can see a double example of usage for a pie graph, one by product category and another by product. In the first one you can easily distinguish which category (*Services*) is more relevant and how much is the relevance (about 60%), while in the second one is really difficult to compare across different products which is more relevant. You can see *Webcam* as the best-selling product but anyway is difficult to compare with *Laptop E5023* that seems to be the second one and then if you try to locate the third, fourth, and fifth seem to be really difficult to identify.

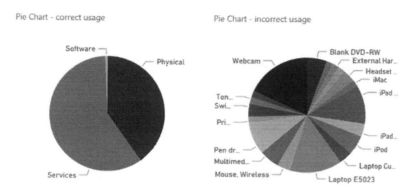

Figure 8-4. *Pie chart examples*

Stacked Bar Graph – Percentage Comparison

If you want to see a percentage of the total when you have multiple values and check it across multiple categories, a good option is stacked bar graph absolute. This graph will have always the same height and then intermediate sections will vary depending on the metric, as you can see in Figure 8-5. In this example we are seeing the percentage of sales for each category across our customer regions. It is easier to distinguish which is the most relevant category in each region and how it is changing the category contribution across regions. You need to take into account in this graph that we are seeing an overall distribution per region, so we fit to 100% the sum of all categories. You need to be clear enough that they are not absolute values; you can see that *Car* in *Central* region is taking more than in *South*, but it affects only to the percentage, maybe in absolute values *South* has higher sales than *Central*.

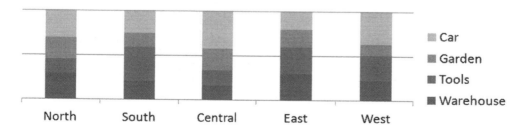

Figure 8-5. *Stacked bar graph example*

Horizontal Bars – Top N Elements Comparison

Another typical graph that we can see in multiple documents is the horizontal bars graph. There are some use cases where we recommend using this graph, and one of them is the Top N elements comparison. In order to easily see which top elements are performing better for a given attribute, it is highly recommended sorting the graph by metric values, so the first bar will be related to the best-selling item, as you can see in Figure 8-6. Here you can see easily that Webcam is the top seller item, that second one is Laptop E5023, third one iPad Mini, etc... much easier to see it than in Figure 8-4 with a pie chart.

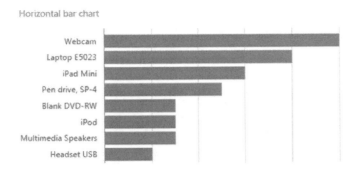

Figure 8-6. *Horizontal bar chart example*

Vertical Lines, Bars, or Areas – Time Evolution

We usually represent the time concept as a horizontal line, advancing from left to right. So in any evolution graph you want to show the evolution of some given KPI across the time you need to use vertical bars, areas, or lines that show time advance horizontally, with a vertical bar next to the other showing how it is evolving. While bars can also be used for comparing another type of attribute, nontemporal, lines, and areas are widely used just for time analysis. We usually relate the line with the evolution so having a line graph to compare just categories or items is not correctly understood by end users. Also we recommend using combined lines and bars when analyzing multiple metrics, a usual graph that we have used in our developments is using the bar to analyze real data and then using a line to mark the objective or the variance regarding the previous year. This last example is the one shown in Figure 8-7, where you can see current year sales, previous year sales, and the ratio of one versus the other.

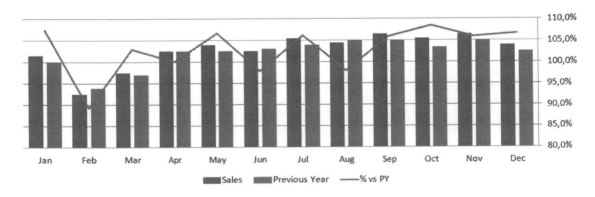

Figure 8-7. *Vertical bars and lines example*

Vertical Bars – Histogram

Using a histogram we try to analyze data dispersion for a metric, trying to see which frequency of repetition of a value of the metric we get. With the histogram analysis, we can see which value appears as the most repeated and which dispersion this value has. The best representation for a histogram analysis is the vertical bar graphic. There are multiple examples of histogram analysis:

- Analysis of qualifications in an exam/test

- Analysis of average pollution index in a city

- Analysis of most used shoe size per shoe model (see Figure 8-8)

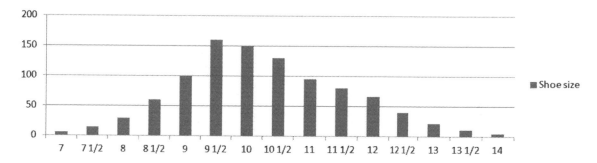

Figure 8-8. *Histogram example for shoe size*

- Analysis of people's age

Bubble Chart – Correlation

A bubble chart allows you to compare multiple metrics among them; depending on tool capabilities, it will allow you to compare up to four metrics, to see the correlation among them. You can compare the relationship between vertical and horizontal position, the relationship between position and size, and the relation between all of them with color. In Figure 8-9 example we are trying to analyze sales inside all states of the United States with the population of each state and the number of seats in the House of Representatives.

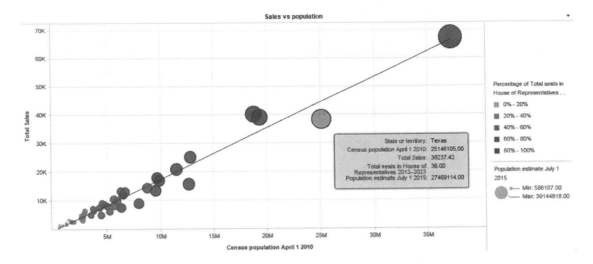

Figure 8-9. *Bubble graph to compare multiple metrics*

In this example we can see in horizontal position the population of each state, in vertical position total sales in the state, colored by number of seats in the House of Representatives and sized by expected population in the next five years. As you can see our sales are quite related with the population because they seem to follow an increasing line. That bubbles over the trend are selling more per capita, the ones below trend have less sales per capita. Also the BI tool used in this example (Microstrategy Desktop) allows us to see a tooltip with the detail of data and we have located the mouse over Texas that seems to be below the trend, to see the related data for the bubble.

BI Tools

We have seen until now some theory about creating dashboards, which graphs fits better for some examples that we expect that you can extrapolate to your needs inside your company. We would like now to have a look at three tools: Microstrategy, Power BI, and Qlik Sense and how to install them and basic instructions to get some dashboard out of them.

Microstrategy Desktop

Microstrategy platform is one of the traditional BI tools that we can find in the market since more than 20 years ago. It is a mature technology that in the enterprise edition has multiple components, starting with the Intelligence Server, the core of the process; Web Server where most of users connect; Mobile Server that provides service to mobile apps for smartphones and tablets; Narrowcast Server and Distribution services to deliver information in email, printers, shared resources; Report Services that provides the ability to define advanced documents and dashboards with pixel perfect technologies; and high interactivity and OLAP services that provide in-memory capabilities to improve the performance of the system.

Design process is an enterprise scenario that can be much more complex than the one we are going to explain for Microstrategy Desktop as far as this is only a small component of the whole architecture, but all the explanations that are written below are completely valid for Dashboards belonging to Report Services inside the platform.

As commented, Microstrategy has also a free edition, named Microstrategy Desktop that provides part of Report Services interactivity powered by some features of the Intelligence Server. It is an installable application for stand-alone computers that allows you to connect to databases; import flat files; connect to network files; connect to Big Data engines; open multiple sources such as Facebook or Twitter; and import files from shared platforms as Dropbox, Google Drive, including import data from the Windows clipboard, as you can see in Figure 8-10 that shows the initial screen when adding data to a dashboard.

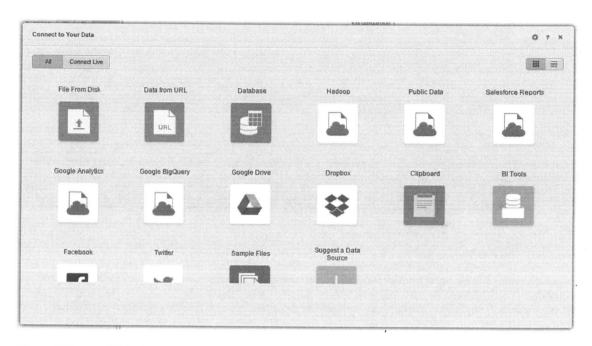

Figure 8-10. *Available data sources in Microstrategy Desktop*

But before using it you will need to install the software in your computer. And in order to install Microstrategy Desktop, the first step, as you can imagine, is downloading the installation package from Microstrategy page. You can find it at `https://www.microstrategy.com/us/desktop` where after filling a small form with your name, company, and email you will be able to download the installation package. You can choose Windows or Mac versions to install; of course it will depend on the computer that you are going to use to work with it.

Once downloaded, just uncompress the downloaded file and launch the installable file, named MicrostrategyDesktop-64bit.exe. You can install it keeping the default installation options (well, of course you can locate the installation wherever you want) and after the installation you will be able to open the tool. First time you open you will see an introduction page with links to videos and different help pages, but in the top you will see the link to Create Dashboard. If you mark that you don't want to see the tips again, next time you open Desktop you will be redirected to the Dashboard Editor that looks like Figure 8-11.

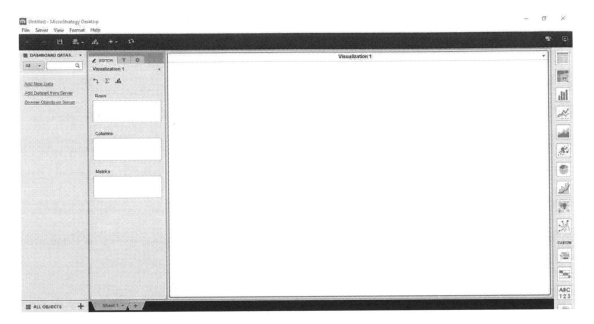

Figure 8-11. *Microstrategy Desktop interface*

In this editor we can find different areas to work with. In the top of the editor we have the main menu to access the different options. The first panel in the left is where we will have our data sources to use them in the dashboard, currently empty until we don't select any dataset. The second panel contains the visualization editor, filter editor, and configuration options for each visualization. The main panel that appears with the title Visualization 1 is the visualization panel, where we can locate one or more visualizations. Finally in the right we have the visualization type selector to choose which visualization we want to use in the visualization panel.

In order to view information from our database we will require having at least the ODBC created that connects to the database. Maybe you have it already created if you are accessing the database from the computer where you have installed the Microstrategy Desktop tool, if not you will need to create it with the proper ODBC driver (you could require some database client installation). In our case we will use our ODBC driver already installed to access to the tables with a read-only user. You can also connect using an ODBC less connection but at the end of the day it means that you won't have available the ODBC for other purposes, the ODBC is embedded into Microstrategy but you need to specify the ODBC library anyway. As a suggestion you can always use MySQL ODBC to connect to MariaDB and vice versa, so if you have one of them installed it should work to connect to the other database.

It is recommended to have a read-only user to retrieve information from the database with access to all tables but only with read permissions in order to ensure that in case of different teams, the BI tool team doesn't modify any object in the database that could affect the responsibility of the ETL process.

When adding new data coming from the database, Microstrategy Desktop will offer you three possibilities: Building a new Query, Writing a new Query, or Select tables. With the first option, the one that we show in Figure 8-12, you can build a query by selecting the tables, defining the joins among them, then selecting which fields of each table you want to add to the dataset, and finally defining if there is any function to apply to some field. Once done you can change to SQL mode to see which SQL query is going to launch in the database that is the same SQL editor that directly opens if you select the second option.

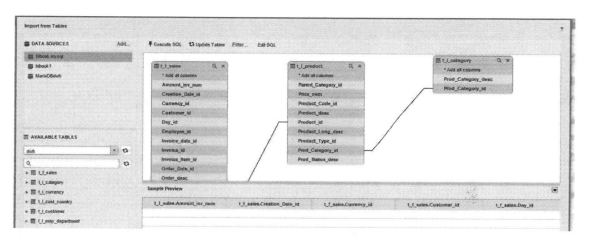

Figure 8-12. Building a query

Once you have selected the tables that you are interested in, you will be able to publish it as it is or click on the Prepare data button. As you can see in Figure 8-13, Microstrategy Desktop automatically identifies based on the data type which fields can be Attributes to analyze the data and which of them can be metrics to be analyzed. If some of them have been incorrectly assigned you can change it just with a drag and drop from Attributes to Metrics or using the right button of the mouse to change it. Also in this screen with the right button you have access to a very powerful configuration where you can define multiple functions for some types of known fields. In case of a date field you can define the whole time hierarchy, Year, Month, Month of the year, Week, etc. In case of geographical attributes such as Country, City or Region, you can define them as geographical and it will allow you to draw it in a map. Once defined all column types you can then import data to be able to analyze the information disconnected from the database with the option *Import as an In-memory dataset,* or you can define the table as a link so anytime that you open Microstrategy Desktop it will run the query again refreshing data from the database with the option *Connect Live.*

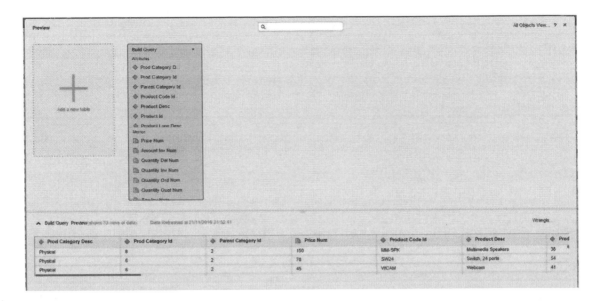

Figure 8-13. *Prepare data options*

With Wrangle button inside the Prepare Data screen you can add functions, transformations, and modifications to the data load in order to adapt the source information to your needs in case that source data has some missing requirement. You can add new derived columns, concatenate columns, or ignore some column if you don't want to include too much information in memory.

It is possible that you have all the information in a single dataset, but it is also possible that you require joining more than one dataset. In this case you will need to define also which attributes will define the relation between datasets just by right-clicking on the proper attribute and selecting the option to link it with the other dataset. You can also link multiple datasets in the Prepare data screen by clicking in the big button in the left of the screen with Add a new table table, as shown in Figure 8-13. Each dataset can provide from different sources but be careful on this because values of the attributes must match across different sources in order to get logical results. You can configure also the join type and define which datasets are primary and which are secondary; in this way you can modify the results of the combination of multiple datasets. If all of them are primary all possible combinations will appear. Imagine that you have a dataset with sales data including the country attribute and another that have just country and population to compare your sales per number of people. If both are primary you can have the population of Uganda but you are not selling in Uganda, so maybe you just want to set the population dataset as secondary to show only data for those countries that you have sales data.

Once we have selected all the source data that is needed for our analysis we can start the analysis itself. Analysis is done using visualizations. You can use a wide set of visualizations specially combined with the options that you have in the configuration tab:

- **Grid**: Single table with rows and columns to analyze data in text format.

- **Heat map**: Visualization area is split in squares through an attribute with the size based in a metric and the color based in another metric.

- **Bar chart**: They can be horizontal or vertical bars, absolute or stacked, showing as height a metric value, split by some attribute with the possibility of changing the width, and the color of the column using metric values.

- **Line chart**: You have similar possibilities than bars, also with the possibility of changing markers.

- **Area chart**: Similar to bars but showing areas instead of bars.

- **Bubble chart**: You can compare up to four metrics split by at least one attribute. One metric defines horizontal location, next one vertical location, other metric shows marks the size of the bubble and another metric the color.

- **Pie chart**: you can define the angle using a metric, color, partitions, horizontal and vertical segmentation based on attributes, and also change the size in case of multiple pies returned.

- **Combo chart**: Combination of bars, lines and areas in horizontal or vertical format.

- **Map**: Geographical localization of metrics based on geographical attributes such as country, region, city, or zip code to draw areas and also using latitude and longitude to locate markers.

- **Network**: Useful to see the relation between two attributes, typical example is number of flights between airports; you can define source and target attributes and then apply color and size based on metrics.

- **Custom**: Finally you have the possibility of adding new advanced visualizations from open source pages that have available hundreds of customized visualizations such as the ones available in:

      https://github.com/d3/d3/wiki/Gallery

Each visualization has its own specifications to work, a minimum number of attributes and metrics that will allow the visualization to work. In Figure 8-14 we show an example of Heat map visualization. By locating the mouse over each visualization button a tooltip will appear with the minimum requirements, as you can see in Figure 8-14, in order to use a Heat map you will require minimum one attribute and one metric. In this same figure we can see the Editor panel where you can drag and drop different attributes for Grouping that will define the number of areas in the visualization, in the Size By area you can drag and drop some metric that will define the size of each area, you can use the Color By to define the basis to color each area, and in the Tooltip area you can drag and drop attributes and metrics that will be shown when you locate the mouse over one of the areas.

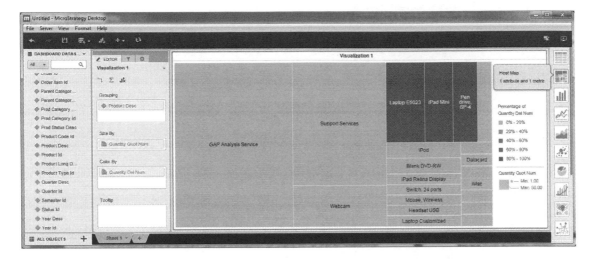

Figure 8-14. *Heat map visualization example*

As commented, each visualization has its own characteristics, so areas opened in the configuration tab will be different in each case. As in the rest of the chapters, this is not a deep analysis on how to use a single tool, so we are not going to see the details that each visualization requires and all the options that you can use in Microstrategy desktop but let's try to summarize main actions that you can do:

- **Multiple visualizations in one sheet**: You can add new visualizations and by default it will split the visualization area in two parts. Then you can drag and drop the visualization to split horizontally or vertically and adjust the size of the visualization.

- **Link visualizations**: You can use visualization to filter data in the rest, so when you click in one area or cell of one graph it will filter the information seen in the rest of the visualizations. In order to do that you need to click on the arrow located in the top right in the visualization and select Use as Filter for that visualization that must act as filter of the rest.

- **Add filter conditions**: Filter conditions will apply to all visualizations in the page; you can filter by metrics or by attributes, defining also how they interact, filtering, excluding or remarking.

- **Add selectors**: In order to filter the information that you are seeing you can add selectors in different formats, button bar, link bar, check box list, slides, etc. In this case you can define which target visualization will be affected by the filter.

- **Use Images and text**: You can also combine images and texts to add titles, comments, explanations, links, and logos.

- **Multiple pages**: You can add new sheets of data by clicking on + sheet shown in the bottom of Figure 8-14.

- **Formatting**: You can change colors, lines, font, and letter size and multiple other options to adapt the dashboard to corporative colors.

- **Thresholds**: You can apply conditional formatting to cells, maps, bars, and markers to most of visualizations.

- **Drilling**: You can define drilling if your attributes have been defined one as parent of other (month as parent of day, for example).

- **Create derived calculations**: You can add formulas that are not directly inside the data, counting attribute elements or defining any other derived formula.

Microsoft Power BI

Microsoft is the most extended software provider so it is quite possible that inside your organization you have already installed Microsoft software and quite feasible that your own computer is using some Windows version, so we are pretty sure that it is not required that we introduce you to Microsoft. But maybe you don't know that Microsoft is offering a free BI tool to create reports and dashboards named Power BI. You have also available Power BI PRO that offers you the possibility of working in a shared environment, whereas in the free edition you will work in your computer using a Power BI Desktop version. PRO version will also provide you with better performance in the dashboard load and bigger data space per user.

In order to install it you can access to `https://powerbi.microsoft.com/en-us/desktop/` and download the installation package. In this page, you will also find some links to tutorials that will provide you deeper knowledge about how to use the tool. Proceeding with the installation itself it is quite simple by clicking on the next button, and finally you can directly open the tool accessing to a screen that should look like as Figure 8-15.

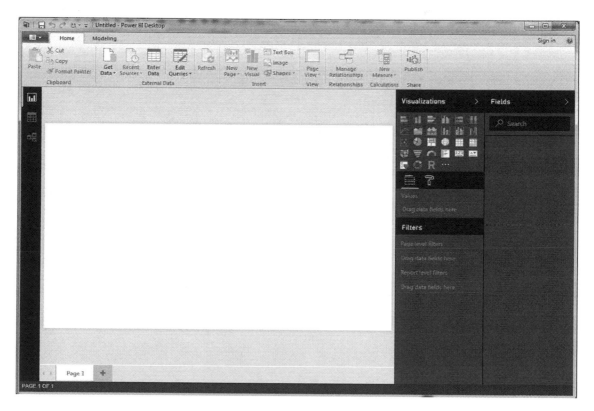

Figure 8-15. *Power BI Desktop interface*

In this panel you can see the button and menus area in the top of the screen, a link buttons vertical bar in the left, the visualization area as the biggest blank area, and then the visualizations panel and the field's panel. Below you can also see the page selector where you can add more than one page if it is required for your analysis.

As a first step, common to all tools, we need to select the data that we want to include in our analysis. In order to do that, we need to use the Get Data option. As we are going to connect to our MariaDB database, you won't see any option directly in the data source list. But if you go to More, then a new menu opens. In this menu we cannot find MariaDB option, but as commented we can use the MySQL driver, and this one is available in the list. If we try to connect, a warning will appear because the NET driver for MySQL is not available, as shown in Figure 8-16. But clicking on Learn More will redirect you to the MySQL download page where you can download the net driver.

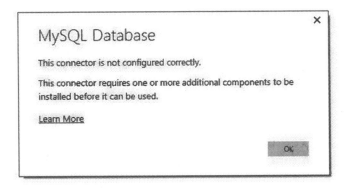

Figure 8-16. Warning message due to driver unavailability

Once installed, (again a simple installation of Next-Next-Next), if you try again to add a Data source from MySQL, then a new window will appear asking you the server name and the database to connect to. In the following next screen, it will ask you about authentication, and finally you access the Navigator menu where we are going to select the same three tables as in the Microstrategy example, t_f_sales, t_l_category, and t_l_product, as shown in Figure 8-17.

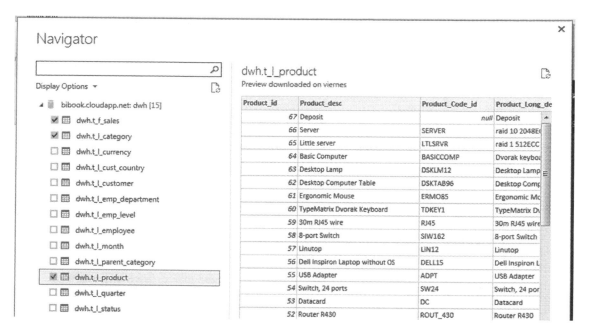

Figure 8-17. *Adding tables to Power BI analysis*

The next step is to define the links among tables. Some of them are linked automatically; in the example we have found that the relation between product and category tables has been correctly defined, but others must be set up, for example, the relation through product_id between sales table and product table has not been created automatically. To manage relationships you need to go Manage Relationships button in the menu. You will see how to create the relationship between Product and Category tables, and by clicking on New button you will be able to create new relationships if there is one missing. In this case we will create a new relationship for Product_id field between sales table and product table.

Another issue that you will see is that by default numeric fields are considered as aggregable as you can see in the right part of Figure 8-18 for Currency, Customer, and Employee identifiers (these are the one shown in the figure, but most of the attributes have been considered as metrics). In order to change this behavior to have them available for analysis, you will be required to change them properly to remove the aggregation.

Figure 8-18. *Modeling options in Power BI*

This can be done in the Modeling menu that is also shown in Figure 8-18 in the Default Summarization option, selecting the required field and choosing in the drop-down menu the option of Do Not Summarize, already shown in this figure. We will do the same for all numeric identifier fields that we see in this situation. Once our mode has been defined with key attributes without the summarize option and linked across tables that contain the same concept, we can start drawing our dashboard by inserting visualizations in the visualization area. In Power BI you have similar visualizations as in Microstrategy, but there are also some other different ones as you can see in following list were we talk about main visualizations available:

- Bar chart: you have horizontal and vertical bars, clustered and stacked, and also 100% stacked. In this case you can fix only value and color.

- Line chart: horizontal lines are available to add to the visualization,

- Area chart: you can have clustered and stacked areas.

- Mixed line and bar chart: bars can be clustered and stacked.

- Waterfall chart: this is new regarding default installation in Microstrategy. You can see how varied a metric it is both in positive and negative for a given attribute.

- Scatter chart: it is similar to a bubble one; you define the location of each point based on two metrics and then you can change color and size. In this case you can also add a play axis that shows you in a moving graph how values vary for each attribute value in the play axis.

- Pie chart: basic pie chart defined by some attribute and some metric.

- Treemap: same concept than Heat map, square areas whose size is based on a metric.

- Map and filled map: location of values over a map based on location, longitude, and latitude. Filled map is showing areas, standard map is showing markers.

- Table and matrix: plain text information organized in rows and columns.

- Funnel: horizontal bar with center bars that allows you to show the relation between elements in percentage.

- Gauge and donut: similar to pie chart but with different shape.

- Slicer: it acts as a selector for the rest of charts.

You have also the possibility of adding customized visualizations. In this tool all visualizations are automatically linked so when you click on one area or shape of a visualization, it shows how it is affecting the rest. As shown in Figure 8-19, when clicking on an India donut segment, the map on the right is zooming in to show just the India part of the map:

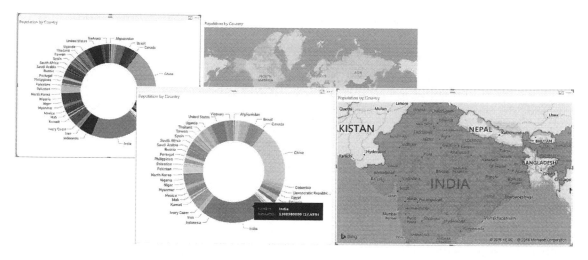

Figure 8-19. *Linked visualizations in Power BI*

You can access all options for each visualization using three main buttons, the ones shown in Figure 8-20. The button with an icon showing two boxes is the one to configure what is shown, which attributes effects to colors, to size, etc. The button with the roller shows options for formatting such as colors, titles, background, or borders. Finally, the button with a graph inside a lens has options to add analytics such as calculating averages, trends, maximum lines, or constant lines. This last button only has options for some of the visualizations.

Figure 8-20. *Visualization configuration buttons*

Regarding options that you can do to create dashboards in Power BI you have available the following:

- **Multiple visualizations in one sheet**: This editor is more flexible to add visualizations as far as you can locate and size them as you want, and it doesn't require you to use the whole space available.

- **Add filter conditions**: You can have filters that apply to all pages or to a single page.

- **Use Images and text**: You can also combine images and texts to add titles, comments, explanations, links, and logos.

- **Multiple pages**: You can add new sheets of data by clicking on + sheet shown in the bottom of Figure 8-15.

- **Edit Interactions**: It allows you to choose how graphs are acting when you select some area in one of them.

- **Create hierarchies and drilling**: You can define hierarchies and use them to drill across data.

- **Edit relationships**: You can modify the relationship among tables also in a graphical editor.

- **Create measures, columns and tables**: Using existing fields you can create new formulas, attributes, and queries considered as tables.

Qlik Sense

Qlikview is one of the first actors that appeared in the Business Intelligence scenario with the idea of data discovery. Since the beginning, this tool has been focused on minimizing development time comparing with classical BI approaches, giving also to the end user the possibility of play with data to discover trends, exceptions, and relevant data in a user-friendly way. We have been working with Qlikview for some years and it is quite a user-friendly tool, but the last version that we have installed for testing purposes related to this book has surprised us in terms of easiness of use and graphical interface. You will see during this section some print screens about the tools that show what we think that is a quite advanced layout.

Qlik Sense Desktop is a free tool developed by Qlikview that has this advanced look and feel and that will allow you to use multiple functionalities of enterprise version in a local way, similar to previous tools analyzed, but then upgrading to the enterprise version you have the possibility of using all the power of the commercial platform. Also without upgrading to a commercial option you will have the possibility of using Qlik Sense Cloud for free sharing your dashboards with up to five users. In order to download Qlik Sense Desktop, you have it available in this URL: http://www.qlik.com/us/products/qlik-sense/desktop. Once downloaded the software installation is again a next-next-next process.

In order to start to work in Qlik Sense, you can just open the tool from the Start menu and begin with application development. The first screen you will ask you if you want to create a new app. This is a welcome screen that you can unmark to not appear when starting up.

Inside Qlik Sense you have the concept hub that can contain multiple applications, the initial screen that you can see if you remove previous initial screen. This hub can be customized by clicking on the button at the top right and selecting Dev Hub, as shown in Figure 8-21. There you can create your own style of application and create Mashups with combined information from multiple applications, entering to configure css styles, html, and so on in order to be integrated in a web environment. We discuss it to show you the option but we won't enter in the detail about how to configure it.

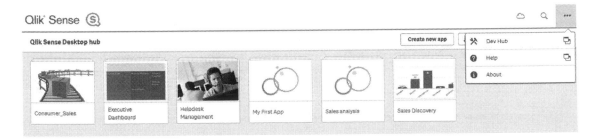

Figure 8-21. *Qlik Sense hub*

An application is similar to the dashboard in Microstrategy and Power BI. Inside the application you can have multiple sheets and inside each sheet multiple visualizations. These visualizations can be created as a stand-alone object in order to be reused in multiple sheets or can be embedded into the sheet and customized differently each time. Also you can organize the Application by creating Stories that can be considered as presentations, which in fact can be used in meetings, as its format is nice enough to do that and you can add comments and descriptions to explain the data that you are showing.

But before arriving there, let's start from the beginning. A first step is to select the data that we want to have available for usage inside our application. Selecting the option of Create new app and setting up the name of the application we will be prompted to Add Data or open the Data load editor, also with multiple buttons and menus that you can open in the top of the screen, such as open App overview, Data Manager, Data Load editor, or go back to the hub. Also in the black band you have two buttons to edit application visualization and modify backgrounds and colors.

Data load editor will open a code editor that will allow you to program loads using some ETL features, transformations, formulas, etc. If you need to do some customization in your load you can find a lot of information inside product documentation and Qlik community. We will use in this example the option of Add data, which will open a menu where you can choose native connectors for multiple databases or, as we do in our case, open an existing ODBC that is pointing to our database.

In order to do a similar analysis for previous tools, we will select the same tables, t_f_sales, t_l_product, and t_l_category. In Figure 8-22 you can see the screen where we select the tables with preview options and also you can check in Metadata button to see which table structure is contained in each table. Once validated and selected you can go to Load data and finish to directly load all tables, or click on Prepare data button to define the table relationships.

Figure 8-22. *Qlik Sense table selector*

As you can guess, we will use this second option in order to ensure that we will have a correct relationship among our tables. Then a curious screen appears: the Associations screen that will show you a bubble for each table. In order to define table links just drag and drop one of the bubbles against the other and a relationship by default will appear if it is able to find matching fields. In Figure 8-23 you can see the Association screen and in this case, which is the relationship between sales and product tables, Product_id field.

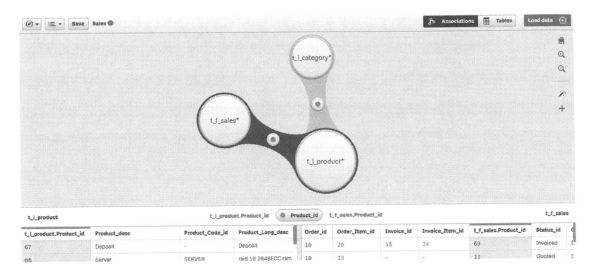

Figure 8-23. *Qlik Sense Associations screen*

Once the associations are defined, you can also go to Tables menu and modify some options on the table definition such as adding new fields or pivoting data from rows into columns. When we have defined required tables and associations we just click on Load data button and we will enter in the design of the sheet to start drawing our application visualization.

We will go then to the sheet editor, shown in Figure 8-24, where we can start defining our graphical interface. In this editor you have different areas: in the top left you have menus to move across Applications, to go back to the Data manager, Data load editor, Data model viewer, or go back to the hub. The next button will allow you to add new data and export visualizations to pdf, duplicate and delete sheets, and access the help. No comments about the Save button as it is pretty obvious.

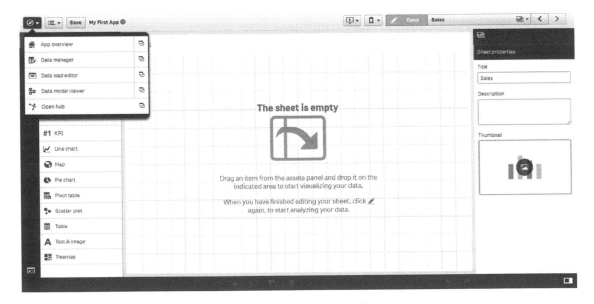

Figure 8-24. *Qlik Sense sheet editor*

In the next subset of buttons, you have access to the Story menu, where you can create new stories that, as commented, can be used (and exported) as a PowerPoint presentation. Next to stories menu, you have the bookmark button where you can save selections of data while navigating through the tool to apply them in any of the sheets.

In the central area, you have first a button bar to add charts to the template. You can see in Figure 8-24 all available visualizations, custom object button to add customized extensions of the tool, the master items button where we can create dimensions, measures and visualizations, and the fields button where you can see all fields for all mapped tables. In order to create graphs and grids, you can directly use fields of the database but we recommend you to use dimensions and measures as an intermediate step as far as it will make the maintenance of the application easier. Center of central area is the template were we drag and drop visualizations, shapes and texts; and finally in the right area we have available the configuration panel, which based on which element of the template you have selected will show different options. In Figure 8-24 we can see sheet configuration options but it changes automatically when you add new visualizations. Also options will vary depending on the visualization that you choose.

In the bottom area you have some buttons to cut, copy, and paste objects; undo and redo actions; and show and hide the configuration panel.

Based on the recommendation of creating an intermediate layer of objects, the next step now is to create all dimensions and measures that you require for the analysis. As you can see with this tool, it recognizes numeric fields used to store dates in format YYYYMMDD as a date and creates automatically multiple date concepts, Year, Quarter, Month, Week, and so on. So in this case the tables t_r_time, t_l_month, t_l_quarter, and t_l_year wouldn't be needed if we are going to work with Qlik Sense.

Dimensions that we create can be Single or Drill-down, depending if we create them with a single attribute or with multiple related ones. We will create now all product-related dimensions, multiple time dimensions for multiple date concepts, etc. Also, we will create all measures that we want to use, by defining the formula. Basic formulas will be the sum of numeric fields and count of dimension fields, but then you can use SetExpressions and other parameters to define metrics filtered, with distinct clauses, etc. You can see options in Figure 8-25; if you are not right about the exact name of a field you can select it from selectors in the right area.

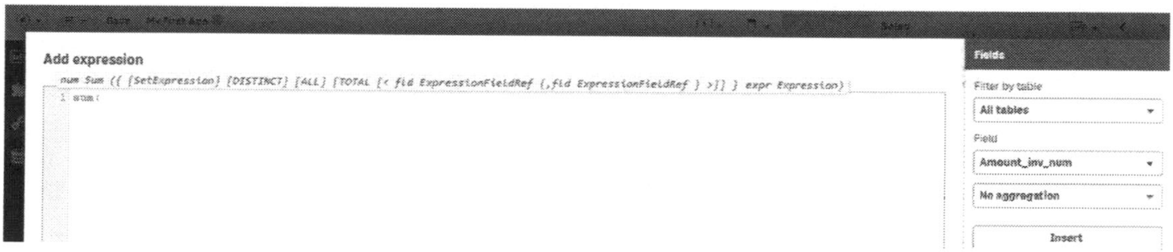

Figure 8-25. *Measure formula definition*

Once the measures and the dimensions are created, we can use them in visualizations. In order to add visualizations in the sheet, just drag and drop them into the template area and then you can start to configure them. When you locate a visualization in the template area, it will appear in the configuration panel visualization options that vary between graphs, but mainly they are grouped into four groups: Data where you define what is inside the visualization and some data options; Sorting where you specify criteria for sorting the chart; Add-ons where you can define added analytics such as trending lines; and Appearance where you can define colors and formats. When you finish the template definition, just click over the Done button in the top right and you will be able to start the analysis.

Qlik Sense automatically links all visualizations so if you click in any region of the graph it is automatically filtering the rest of graphs. In order to filter, when you are in visualization mode you just need to click on a bar of the bar graph, a cell of the table, or any section of any graph and then it will show you the filter buttons, as shown in Figure 8-26. You will be able to click one by one to select and unselect elements, but you can also use the *laso* option to round multiple values, as you can also see in Figure 8-26.

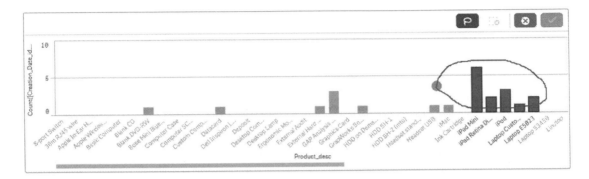

Figure 8-26. *Graphical filtering options*

Anything that you are filtering is located in the top area; just click on the X to remove the filter applied in all visualizations. In Figure 8-27 you can see filters applied to previous graphs through Product_desc field. Just clicking on the X will reset the visualization to the previous screen.

Figure 8-27. *Filters applied*

You can also take snapshots of the visualization by clicking on the camera icon that appears when you move the mouse over a visualization and open the visualization in full screen mode to drill and analyze data inside the visualization.

Once you have all your sheets defined, it is the right moment to create a Story. If you do that through the story button that you have in the top right of the screen, you will enter to the Story editor, shown in Figure 8-28. Here you can see the thumbnails of slides on the left, with the bottom button where you can add new slides; also you can see the tools button bar on the right, where you can select the objects that you want to add, pictures, text, shapes, effects, media objects, and Qlik sense sheets. With all slides prepared, you can visualize them in the whole screen with the green Play button.

Figure 8-28. *Story Editor*

Conclusion

Design is playing an important role in dashboard definition, and you need to be crystal clear that BI interface is a graphical tool and if you want a successful BI project, it needs to be useful, user friendly, and nice for your end users. Because of that, in this chapter, we have started reviewing different BI tool approaches and seeing multiple design recommendations for our dashboards and documents that will allow us to define these useful, user-friendly, and nice dashboards for our customers that we want to develop, also seeing with multiple examples best practices of graph usages. Following these recommendations it is a must to ensure success in your BI implementation.

But also, you need to take into account that functionalities that your model and your BI tool are offering to your users must meet with their requirements and expectations. Only with the presented best practices, you cannot be successful. We have reviewed just some of the options available in the market that we have considered interesting to add in this book, but there are plenty of tools with different versions, free, commercial, enterprise with a wide range of prices, functionalities, configurations, and capabilities.

Regarding the three tools chosen for analysis, Microstrategy, Power BI, and Qlik Sense, we have shown an overview of functionalities but they have much more options, especially if you move to their commercial or enterprise editions. We consider functionalities of free versions quite similar. Maybe Qlik Sense with the possibility of creating stories gives you an added value but they are doing mainly the same just with some different options. If you have in mind to use them to create a pilot project as a previous step of a commercial installation, then you need to take into account multiple other factors, number of users, expected response, expected functionalities, usage of cloud, server layout or data volume, and evaluate which tool, inside these three or outside them, fits better with your needs. In our humble opinion, Power BI is cheaper and can be considered an option for small and medium projects, Qlikview is the intermediate option in terms of data volume and pricing, and Microstrategy offers the platform most robust and with wider functionalities from administration and data delivery of these three but also the most expensive in terms of licensing. But this is just an opinion.

Ending this chapter, if you have followed the instructions from the entire book, you should have a BI platform with its main components, from your transactional, your database, an ETL process to fill it up, and a BI tool accessing the data model. Of course the entire project should have followed Agile methodologies and your database and your ETL will be optimum. Let's go ahead to the next chapters that will allow us to simulate scenarios with a MOLAP tool, to see how to maintain all these solutions, how to orchestrate all processes in an automated way, and finally options for moving the solution to a cloud environment. We hope you like it!

CHAPTER 9

■ ■ ■

MOLAP Tools for Budgeting

This is time for an extra bonus track! We consider MOLAP tools as part of the whole BI platform but it is usually out of the scope of usual consideration of BI. But as explained during Chapter 1, we consider this tool as the final step of the BI process inside a company mainly for two reasons: they can be used anyway to analyze data by end users, so they can be considered BI; and the definition of budget and objectives for the following periods helps the company to focus in its daily operations and it can be an input for the BI systems, when for example we want to compare actual and target scenarios. We are working in our operational system saving the daily activity of the company; then we extract it using an ETL tool, loading into our database, information that will be analyzed using a BI tool. In this moment we extract some conclusions from the data analyzed and we think of some actions to do in order to improve our company performance, but we want to know what each action implies in terms of net revenue improvement before applying it.

Apart from being a bonus track, this component is usually deployed separately from the rest of your BI environment. It is possible that this component already exists in your company when you start the BI process, depending on the process maturity of each department. As far as these kind of tools are used many times for financial analysis, it is possible that your finance department is already using some tool for budgeting and simulation of different scenarios. On the other hand, you may prefer having your BI process consolidated and once your system is fully working for gathering real data, then you start with the capability of making "what-if" simulations to see how they will affect the rest of KPIs based on some supposition. So if your intention is to follow this book step by step to play with proposed tools, just go ahead with this chapter and try our proposals. But if you are following this book as a guide to implement your productive BI system, ensure first that there is no other department using this kind of tools and if not, maybe it is better that you leave this chapter in freeze mode until you consolidate the rest of the BI process. First you will need to analyze your company's actual data defining which KPIs will be the focus of your company, gathering historical trends to analyze data, and then you can start thinking of simulating scenarios for the future.

Let's go ahead anyway with some examples regarding the usage of MOLAP tools. Imagine that you are back in the example of the hardware store chain. You, as general manager, can be thinking on opening a new store to improve your sales, and this can be possibly true if you open it, but what can happen with your Net revenue results? You can try to simulate the result based on multiple scenarios, playing with multiple variables and comparing the results. You can evaluate what happens if the three stores that are closer to the new one decrease their sales because of the proximity of the new store opening. You can evaluate how this affects the rappel that you get from your providers by increasing the overall sales of the company. You can also evaluate the impact in net revenue of the need of filling up the new store by reducing the cost of stock in the rest of the stores. You can check what happens if the location of the store is not good enough and you are selling just half of what is expected or how you can press your Garden goods providers to increase discounts because this store has the expectancy to double the sales for Garden products because it is located in a region with plenty of houses with gardens.

© Albert Nogués and Juan Valladares 2017

A. Nogués and J. Valladares, *Business Intelligence Tools for Small Companies*,
DOI 10.1007/978-1-4842-2568-4_9

For sure you will be able to simulate sales scenarios just comparing Excel sheets, and there are also multiple budgeting tools that help you with budgeting processes without deploying a MOLAP system. But using MOLAP tools will give you the capability to save the information in a single repository shared across users, ensuring that there are no errors inside complex Excel files that use hundreds of formulas and using complex calculations in a prepared mode to analyze data at aggregated levels with a high speed. MOLAP can use two main strategies: precalculated models that contain data already precalculated at multiple levels that returns data immediately, or online models that quickly aggregate data to return the results. And this is one of the main benefits of Multidimensional databases: information can be queried at multiple levels of each dimension of the cube or database giving a very quick performance. But before entering in the detail of how to use a multidimensional tool to analyze data, let's see main concepts of MOLAP. After analyzing general concepts we will see an overview of a MOLAP tool because in this case it is difficult to find open source tools or free versions of commercial ones, MOLAP tools are not so widely used as BI ones. We have lots of experience with Hyperion Essbase provided by Oracle but there is no free version for this product so we have searched through the market and selected a commercial tool that offers a free version of its application, PowerOLAP, to go ahead with some examples about MOLAP capabilities. We hadn't any previous experience with this tool but we have evaluated it for this book, and the results and learnings have been good enough to go ahead with some MOLAP functionalities examples. But anyway, let's start as commented with the general concepts.

Multidimensional Databases

Multidimensional databases can be used for budgeting purposes but they can also be used for data analysis itself. We are focusing now on the budgeting process but there are also some other utilities for them. We are using them as budgeting tools because they will usually provide a user interface that is able to insert data while typical BI tools only allow you to retrieve data, and you need some kind of ETL process to add data to the database. There will be applications were we can use both relational or multidimensional databases, such as basic data analysis, but then there will be some applications where we can take profit of benefits of a multidimensional database and other scenarios where disadvantages of a multidimensional database will force us to use a relational one.

A basic characteristic of a dimensional database is that we can have precalculated the information across all possible levels of any combination of hierarchies, so they have a very good performance for queries, as far as data is already precalculated in the database at the desired level. In a relational OLAP (ROLAP) database, you can have the same behavior but it will needed most of the times to do some aggregation of data at the desired level used in the query.

Multidimensional databases are commonly known as Cubes. If you hear about the concept Cube, you can visualize an object with three axes but of course, speaking about multidimensionality we will have more than three dimensions. Anyway, the idea of the cube is quite powerful to understanding how they work. Imagine that you have a Rubik cube so let's try to build a database with three dimensions of three elements each. In one axis you have the time dimension, so you will have data for 2014, 2015, and 2016; in other axis you can have the product dimension, inside our limited example we will have just three categories, Garden, Tools, and Car; and finally in the third dimension we can have a Metrics dimension, with Revenue, Cost, and Profit. Then inside each piece of the Rubik cube you will contain the related value, so in the first piece you will have, as you can see in the example of Figure 9-1, $5,000, corresponding to the value of Revenue for Garden in 2014. We know that adding more than three dimensions is difficult to visualize but based on this image, you can now start developing it in your mind by adding dimensions and elements to the model.

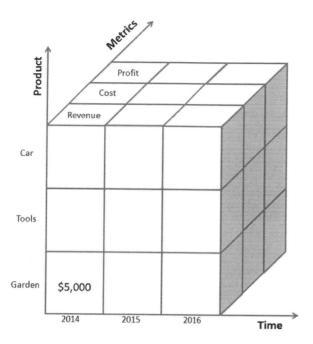

Figure 9-1. *Multidimensional Database example*

You can see the multidimensional database as a database with a single table, where each column is related to a dimension, but taking into account that in each dimension column you can have values for multiple levels of the dimension. You will have a row with data for total year, one row for each quarter, and one row for each month. Then you can imagine that metrics or measures dimension as another standard dimension, so you will have a row for each metric (revenue, benefit, percentage of increase of benefit vs. last year, etc) with a single field for all values, which would be a normalized way of imagining it or a column for each metric, considering metrics dimension as a particular one. The way that each multidimensional vendor is storing data will depend on each vendor, and also there are some vendors that will allow you to configure each dimension to save the information in a sparse mode or dense mode, which could be identified with rows and columns. Sparse mode will create new rows only when a combination of the key exists, and dense mode will create the cell just by creating the member, so they can be considered as columns.

Dimensions or Axes for Analysis

As you can suppose, inside the Multidimensional database definition the main topic to define is which dimensions we will use in the analysis. In this case we can do a similar analysis than the one done in Chapter 5 where we saw how to model our datawarehouse. It cannot be the same as far as in your multidimensional database you won't have usually the same level of detailed information than in the datawarehouse; maybe you have a model related to a datamart inside your datawarehouse that can be directly used as your multidimensional model but most of the times you will need to redo this analysis for the multidimensional database. There will be some attributes that won't make any sense to be used in a multidimensional database, such as order and invoice number, as far as they will have only information for a year, a customer, a product or a scenario, so it won't make sense to compare how an invoice number varies depending on customer or time dimension because we will have this invoice number just for a single customer and period. In the concrete case of this chapter where we want to use a multidimensional database for budgeting, there will be also some attributes that should be possibly skipped; you will have in your database information at level product-customer and unless you have few customers and few products in your business it won't make any sense to try to define a budget for next year's sales at product-customer level. Most likely your budget for next year will be done at higher levels of product hierarchy such as category or family and by customer type or customer groups.

The concept of dimension in this area is quite similar also to the one seen also in Chapter 1; a dimension is an axis that allows us to split the information across different dimension levels in order to see how metrics perform through each dimension member. The main difference is that in this case the dimension contains members from multiple levels in a defined hierarchy and in a pure BI dimension we consider it as a group of related attributes.

Another difference in the usage of dimensions between relational and multidimensional databases is that while in a relational database you can select any field that you want from any table, in the multidimensional database you will be required to use at least one member for each dimension in your queries. If you don't select a member of a given dimension it will return you the value for dimension name, usually containing the aggregation of the whole dimension, taking anyway a member of the dimension.

Usually in most of the tools you can distinguish three main dimension types:

- **Metrics dimension**: It contains the metric definition for all metrics that we will use in the model. It is mandatory as far as without metrics you won't have anything to analyze. Some tools don't consider Metrics or Measures as a dimension itself but as fields of the cube, so visualizing them in columns rather than in rows. But in all applications you will use metrics to be able to work with cubes.

- **Time dimension**: Time has usually some specific properties and functions available to facilitate dimension management such as Year to Date, Year to Go accumulations, or Previous Period that will allow you to easily calculate these types of aggregations and time view changes. This kind of time view changes can be related with time transformations seen in Chapter 5.

- **Standard dimension**: The rest of the dimensions can be considered as standard.

■ **Note**   All metrics will be numeric values usually, but some applications allow saving text and date metrics but they don't allow having free text, just a selector of values that are mapped to a number and dates are saved as a number, so at the end of the day they are saving numeric values.

A dimension can be basic, having just some members that allow you to see multiple elements or hierarchical, organized in a parent-child mode where each parent is the sum of its children. Inside a hierarchical dimension we can have multiple levels of hierarchy and grouping with a homogeneous structure, where all branches of the dimension are completed or with a heterogeneous structure, where you will have different numbers of levels in each branch. An example of a homogeneous structure would be the time dimension where you have always the same number of levels, Year, Quarter, and Month. It would be very strange to have a time dimension where in June you arrive to a daily detail and in April you arrive just to a monthly detail. Typical heterogeneous dimensions could be metrics where you can group metrics based on the aggregations that make sense to perform; in case of a financial analysis you will have some costs that will be grouped in cost types; then you will have gross sales, net sales, net revenue, and all these metrics will have different levels of metrics defined as children and descendants. Structure type for each dimension will vary depending on your needs and it will have more or less implications depending on the tool that you choose and function availability for level references.

In order to go ahead with the dimension definition explanation, let's try to propose for you a model based on the example that we have used for the rest of the book. In this case we will use our budgeting model to define our targets in terms of sales and net revenue, so it will be a mix of sales and finance analysis, as you can see in Figure 9-2.

	Type	Structure	Hierarchy	Description
Metrics	Metric	Basic	Homogeneous	We will define some sales metrics and some financial metrics
Time	Time	Hierarchical	Homogeneous	This dimension will contain three levels, Total year, quarter and month
Product	Standard	Hierarchical	Homogeneous	This dimension will contain four levels, Total product, Category, Subcategory and Family
Customer	Standard	Hierarchical	Heterogeneous	We will have some important customers with detailed information by region and some small customers grouped into aggregated members
Scenario	Standard	Basic	Homogeneous	It will define the type of data, if it is actual, budget, forecast and comparisons among them
Year	Standard	Basic	Homogeneous	Separated dimension from time in order to easily cross information of different years

Figure 9-2. *Dimension definition*

While most of the dimensions are present in the datawarehouse, there is a specific one that is related to the budgeting process, the scenario one. Scenario will allow you to have multiple versions of the same information in order to have the possibility to compare data among them; in fact, we will have already a dimension member that calculates the comparison across them. Another comment regarding dimensions chosen in our model is related to the Year one. We have decided to have a separate dimension for the year regarding Time dimension in order to easily compare different years, by selecting months in columns and years in rows or vice versa, and also to simplify time dimension maintenance, as far as each year you will need to add just a member (the year number) and not 17 (1 year, 4 quarters, and 12 months).

Depending on the tool, you will refer to each level of the hierarchy as a numeric level, starting on zero for the ones that have no children, you will have the possibility of referring to them as generations, starting in generation zero from the top of the hierarchy and if your tool allows that, you will be able to define the levels of a hierarchy with names, as shown in Figure 9-3 for time dimension.

Time member	Level	Generation	Named level
Total Year	2	0	Year
Quarter1	1	1	Quarter
January	0	2	Month

Figure 9-3. *Time level example*

Dimension Members

All dimensions will be compounded by multiple members. A member is a dimension value, in the dimension year we will have 2014, 2015, 2016, each member of customer hierarchy that we want to have in the cube, each product element, each metric, etc. Depending on the tool that we use, you will be able to define multiple types of members: some of them fixed members that store data and others that are just formulas defined in the dimension. You can have also pointers to other members also named shared members, other members that can be just labels without any data inside. It will depend on the tool that you choose that you can have different types of members. You can also classify members as base or aggregated elements, taking into account that base elements will be the ones that don't have any child, also known as level zero in some tools.

In our example we will create a sample model with a few elements, as shown in Figure 9-4, by defining just a small cube to show you how it works, but that will have enough data to understand the options that you will have available.

Metrics	Time	Product	Customer	Scenario	Year
Gross Sales	Total Year	Total Products	Total Customers	Actual vs Forecast	2014
Invoice Costs	Quarter1	Tools	Customer Group 1	Actual	2015
Trade Deals	January	Manual Tools	Customer 1	Forecast	2016
Net Revenue	February	Electrical	Customer 2	Mid Year Review	2017
% Margin	March	Warehouse	Customer 3		
	Quarter2	Boxes	Customer Group 2		
	April	Closets			
	May	Car			
	June	Spare			
	Quarter3	Car Cleaning			
	July	Garden			
	August	Garden Tools			
	September	Plants			
	Quarter4				
	October				
	November				
	December				

Figure 9-4. *Dimension members table*

A **tuple** is defined as the intersection of a single member of each dimension. Using the example of the Rubik cube, a tuple would be each piece of the cube, and the one uniquely identified by Garden, 2014 and Revenue has a value of $5.000. The cube defined by dimensions shown in Figure 9-3 is an small cube but takes into account that if it is completely filled up you will have more than 100,000 combinations of data or tuples, just with these small dimensions, in order to load the whole cube you will have just with this dummy cube a file of about 20,000 rows if you set metrics in the columns. It will depend on how it is managing the tool and the data below but usually not all combinations will be created automatically, just that ones with data. But in case of hierarchies if you add data for example in January, it will create data for Quarter1 and Total Year, so if you have a hierarchy with 10 levels and you add data in the bottom level, it will create up to 10 times the amount of data inserted. So be careful when you are defining your cube and evaluate implications on disk space and performance for data loads before adding a huge number of members in a dimension. You shouldn't consider a multidimensional database a datawarehouse; just use it to store a limited and aggregated number of elements.

■ **Note**   It is important to remark that while in a relational database you can remove an element of a dimension table and reload it without losing fact data, MOLAP databases are defined by the dimension members and data that are located in the intersections of members (tuples), so if in a reload process we remove a member of a dimension, data related to this member is completely lost, for all the combinations with the rest of dimensions. So be careful with your dimension load strategy ensuring that you are not removing needed elements from the dimensions. Trying to clean unused elements could cause losses of information for past scenarios.

Sharing Properties across Cubes

In a MOLAP tool, we will usually see some way of grouping cubes that will allow us to share some property. In order to explain you the sense of using these groups, let us show some particular exampled based on existing tools and how they name them. In Essbase, they refer to database for the cube and application for the set of cubes and you can define access, parameters, limits, and service commands such as start and stop. In PowerOLAP they name them as cube and database respectively and you can share dimensions across different databases. We will see deeper detail of how to use them in next sections, when analyzing how to use them, especially PowerOLAP that has available a free version.

MDX Language

MDX is a language that will be used from reporting interface tools to connect to the cubes to retrieve information in order to show it for analysis purposes. The MDX language, acronym of MultiDimensional expressions, is similar to the SQL language seen in Chapter 3, but while SQL was used to speak with a relational database, MDX is used to speak with a multidimensional database. They have some similarities, but MDX is just restricted to read data with the possibility of applying calculations and formulas to the source data, while in SQL you can also create database objects. There are MOLAP tools that provide you a user-friendly interface that helps you avoid the usage of MDX as far as they directly translate your requests into MDX and show you the results. Other tools don't have this utility and you need to interact with the application with MDX and other tools allowing you a mixed usage; they provide a basic MDX based on your requests and then you can modify and tune this MDX, but for this you should read much more documentation about MDX than the one that we are going to explain in this section.

While we have dedicated a whole chapter (Chapter 3) to talk about SQL, we don't want to enter in the detail of how MDX works because we consider MOLAP as an extra feature and it is possible that you don't need it, so we will try to give you a brief overview.

Main operation in MDX is to perform selects organizing in columns and rows the way that you are receiving the information, with the possibility of adding axis to the query, such as pages, chapters, and sections, but with the limitation that you cannot have missing axes. So you cannot do a select organized in chapters if it doesn't contain something in columns, rows, and pages axes.

In select statements you will use the FROM clause to define the cube that you are using for retrieving information and you can add WHERE clauses similar to the ones seen in SQL to filter the information that you are retrieving, generating a **slice** of data. This could be a select statement from the cube Forecast that we will create to analyze our sales prevision for future years:

```
SELECT
    [Metrics].Members ON COLUMNS,
    [Time].[Month].Members ON ROWS
    [Product].[Category].Members ON PAGES
FROM
    [Forecast]
WHERE
    ([Year].[2016],[Scenario].[Actual])
```

In this statement we are selecting data for year 2016 and Actual scenario, as you can see in the **WHERE** clause of the select statement. Result will be organized in five columns, one for each metric defined by the clause **ON COLUMNS**, in twelve rows, one for each month defined in the clause **ON ROWS** and with four pages, and one for each category defined in the clause **ON PAGES**. There are some dimensions that can be skipped in the select statement, in this example we are not fixing for which Customer we want to extract data or if we want to have a page per Customer. When we don't have any reference to an existing dimension on the cube, it will return the default value, usually linked to the total, so in this case it will return data related to Total Customer.

Also in select statements you can add calculated members in the retrieve time or saving them as a member, adding complexity and multiple functions to be able to get data derived from the one that you have saved in your database. Added to this, you will be able to use sub-select statements to be able to perform complex queries.

■ **Note** As commented in the beginning of the section, we don't want to do an exhaustive analysis of MDX possibilities because it is quite feasible that you don't need them in your MOLAP implementation, but if you want to analyze in a deeper detail how MDX works, we recommend you to have a look on the documentation of a reporting tool based on MOLAP cubes, icCube, where you have an MDX reference quite friendly (to be MDX), available in the following link: https://www.iccube.com/support/documentation/mdx/index.php

---

MOLAP Data Interfaces

It is quite common using MOLAP tools (PowerOLAP will work in this way, also Essbase does) that you get some plug-in installed in your Microsoft Excel application that allows you to connect, retrieve, and load data in the MOLAP database using Excel, so the main user interface for this set of tools will be your already known and useful software provided by the most popular software vendor in the world. Using Excel as an interface will allow you to select a data portion or slice from the whole database and read and write data in an easy way. In order to do that you usually locate some members from one or multiple dimensions in columns (one dimension per column), some members of one or multiple dimensions in rows (one dimension per row), and some members of the rest of dimensions in the initial rows of the Excel sheet acting as pages of your data, and then you will be able to launch, retrieve, or insert commands using the interface plug-in menu. A logical restriction to read and save data is that you cannot locate members from the same dimension at the same time in more than one axis, rows and columns, columns and pages, or rows and pages. In Figure 9-5 we show some example about how it looks like a correct Excel data template and an incorrect one.

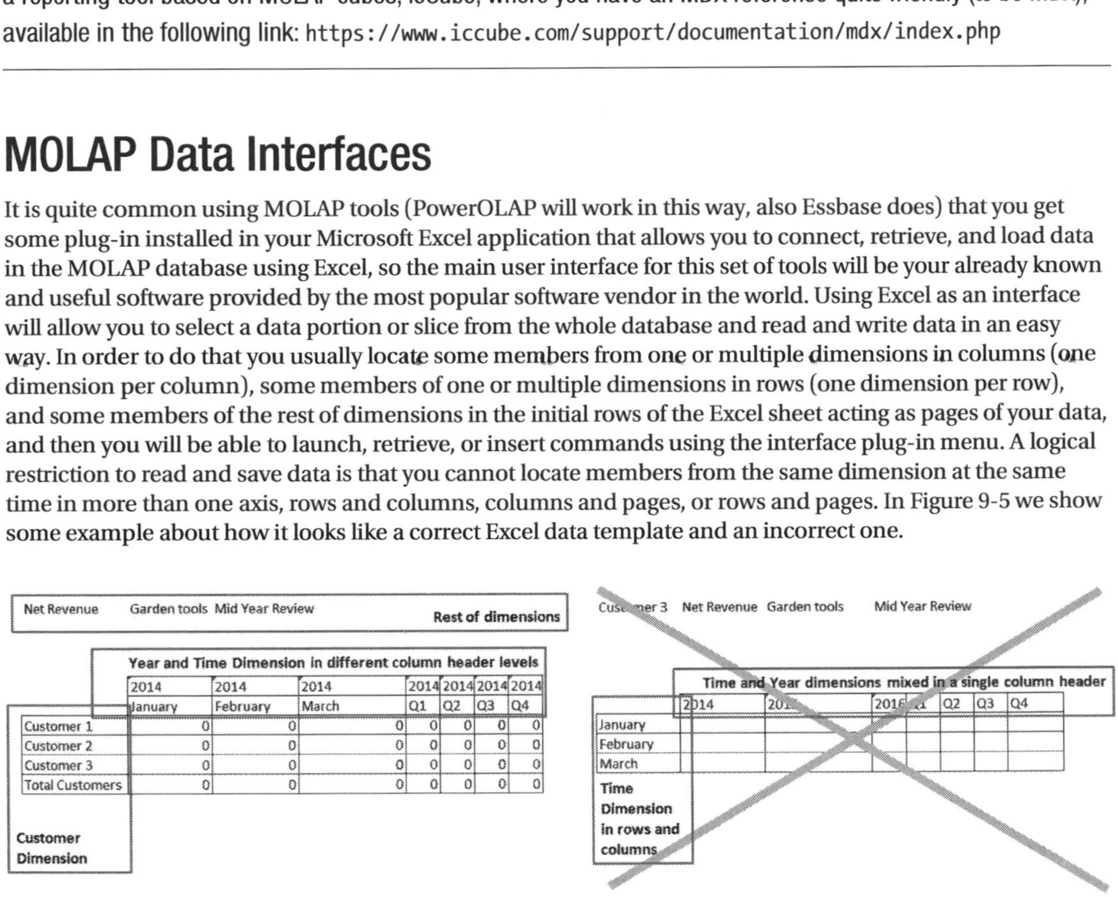

Figure 9-5. *Correct and incorrect Excel templates*

On the other hand, there are tools that have its own data access interface both for loading and analysis. In such cases, each interface will have its own restrictions and rules.

■ **Note** There are some BI tools that allow you to connect to multidimensional databases offering you all BI capabilities also for data inside MOLAP databases with the possibility of comparing data from your datawarehouse with data inside MOLAP, also useful to check data integrity across systems.

Data Import

Having a structure of the database created with all dimensions and members in place is a good starting point but once you have created your database structure you will need to fill it up with data, otherwise all this process would be useless. In order to load information into MOLAP databases we will have mainly two possibilities: fill it manually through application interface (either Excel or own interface as seen in previous section); or automate data load from some different datasources such as flat files, Excel files, or relational databases by configuring ODBC or JDBC connections.

Data import will accelerate and facilitate the load of data into the database by establishing a direct connection between the database and the data source. It can require some bigger effort to initialize the process than directly writing and saving data into the database in a manual way but it will enhance the process of data load in case of big amount of data. Data import is widely used for Actual scenario loading as far as it comes directly from an existing system but when saving forecast, budget, and prevision scenarios it is possible that the data import feature is not so useful. It can be used anyway during the initial load for budget and then you can edit and save data from the manual interface to adjust each metric based on your needs.

Data import will allow you also to facilitate dimension updates based on relational tables or files by enabling the possibility of dynamically changing the database structure adding members, updating descriptions, or changing parent-child relationships.

Another possible benefit of data import when the MOLAP tool has command-line interface is the possibility of automating the process of data import using some workflow tool for process synchronization. We will talk about this topic within Chapter 10.

Source Data Preparation

In case you are using a manual interface to load data, you will just need to prepare some Excel sheet with the information prepared to be loaded in a defined level, or you just will need to connect to the database and send information there. But if you are thinking of some kind of automation to load data into your MOLAP database it is quite possible that you need some data manipulation to achieve the defined level. You will probably use one of two strategies: create a view with a select that returns data aggregated to the desired level, or create a table and a related process to fill it up for the same purpose. Most of the time, each of them will require an aggregation select as far as will be usual that you have a smaller level of detail in the MOLAP database compared with your datawarehouse.

If we try to link a proposed data model with the one that we have used in our relational database, we will see that we had a fact table, t_f_sales defined at a customer, day, product, invoice level among others and you will want to summarize it up to product family through t_l_product table (that in fact in the source data didn't exist directly in the product dimension table, let's suppose for this example that it is related to the Product category and in the cube we will have the category description), you will be required to summarize it up to a month using t_r_time table and you will need a custom aggregation for customer hierarchy, as far as some customers will be inserted directly but other minor customers will be joined in groups. So the query required to create the view or to fill up the table would be:

```
select  date_format(s.day_id, '%M') month,
        cat.prod_category_desc category,
        'Actual' scenario,
        t.year_id,
        case when c.customer_desc in ('Customer 1', 'Customer 2','Customer 3')
            then c.Customer_desc
            else 'Customer Group 2'
        end customer,
        sum(Amount_inv_num) Gross_sales
from t_f_sales s, t_r_time t, t_l_product p, t_l_month m, t_l_category cat, t_l_customer c
where s.day_id=t.day_id and s.product_id=p.product_id and t.month_id=m.month_id and
p.prod_category_id=cat.prod_category_id and s.customer_id=c.customer_id
```

Returning some data at the desired level as shown in Figure 9-6, so, at the desired level of input for our cube.

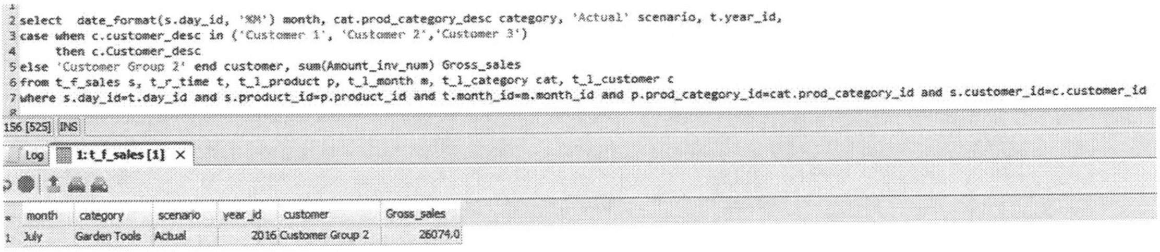

```
1
2 select  date_format(s.day_id, '%M') month, cat.prod_category_desc category, 'Actual' scenario, t.year_id,
3 case when c.customer_desc in ('Customer 1', 'Customer 2','Customer 3')
4     then c.Customer_desc
5 else 'Customer Group 2' end customer, sum(Amount_inv_num) Gross_sales
6 from t_f_sales s, t_r_time t, t_l_product p, t_l_month m, t_l_category cat, t_l_customer c
7 where s.day_id=t.day_id and s.product_id=p.product_id and t.month_id=m.month_id and p.prod_category_id=cat.prod_category_id and s.customer_id=c.customer_id
8
156 [525] INS
```

Log	1: t_f_sales [1] ×

.	month	category	scenario	year_id	customer	Gross_sales
1	July	Garden Tools	Actual	2016	Customer Group 2	26074.0

Figure 9-6. *Data preparation example*

Data Export

In the same way that you can configure data import processes, you can configure data export processes that allow you to bulk export the information inside the database to flat files or relational databases. An important utility of this data export would be to finalize the BI life cycle. Once we have validated our budget doing multiple simulations and choosing one scenario as the result of our analysis, you can export this data to be loaded into the datawarehouse to track how it performs the year results compared with the expected KPIs, especially if your BI application doesn't allow you the possibility of connecting it to the MOLAP database. Data export can be used for any other purpose that would require you to have the information in text files or in a database, such as saving it in the ERP system for any requirement that you have, to transport data across multiple MOLAP environments, to send it to customers, or any other data flow requirement. Again you can export also the database structure, so it would be useful to configure a backup of the structure, to load this structure in another MOLAP environment, or to export the structure as a sample and then use it as template to add more elements into your MOLAP database.

Calculations

Data is usually imported at the bottom level of each dimension and then aggregated as far as most of the calculations required are just sums of the values related to children elements. But this calculation can be any other formula that you could think as ratios, differences, multiplications, percentages, or any other formula required for your business. Both basic aggregation and other calculations must be defined in some way in the MOLAP tool in order to allow your system to be able to know how to calculate data related to each member that has not been directly uploaded manually or automatically.

But calculations are not only done bottom-up. Imagine that you are doing the budget for next year and your product hierarchy has six levels: Total product, Category, Subcategory, Family, Subfamily, and Product; and the whole hierarchy has 4,000 products. Defining a budget for next year sales at product level could be a nightmare, including inconvenient such as if you have 4,000 products it is quite possible that during a year you will have new products in your hierarchy and some discontinued ones. So in this situation it is quite possible that you prepare a budget at category or subcategory level, defining the budget for 20, 30, or 50 elements, but not for 4,000 elements. This point can be interesting for you to have a top-down calculation that spreads the total amount of expected sales at subcategory level down to the product level, using some metric as a spread driver to arrive to the product. This capacity is one of the most useful benefits of MOLAP tools. You can achieve these calculations by using some compiled functions in all MOLAP tools that allow you (with different syntax in each tool) to refer to the data from related elements through hierarchy or through position. You will have the possibility to refer to PREVIOUS elements based on time hierarchy to get data from previous months or years, needed in case of the spread because you will spread Budget for 2017 based on Actual of 2016. You will have the possibility of referring to PARENT value to multiply current member value by its actual data and divide it by actual data of the parent member. You will have the possibility of referring to the top of the hierarchy member to get the value of the percentage of increase of sales regarding current year that you want to achieve for the next exercise and multiple other functions that can be useful to perform MOLAP calculations.

MOLAP functions available and capabilities are very similar to the ones offered by MDX language but sometimes scripting language is particular for the tool you are using, so refer to the tool documentation in order to know which functions are available for your case and the specifications about how to use them.

Now that we have seen the justification of using a MOLAP database for our budgeting process and we have reviewed some relevant topics to understand MOLAP concepts, let's go to analyze them using PowerOLAP, one of the options that we have available in the market.

PowerOLAP

PowerOLAP is a product offered by PARIS technologies that allows MOLAP functionalities in a user-friendly way with a simple interface. There are mainly two options, the commercial one that includes the server component as the core of the system that gives service to the users, and the personal version that is available for free. This will be the one that we will use in order to show you how to implement a MOLAP system, as far as it will give us the same design functionality but just applying some limits in the size of the cubes that we will be able to use: 10 cubes, 10 dimensions, and 10,000 elements; and some administrative tasks as security management. In order to fill up our database we will have mainly two possibilities: use the own PowerOLAP tool or a PowerOLAP plug-in added to Excel, both of them installed during the installation process.

PowerOLAP engine follows the strategy of using an online model, so data is stored at the base level and then aggregated in a very quick strategy when you query it. This is useful to have quick refresh when base data is modified, as far as precalculated models usually have big-time consumption for the calculation using batch processes, while online models recalculate the information in a short time. Retrieve won't be immediate as in precalculated models but the idea is to have a balanced scenario between loading time and retrieve time.

Starting with PowerOLAP

As the first step to using PowerOLAP, as usual we will need to install it. For that go to Paristech page in this URL http://paristech.com/products/powerolap and click on download PowerOLAP Personal. It will ask you to register and they will send you a link to access the download, and it is quite feasible that you are contacted by somebody from PowerOLAP sales team. It will start the download of a zip file named PowerOLAP-Personal-16044.zip or similar, depending on the available version when you are reading these lines. Just unzip and it contains an executable file. Installation is quite easy, Next-Next-Next and finally you will have installed PowerOLAP Personal into your laptop both components already commented, application and add-in for Excel. You can find the application on Start Menu ➤ PowerOLAP folder, or when accessing to Microsoft Excel you will see the Add-in and from there you can launch also the application using the first button of the ribbon installed in your Excel, as shown in Figure 9-7.

Figure 9-7. *PowerOLAP ribbon*

When you start PowerOLAP you will see an interface similar to the one shown in Figure 9-8, with five main menus. Home is where you can find some general options, such as cut, copy, and paste buttons, options to configure the database and other visualization options, including a link also for Help and Genius Center that contains links for support, videos, and trials. In the Model menu we will find access to the main components for the database design, dimensions, and cubes and also access to the check syntax button (well and if you have the server version you will have also access to the security tabs). Next menu is Slice menu that contains all the required buttons to manage Slices, and we will see in a few lines what they are and the main options. We have also Data menu that we can use to connect our cube with external sources, mainly flat files and databases in a bidirectional way, with the possibility of reading from the database to load data or to build dimensions and with the possibility of exporting both data and dimensions. Finally we have the Tools menu where we can access tservers and configure multiple options, but available only with the commercial version.

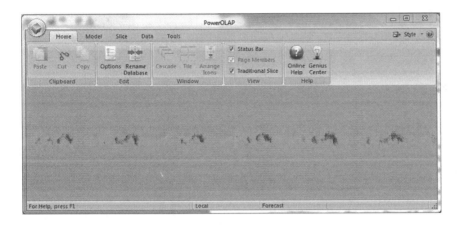

Figure 9-8. *PowerOLAP interface*

Once we have presented you the interface, let's see how it works. PowerOLAP uses dimensions that act as Axis to analyze information, as already explained. In case of PowerOLAP, dimensions are isolated objects, in other databases they are part of a cube; but here we have cubes and dimensions independent of each other, so you can have cubes that are not using some dimensions of the database and dimensions used in multiple cubes.

■ **Note** Dimension member management must be done taking into account that a dimension can be used on multiple cubes, so removing a member for a dimension can affect other cubes that we are not working with in that moment. In order to avoid that you can use the Persistency characteristic in each member that you want to keep untouched.

In order to start working with PowerOLAP, the first step will be the creation of the database, which at the end will be just a file in your computer that will contain all required infrastructure and data, with .olp extension. You can do it by creating in the top left button with an orange cube and selecting New. There are some configuration options accessible from File menu that are defined at database level such as Calculation mode, Autosaving options, Caching level, or some formatting options. You can see them in Figure 9-9. From this screen calculation mode is the most important parameter as far as it will define if aggregated data in your database will be calculated just by sending data or if you need to do a manual update of data.

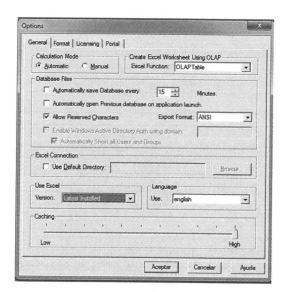

Figure 9-9. Database configuration options

Creating Dimensions in PowerOLAP

Once the database is created the next step will be create the desired dimensions. We need to go to the Model menu and click the Dimension button to enter in the dimension catalog, were we will add all our dimensions, by typing a name and clicking on Add button. You can see the result in Figure 9-10, once we have created all our dimensions.

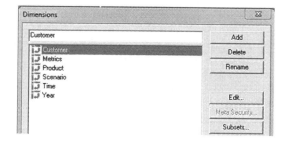

Figure 9-10. Dimensions tab

After adding the dimension we will need to add dimension members and define the hierarchy. For that we press on the Edit button shown also in Figure 9-10 and we will enter into the dimension Editor, which you can see in Figure 9-11.

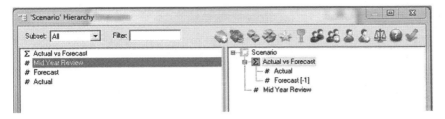

Figure 9-11. *Dimension editor*

We have chosen this dimension to show you because here you can see multiple options and particularities. On the left side you have the available members to use in the hierarchy and in the right area you have the hierarchy build. With the first button you will be able to add new elements to the list, which doesn't mean it has to be considered in the hierarchy, and once created you just drag and drop the element locating it at the desired level with the desired parent-child relationships. By default locating the element below another will imply two things: new element data will be consolidated to the top level, by default aggregating 100% of its value; and top level is not considered anymore as a base element and its icon changes from # to Σ. We have said aggregating by default 100% because this can be changed with the balance icon changing the weight of the element value. In this example we have changed the value to -1 in order to have the element Actual vs. Forecast as the difference between Actual and Forecast, but you could also create some calculations such as weighted average by using the required weight.

■ **Note**   Due to the structure of PowerOLAP cubes, you cannot save data directly in aggregated members; you can only save data in base elements. In order to perform calculations such as cost spread you need to have a plain hierarchy without aggregated elements and create a calculation to perform the split. So in this case it wouldn't be the ideal tool.

There are more things to do in a dimension definition. For example you can define Aliases to the members in order to have different labels to refer to the same object, something that could help you to differentiate what you want to see in the slice versus what you need to use in your data load. Imagine that you are loading fact data related to products using a public EAN code but you want to see in your retrieve a description of the product, you could create both Aliases and use one in each moment. You can access there using the third button shown in Figure 9-11. You can modify also some properties for each element that can be then used in calculations, accessing to the fifth (star) button and mark elements as Persistent in order to not lose data related with the element in case of dimension automatic load, as already commented in a previous note. Finally you have available some buttons that can facilitate the hierarchy implementation by adding members as sibling and children.

You can also define subsets of elements in each dimension that can facilitate hierarchy creation and also they can be used in calculations.

Cube Definition

When you have all dimensions defined it is quite easy to define a cube, you just need to access to the cube catalog by going to the Model menu and clicking on Cube seeing a tab like the one shown in Figure 9-12. Then you need to define a name and click on Add, opening a new screen were you select the dimensions that you want to use in your cube from all available dimensions. At cube level you can define a set of options, mainly related to caching in the Personal edition, by clicking in button Properties shown in Figure 9-12 and you can also define the formulas related to the cube by clicking on the Formulas button. Security options are only enabled in the commercial version.

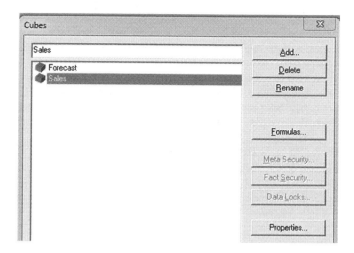

Figure 9-12. *Cube catalog*

Slices

As seen during the introduction we can have thousands, even millions of tuples (or cells) just with few elements in the dimensions as far as you have a tuple for each combination of dimension members. In order to facilitate data maintenance we will need to fix portions of the database where we want to load and analyze data; otherwise we would need a cinema screen with super-very-ultra-high resolution to see all the cells in a single screen. The way of segmenting the database to be able to manage data is through slices, portions of information defined by columns, rows and pages. Slices can be created and used both from Excel and from PowerOLAP applications. In order to access to slices from PowerOLAP just go for Slice menu and there you can manage them, by creating new, opening, closing, copying and managing Slice buttons. Let's go for New Slice and then a menu with a cube selector will appear. We will select one of the recently created cubes and when we click on Ok a new slice will appear with an aspect shown in Figure 9-13.

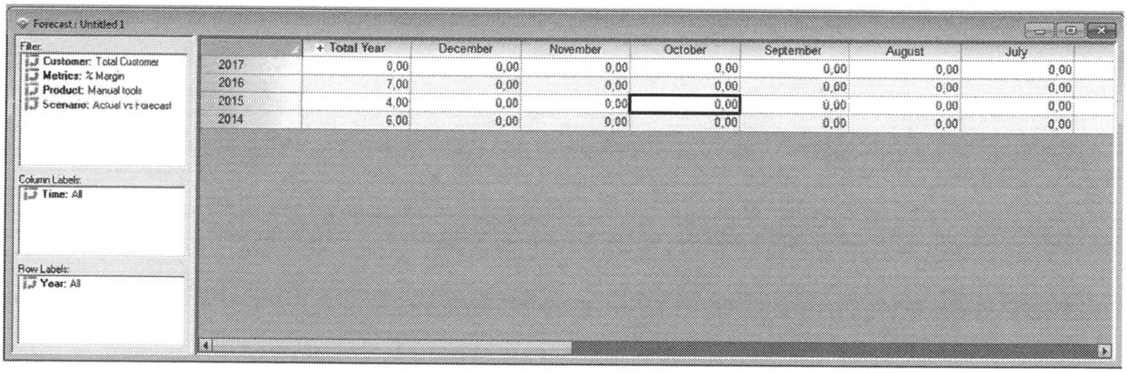

Forecast : Untitled 1								
Filter:		+ Total Year	December	November	October	September	August	July
Customer: Total Customer	2017	0,00	0,00	0,00	0,00	0,00	0,00	0,00
Metrics: % Margin	2016	7,00	0,00	0,00	0,00	0,00	0,00	0,00
Product: Manual tools	2015	4,00	0,00	0,00	0,00	0,00	0,00	0,00
Scenario: Actual vs Forecast	2014	6,00	0,00	0,00	0,00	0,00	0,00	0,00

Figure 9-13. *New slice example*

In order to adjust what we want to use and what we want to show, we can move the dimensions across filter, column labels, and row labels areas and by clicking F9 to refresh we will see what it looks like. By double-clicking in a dimension we will be able to select which members of each dimension will be used in columns, rows, and filters. Once saved from the Manage Slice button you can perform some actions like adding constraints to the data that can be inserted, to try to minimize data entry errors. In here you can also mark to hide empty and zero rows. While it can be interesting for reading purposes, it won't be useful for inserting data as far as it will be quite possible that the full row is empty when you are filling up the cube, so you won't visualize it and you won't be able to add data. When you have finished Slice definition, you can open it in Excel by clicking on the Worksheet button from Slice menu. Also we would like to comment that you can see data in a graphical format by selecting one of the graphs from Graph button.

■ **Note** If you have defined subsets during dimension definition, you will be able to use them in this moment in order to select the members that you want to use in the slice.

Formulas

In the button Formulas that we have commented in the cube definition part, we access to a formula editor that allows us to define which will be the calculation applied to the cube. If you don't define any formula, data will be aggregated summarizing base values up through the different hierarchies. This is what is considered **Pure Aggregation**. This aggregation is the most optimal possible as far as it only takes into consideration those cells that contain information, skipping empty rows as far as the tool is able to detect that there is no data before querying the real value. But we can also define each calculation that we can think about. We can have ratios, percentages, differences, averages, maximum, minimum, or any other formula that we need for our analysis. To add a formula we will just click on Formulas button from the cube catalog, shown in Figure 9-12. Then we will see a formula editor as the one shown in Figure 9-14 that contains also an example of calculating a metric based on other two metrics, *% Margin* defined as *Net Revenue / Gross Sales*. We don't want to enter into deep detail about programming language for PowerOLAP, but just a few comments. With the first part of the statement we fix the level of members where formula will be applied, in this case All means all the cube (for the rest of dimensions a part from Metrics that is the member that we are calculating), but we can select also Aggregates and Details, to fix aggregated or base members, then the word and after that the element (or elements, you can specify more dimensions separated by ands) that we want to calculate. On the other side of the equal we set the metrics that define the formula, in this case Net Revenue and Gross Sales divided. The statements must finish with ";".

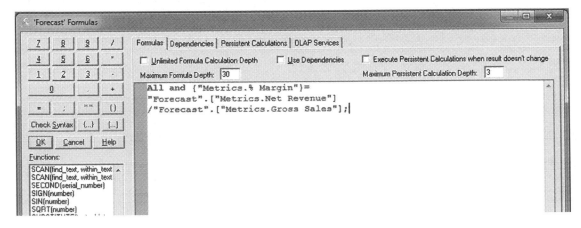

Figure 9-14. *Formula editor*

From this editor re-mark some buttons:

- {...}: it is used to select the target area and members of the calculation.
- [...]: it is used to select the members included in the formula.
- Check Syntax: to validate that what we are writing is correct.

■ **Note** By using the dependencies feature, you can save much execution time during calculation. It is used to only calculate values for the result metric when the metric used in the formula is not empty. To use it, you must mark Use Dependencies flag and define the dependence in Dependencies tab. See the product documentation for more information about formulas and dependencies.

Caching

With the caching configuration you can save time especially for those costly calculations that are using all the cells because dependencies are not an option. Cache is saving some aggregated values already calculated using an algorithm to choose which values are the best to be saved based on the amount of operations required to get the data. You can define caching configuration at cube level by enabling or disabling it in the cube properties, and also at database level where you can define the level of caching, which marks the number of calculations required for a cell to be considered as a cacheable cell. In Low caching approach, it will consider a higher number of operations so it will save fewer cells as far as in the lowest case it will save only the cells that require thousands of calculations. In the highest level it saves a cell each 500 calculation units, so the cache file is bigger. We can tune the cache level by iterating tests and getting what we consider the best ratio *space used/retrieve time*.

Data Import and Export

Using slices to fill up a cube can be useful for small cubes or budgeting processes where we want to create multiple scenarios and comparisons among them, but as commented in previous general sections it can be useless if we want to fill up big amounts of data into our database. As we have seen, there are usually multiple ways to import data with tool that can allow you to easily update dimensions and fact data, located in the menu Data shown in Figure 9-15. From here we can Import and Export data from files, Transfer data across cubes, launch Persistent calculations, and create cubes from zero from a relational database using OLAPExchange options. Metadata option is used to export/import the cube structure and formulas, Factdata option is used to export/import numeric data into the cube.

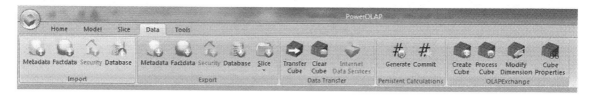

Figure 9-15. *Data menu options*

In order to load metadata and fact data we strongly recommend you to first export a file to know the format that you need to implement to correctly load the files. Here below is an example for our model with just some data rows with fields separated by tabs (this can be chosen at the import time), but as you can see it requires a header to correctly understand the fields below. Then with this sample you can generate and load the full cube file.

```
SCHEMA   TRANSFER
COLUMN   M       Customer
COLUMN   M       Metrics
COLUMN   M       Product
COLUMN   M       Scenario
COLUMN   M       Time
COLUMN   M       Year
STARTINLINE
Customer 3       Net Revenue     Garden tools    Mid Year Review February 2014    21,00
Customer 3       Net Revenue     Garden tools    Mid Year Review January 2014     12,00
Customer 3       Trade Deals     Garden tools    Mid Year Review January 2014      2,00
...
ENDINLINE
```

Exporting fact data allows you also to save the data into a relational database, allowing you to access to it from your BI tool or do whatever management that you need to do. At the export time you can define the portion of the cube that you want to export, by filtering by any dimension or using subsets of data to get the desired portion of data into the relational database. Export configuration is saved before being executed, so after that you can execute it as many times as you need.

With the option of Transfer Cube inside Data Transfer menu, you can copy data across cubes or inside a cube through some field. It will be very common to do a copy from one scenario to another to save versions of your data to have the possibility to compare after any simulation of conditions that you can try to do. Remember the introduction of this chapter: you can want to compare what could happen with your company Net Revenue if a new store implies a reduction on total costs or a reduction of total sales for the rest of them.

We have finally a section of options to connect to a relational database and create from zero cubes and dimensions. By selecting the desired tables for dimension creation and fact load, the cube will be automatically created, defining members, member aliases, member properties, drill through attributes, rollup attributes, member order, etc. Once you have saved the cube definition you can update it fully or incrementally to have a better update performance, also with the possibility of updating single dimensions of the cube.

Conclusion

In this bonus chapter you have learned the main theoretical concepts of OLAP databases that allow you to prepare next period objectives and create and compare multiple data scenarios for budgeting that purpose in an easy way by centralizing all data into a server accessible by multiple users. We have seen also main OLAP functionalities examples with a free version of PowerOLAP installed in a laptop, using a single user installation. Maybe it doesn't fit your needs and you require a bigger installation with bigger cubes and more users but for that, we recommend you to go for a commercial tool, as far as free versions are not powerful enough and some open source that we have evaluated are quite complicated to use, requiring programming skills in Java.

If you have arrived until here without skipping any chapter you have now an overview of all BI components that are required to have a complete BI solution. Congratulations for finalizing the whole BI platform infrastructure!

But we haven't finished yet; we still have some other areas to see. You should have installed multiple components in multiple servers in this moment, with multiple processes for ETL, reporting and load into your MOLAP tool, which are related to each other in a logical way but we will need to orchestrate all these processes adding dependencies among them with some coordination tools.

In a real scenario you will have more than one environment for developing new processes and testing them before going to production with that, and there are some best practices that we want to show you. Finally, it can be recommended based on your needs the possibility of going to cloud solutions for all the components and in this area there are also some considerations to think about.

All these topics will be analyzed in further chapters, so if you are interested in all these points don't hesitate and go ahead with them!

■ ■ ■

## CHAPTER 10

BI Process Scheduling: How to Orchestrate and Update Running Processes

So far, we have built our solution, from the database until the reporting layer and the budgeting system. Maybe you have built the best dashboard ever seen with all the KPIs that your users need to analyze but your users will want fresh data periodically. This means that we must put a system in place that runs every day; or with the frequency we need; and in case of any error or issue during the load process, we get at least an alert, so whoever is in charge can review and analyze what has happened. In this chapter, we will see first, how to finish our ETL project by designing a final job that will launch the transformation, gather information, and control the flow from its execution; and take care if something goes wrong. Then, we will move on to see what mechanisms PDI has to trigger jobs and transformations, and then we will move on to see how to schedule them.

By the end of the chapter we will also see how to perform some maintenance tasks in the database to ensure that our processes run well after some time, by gathering fresh statistics and performing consistency checks on the data.

Finishing the ETL

The first step before starting the scheduling stage of our processes is to finish our ETL. By finishing we mean to encapsulate our transformation inside a job. Whereas it is possible to launch Kettle transformations directly from the command line or any other task scheduler, usually it is not a good option. If we want to add flow control, email notifications in case of failure and other checks, we need to bundle these inside a job.

Also, it is very common to have more than one transformation. To synchronize, put dependencies, and organize the flow of the data and actions to be performed on it, we need a controller. This controller is the job. Our first task for this chapter is, then, to clean up the ETL and create the job containing our transformation(s). Let's start!

Creating the Job in PDI

We had two transformations, one that grabs the data from the Odoo database and stores it in the staging, applying the first batch of transformations; and another one, which grabs the data from the ODS area, and stores it in the final datawarehouses tables. To create a new job, we need to open the Spoon editor, click on File, then New, and finally select Job.

© Albert Nogués and Juan Valladares 2017

A. Nogués and J. Valladares, *Business Intelligence Tools for Small Companies*,
DOI 10.1007/978-1-4842-2568-4_10

In contrast with transformations, where we start reading the data source (usually, not always as sometimes we read parameters or other variables like system variables), in a PDI job, the starting point is the Start step, a typical green arrow that we can find in the general section. Drag and drop it into the canvas, and then, after it find two transformation steps inside the same general category and drag both to the canvas as well.

Finally, find a step called Mail, inside the Mail folder, and drag and drop two of these inside the canvas as well. To create the appropriate flow, make sure you link the objects as seen in Figure 10-1.

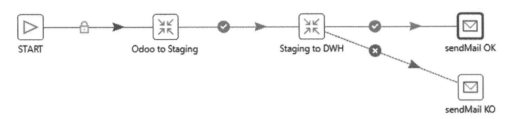

Figure 10-1. *Connecting the steps in the canvas to create the job that will launch our both transformations*

Then, to tell PDI which transformation must run in each step, you need to double-click the transformation step, and depending what are you using (plain files or repository), you select the appropriate option to locate the transformation you will use. In my case, since I am working on my computer, and I have not yet moved these files to the server, and only for testing, I am selecting the file manually on my filesystem. But if on the server, we do plan to use a repository, which should be the case, then we will need to change it. We will see how to do it later, but at this point simply select the file in your computer, and this will let us to try the full job. Check Figure 10-2 for more help.

✕ Job entry details for this transformation:	— □ ✕

Name of job entry: [Odoo to Staging]

| Transformation specification | Advanced | Logging settings | Argument | Parameters |

⦿ Transformation filename: [C:\Users\Albert\Desktop\OdooToStg.ktr]

○ Specify by name and directory

○ Specify by reference

Figure 10-2. *Specifying the transformation to run in the Transformation step*

The last step is to configure both email sending steps. These will send us a message when the job is done, specifying if all has gone ok (the first one) or another message specifying if something has failed, along with the log of the error, so we can quickly check from our mail inbox and go wherever we need to go to fix the error: a hung database, a problem with the ETL, a problem with the quality of the data …

■ **Note**   To set up the email step we need an SMTP server to be able to relay the mails to the specified destinations. Most companies have an internal SMTP server, but if it is not your case, do not worry. Many companies provide SMTP emails for free. Check out the conditions of each one, but I recommend `https://www.mailjet.com/` that offers up to 6.000 emails / month for free using their SMTP servers.

The easiest way to configure both email delivery steps is to create first one, and then create the second one by copying the first, renaming it, and adjusting the new properties. They won't be exactly equal, as in the error step, we decided to ship also the log files and the line of the error on the email step, so when the error is raised, whoever gets the email will know exactly which error has been produced without even needing to check the ETL process.

These emails will be targeted to the process controller, but it is a good option also to add another different email for business users in order that if the load has worked fine they know that they have fresh data available. On the other hand, it can be useful to have some control table that contains the last execution date and status of each load step and add to your end user dashboard a small grid including this information in order for them to know in the same screen that they are analyzing data, and when the last refresh process was run. Also proposed free versions for the BI component don't let you schedule emails and an automatic refresh of data, but if you decide to move to a tool that allows you that automatic execution from here, we can run a command that uses the BI tool command line to trigger the update of data inside the tool or the automatic sending of information attached via email.

To set up the SMTP step we need to know the details of your SMTP provider. You may use an internal SMTP server or an external one. In this case, we opted for using mailjet SMTP service, and the details we need to set the step up is basically the hostname of the SMTP server, the port, the username, and a password. Once we have these details (check your sysadmin or the tech guy if you're using an internal SMTP), we are good to go.

Configuring the SMTP step is easy. Just double-click on the step and in the first tab we only need to configure the source email account, the destination email, and the sender name. Have a look at Figure 10-3 for an example:

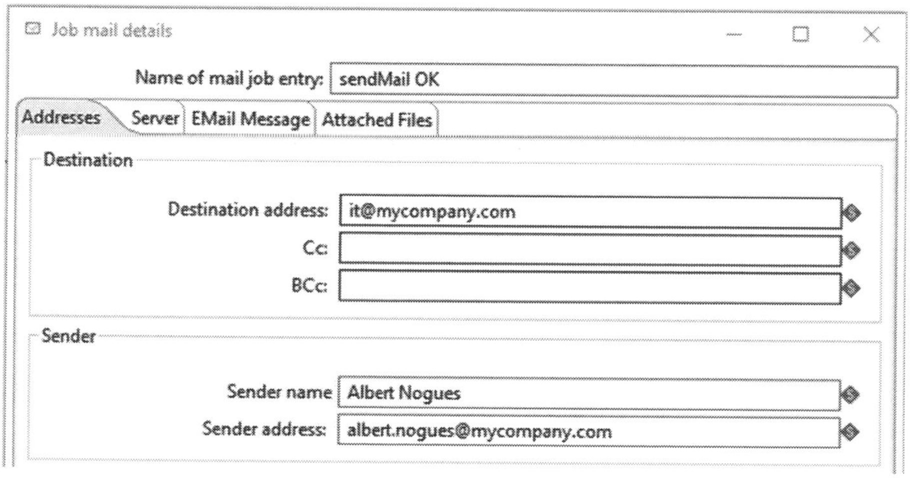

Figure 10-3. *Configuring the Sender address and the destination address of the email step*

The server step is a bit more complex. Here we need to configure all the details related to the server, the user, and password, as well as the port. If you use mailjet like me, be sure to specify port 587, as we have had issues using the standard 21. As a login either use your username, or the API KEY if you use mailjet, and as a password the provided password for that account or API KEY. Then you can decide to choose if you want to use secure authentication or not. If you are sending sensitive data, this should be a must. You can choose between TLS or SSL. Check out the details sometimes by choosing authentication so that you need to use a different port. In our case we have not checked the authentication box, but if you use mailjet you can choose TLS without any problem. See Figure 10-4.

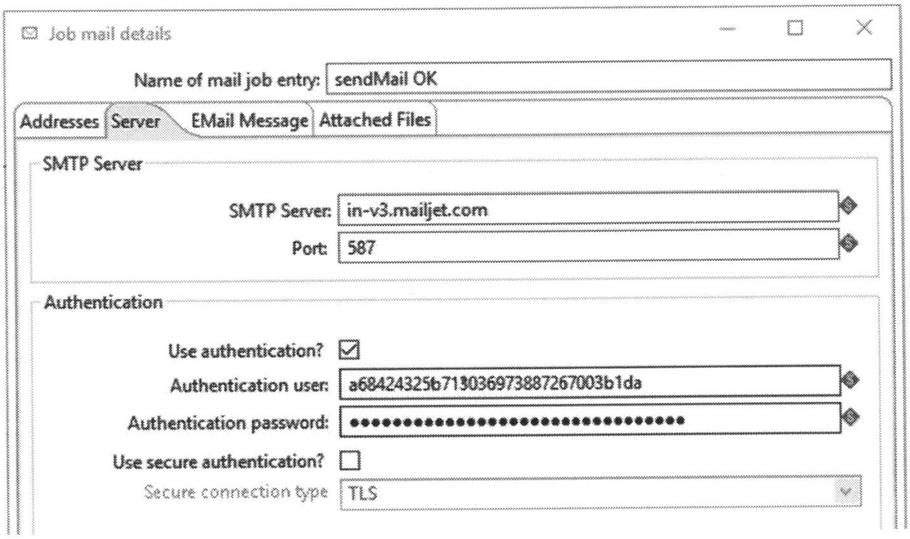

Figure 10-4. *Adding the server, port and password*

In the third tab, the Email message, we leave the body of the email empty but on the subject line we write: The ETL finished OK. And we skip the last tab, as if the process finished ok, we don't want to attach any file. After this, we are ready to run the job. As we are still testing it, we can launch it directly from spoon. Once it finishes, we will receive an email like the following one in our destination box, with subject "The ETL finished OK":

```
Job:
-----
JobName    : KettleJob
Directory  : /
JobEntry   : sendMail OK

Message date: 2017/01/21 12:26:03.005

Previous results:
-----------------
Job entry Nr      : 2
Errors            : 0
Lines read        : 0
Lines written     : 0
Lines input       : 0
```

```
Lines output        : 0
Lines updated       : 0
Lines rejected      : 0
Script exist status : 0
Result              : true

Path to this job entry:
------------------------
KettleJob
  KettleJob :  : start : Start of job execution (2017/01/21 12:17:45.716)
  KettleJob :  : START : start : Start of job execution (2017/01/21 12:17:45.717)
  KettleJob :  : START : [nr=0, errors=0, exit_status=0, result=true] : Job execution
  finished (2017/01/21 12:17:45.717)
  KettleJob :  : Odoo to Staging : Followed unconditional link : Start of job execution
  (2017/01/21 12:17:45.718)
  KettleJob :  : Odoo to Staging : [nr=1, errors=0, exit_status=0, result=true] : Job
  execution finished (2017/01/21 12:21:55.690)
  KettleJob :  : Staging to DWH : Followed link after success : Start of job execution
  (2017/01/21 12:21:55.690)
  KettleJob :  : Staging to DWH : [nr=2, errors=0, exit_status=0, result=true] : Job
  execution finished (2017/01/21 12:26:02.996)
  KettleJob :  : sendMail OK : Followed link after success : Start of job execution
  (2017/01/21 12:26:02.996)
```

Time to configure now the error step. In that case, we will keep basically the same configuration, but we will also include a log in the email to track the error that has been raised. For this, we double-click in the sendMail KO step, and in the fourth tab, we enable the checkbox Attach file(s) to message and we select the Log, Error Line, and Error options.

That's not all; we need to do a slight change in our ETL process. To be able to trap any error in any step, we have to configure an error exit in each step. Since we are not very concerned at this point in dealing in a different manner with our two transformations, we can link both error outputs to the same send mail step. For this, create a new link by dragging and dropping the output step of the first transformation in the canvas, the one that moves the data between the Odoo database and the staging area in our datawarehouse database and connect it to our sendMail KO step, like in Figure 10-5:

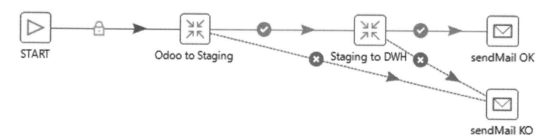

Figure 10-5. *Linking both error outputs of the two transformations to the same sendMail step*

After this, a new slight change is required in each transformation step. We need to enable the log tracing in both transformation steps. Double-click on the first one, head to the logging settings tab, and click on the Specify Logfile? Check the box and write the name of the logfile you want in the Name of Logfile box, and then leave the txt extension in the Extension of logfile textbox. The Log Level drop-down box lets you choose the verbosity of the output log. We will leave as it is (Basic) for the moment, but you can configure here the detail of the error log you want to receive. The result should be the same as that in Figure 10-6:

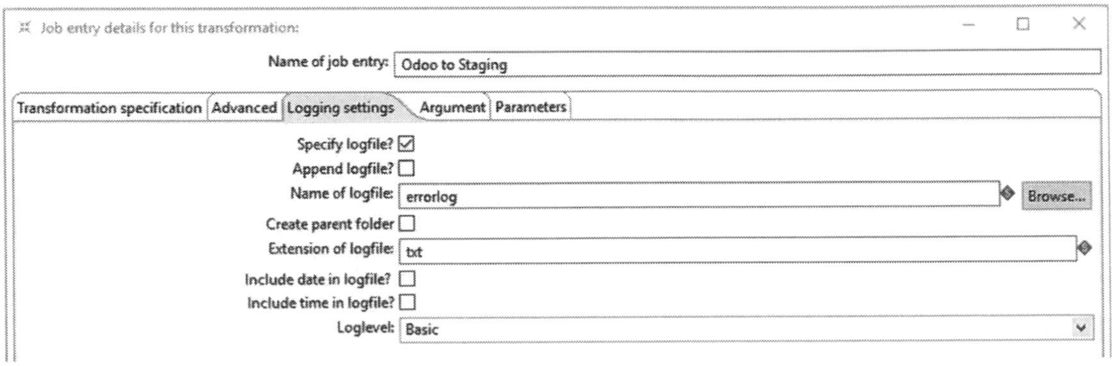

Figure 10-6. *Configuring the log generation in each transformation step*

Now imagine that an error has been triggered in the ETL process. In our case, it is a problem connecting to the database. But it can be many other problems, like duplicated values, wrong data, or wrong data format, a missing table, a connection issue with the database, or any other problem you can imagine. If the ETL has been designed properly, we will get that error in our mailbox. This time the subject of the email will be "The ETL finished KO" and the content will be something like the following:

```
Job:
-----
JobName    : KettleJob
Directory  : /
JobEntry   : sendMail KO

Message date: 2017/01/21 12:50:29.672

Previous results:
-----------------
Job entry Nr        : 1
Errors              : 0
Lines read          : 0
Lines written       : 0
Lines input         : 0
Lines output        : 0
Lines updated       : 0
Lines rejected      : 0
Script exist status : 0
Result              : false

Path to this job entry:
-----------------------
KettleJob
  KettleJob :  : start : Start of job execution (2017/01/21 12:50:23.022)
  KettleJob :  : START : start : Start of job execution (2017/01/21 12:50:23.022)
```

```
KettleJob :   : START : [nr=0, errors=0, exit_status=0, result=true] : Job execution
finished (2017/01/21 12:50:23.022)
KettleJob :   : Odoo to Staging : Followed unconditional link : Start of job execution
(2017/01/21 12:50:23.023)
KettleJob :   : Odoo to Staging : [nr=1, errors=1, exit_status=0, result=false] : Job
execution finished (2017/01/21 12:50:29.653)
KettleJob :   : sendMail KO : Followed link after failure : Start of job execution
(2017/01/21 12:50:29.654)
```

Note that the result value now is false. This means that the job hasn't completed all right. This line is also interesting: "Odoo to Staging: [nr=1, errors=1, exit_status=0, result=false]" as explains in which transformation the error has been produced. In our case, it has been in the first one. But that is not all! We still have an attached file to review for further information, the error log file. If we double-check the email we just received, we should find an attached file, called error log as it is the name we added previously. If we check the content of the file, we will get more information about the error, with the first lines being very informative:

```
2017/01/21 12:50:23 - Odoo to Staging - Loading transformation from XML file [OdooToStg.ktr]
2017/01/21 12:50:23 - KettleTransf - Dispatching started for transformation [KettleTransf]
2017/01/21 12:50:23 - sale_order.0 - ERROR (version 6.1.0.1-196, build 1 from 2016-04-07
12.08.49 by buildguy) : An error occurred, processing will be stopped:
2017/01/21 12:50:23 - sale_order.0 - Error occurred while trying to connect to the database
2017/01/21 12:50:23 - sale_order.0 -
2017/01/21 12:50:23 - sale_order.0 - Error connecting to database: (using class org.
postgresql.Driver)
2017/01/21 12:50:23 - sale_order.0 - Connection refused.
```

Overview at PDI Command-Line Tools

So far so good. Until this point we have seen how to run a PDI job (or transformation) and send the status of its execution to us. This is very interesting but there is still one supposition that we assumed until this point. We are launching the tasks directly from spoon, but usually the server won't have a graphical interface and even more, we want to schedule the tasks to run automatically, so having somebody to launch the tasks is not an option. To solve this issue PDI has some command-line tools that help us to launch jobs from the shell (Windows or Unix systems alike). These two tools are called Pan and Kitchen. Pan is the tool to launch stand-alone transformations (without any job) and Kitchen is used to launch jobs (which may have several transformations inside). We will see a sample of both, by using first Kitchen to launch our first transformation and, later, Pan to launch the entire job we designed at the beginning of this chapter.

Launching Transformations from the Command Line with Pan

There are several parameters you can pass to the Pan executable to launch a transformation. These parameters can be easily visualized if you call the executable without any parameter. The most important ones are the following:

```
Options:
  /rep        : Repository name
  /user       : Repository username
  /pass       : Repository password
  /trans      : The name of the transformation to launch
```

```
/dir         : The directory (dont forget the leading /)
/file        : The filename (Transformation in XML) to launch
/level       : The logging level (Basic, Detailed, Debug, Rowlevel, Error, Minimal,
               Nothing)
/logfile     : The logging file to write to
```

■ **Note**   You can check all options along with the explanation either by calling the tool or by reading Pan documentation on the PDI website here: https://help.pentaho.com/Documentation/6.0/0L0/0Y0/070/000

Depending if we use a repository or not, we need to call the Pan executable with a set of specific parameters. Let's have a look at the easier option, which is calling a ktr (PDI transformation file) directly from the executable. The transformation is named OdooToStg.ktr in our case, and it is the first transformation of our job, the one that copies data from Odoo to the staging area. Since we don't need to pass any parameter to our transformation, we will be only specifying an output log, just to check if the transformation ran successfully. This can be accomplished with the switch /logfile. We will keep the default logging level.

Take account if you are on Windows or Unix as you need to call the appropriate version of Pan (Pan.bat or pan.sh). So, for Windows the call should be:

```
Pan.bat /file OdooToStg.ktr /logfile OdooToStg.log
```

And the output should be something like this (it has been truncated to show only the most important part):

```
...
2017/01/21 16:07:37 - Pan - Start of run.
2017/01/21 16:07:38 - KettleTransf - Dispatching started for transformation [KettleTransf]
2017/01/21 16:07:39 - stg_account_tax.0 - Connected to database [bibook staging]
(commit=1000)
2017/01/21 16:07:39 - stg_hr_employee.0 - Connected to database [bibook staging]
(commit=1000)
2017/01/21 16:07:39 - Table output.0 - Connected to database [bibook staging] (commit=1000)
2017/01/21 16:07:39 - stg_product_template.0 - Connected to database [bibook staging]
(commit=1000)
...
2017/01/21 16:11:50 - Table output.0 - Finished processing (I=0, O=3653, R=3653, W=3653,
U=0, E=0)
2017/01/21 16:11:50 - Pan - Finished!
2017/01/21 16:11:50 - Pan - Start=2017/01/21 16:07:37.716, Stop=2017/01/21 16:11:50.629
2017/01/21 16:11:50 - Pan - Processing ended after 4 minutes and 12 seconds (252 seconds
total).
2017/01/21 16:11:50 - KettleTransf -
2017/01/21 16:11:50 - KettleTransf - Step Generate Day Rows.0 ended successfully, processed
3653 lines. ( 14 lines/s)
...
```

If we set up a repository at some point (no matter if we set up a file repository or a database repository) or even a Pentaho repository (newer versions only), we need to supply the parameters of the repository to launch our transformation. At this point it is like launching a stand-alone file, with a minor variation. The command would be:

```
Pan.bat /rep pdirepo /user admin /trans OdooToStg
```

And this will launch our Transformation from the repository. Note that we did not specify any password for the repository. If you specified one, you should add the /password switch and provide the password for the repository; otherwise Pan will throw an error. Having this clear, let's move on to launching jobs with Kitchen!

Launching Jobs from the Command Line with Kitchen

Launching jobs with Kitchen is very similar to launching transformations with Pan. The options are basically the same as those in Pan. To launch a job in our repository we will use the following command:

```
Kitchen.bat /rep pdirepo /user admin /job KettleJob
```

In case we get a message like this: "ERROR: Kitchen can't continue because the job couldn't be loaded," this probably means we are launching a job that does not exist or more probably, we have misspelled the job name, or we are specifying an incorrect repository. To list the repositories we have, we can use the following command:

```
Kitchen.bat /listrep
```

Which will throw us the following output:

```
List of repositories:
#1 : pdirepo [pdirepo]  id=KettleFileRepository
```

We checked that the repository is ok. To check the directory and the job name we can use another two set of commands:

```
Kitchen.bat /rep pdirepo /user admin /listdir
Kitchen.bat /rep pdirepo /user admin /dir / /listjobs
```

And the latest command will have shown that we have only one job called KettleJob. And relaunching it again with the correct directory:

```
Kitchen /rep pdirepo /user admin /dir / /job KettleJob
```

Which in fact it is the same as the first code snippet we saw, as if the Job is placed in the root directory "/", it is not necessary to explicitly tell it to the program. After a while we will see:

```
2017/01/22 11:09:27 - KettleJob - Job execution finished
2017/01/22 11:09:27 - Kitchen - Finished!
2017/01/22 11:09:27 - Kitchen - Start=2017/01/22 11:01:00.126, Stop=2017/01/22 11:09:27.297
2017/01/22 11:09:27 - Kitchen - Processing ended after 8 minutes and 27 seconds (507 seconds total).
```

Which means that our job has finished successfully.

■ **Note**   You can browse the content of a repository from Spoon by going to the Tools menu, then to Repository, and then clicking on Explore. From there you should be able to view all your objects in the repository and the hierarchies of files and folders. Moreover, from the Repository submenu you can import stand-alone files (or export them) in and out of your repository.

We have seen how to interact with the command-line tools of PDI. This is important to know as it will enable you to efficiently schedule jobs from any single task scheduler we may think of. For the rest of the chapter, we will see how to schedule tasks with the two most common task managers available: Cron for Unix systems and the Windows Task Scheduler for Windows systems.

Scheduling Jobs in the Task Scheduler

It's time to see how to integrate our jobs within a flow. Usually you will have several processes running at some point in the day, usually at off-peak hours. So, the BI part needs to go in some sort of harmony between these. Since every company (small or big) has different needs, we will present an easy and free way to launch your jobs. Obviously, this has some limitations so if you need to establish complex dependencies between jobs, or you need a tighter control and integration between all processes in your organization, you may need to look further for more specialized tools. But for our needs we will stick to the task scheduler bundled with the operating system.

Scheduling a PDI Job in Windows

In Windows, we have the task scheduler. To create a new task, click on "Create Basic Task." Name the task as you want and enter any meaningful description. After that window, another one will come asking for the recurrence of the task. In our case, we will select Daily. The start time and the recurrence will be set in the following dialog. Since we want the data to be loaded off hours, we will set the start date at 00:00 AM, which is basically at midnight, and we will leave the recurrence to every 1 day, as filled by default. In the next step, we are given three choices about what type of task we want to schedule. The first one is starting a program, the second one is sending an email and the third is displaying a message. We will leave the first one as selected, and we will click Next again. The last step is the most important one. The scheduler basically asks us which program to launch, which parameters to pass to the executable, and from where it should be run. Misconfiguring this will lead to problems in starting our job, so be sure the information is filled in correctly.
We will fill the boxes with the following information:

- In the Program/script box just add the path plus the name of your Kitchen executable. In Windows this will be something similar to this: D:\data-integration\Kitchen.bat

- In the arguments box, we will put all the switches we used when calling Kitchen manually. These are: /rep pdirepo /user admin /dir / /job KettleJob

- In the start in section, we will put the directory where PDI is uncompressed. In our case this is: D:\data-integration

Click now next to finish adding the task.

Before being able to run the task, we want to make sure the task will we launched regardless of if we are connected to the system. For this, double-click the task name in the Task Scheduler, and go to the first tab, called General. Make sure you select Run whether the user is logged in or not. The result should be the same as what you can observe in Figure 10-7:

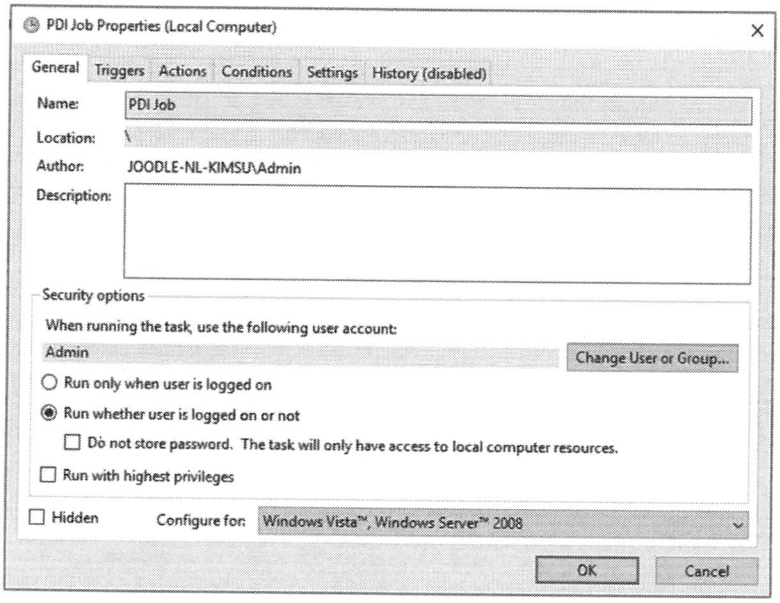

Figure 10-7. Run the task even if the user is logged out of the system

If your task needs administrative privileges, make sure you select the check box named Run with highest privileges. After that click OK, and you will be prompted with the credentials to run this task. We recommend that you create an account for automated tasks with only the required system permissions. With that you have your job scheduled. In the Last run result, you can see the result of the latest execution. Instead of a description it is shown in a code format being 0x0 the operation completed successfully, 0x1 (Error), and 0x41301 the task is currently running, the most common results.

▪ **Note**   If you want a powerful task scheduler in Windows, check out nncron. Nncron is compatible with crontab and it has a lite version that is powerful enough, and it is free. The website of the authors is: `http://www.nncron.ru/index.shtml`

Scheduling a PDI Job in Unix/Linux

To schedule a job in Linux we have choose cron. There are others, but all Unix systems come with cron shipped. While it may be required to learn a few configuration parameters to understand it, it is a very powerful tool with a very small footprint.

To list the content of your cron, you can use the following command:

```
crontab -l
```

If you don't have anything scheduled it is quite possible that you will see a message like the following one:

```
no crontab for bibook
```

The crontab file has a specific format. Basically, the file works in lines. Each line specifies a task to run, and then each line is split in six columns. Each column can be separated from the following one either by a space or a tab. We recommend using tab as a separator for easier comprehension.

The columns are used for the following:

- The first column specifies the minute when the task will run.

- The second column specifies the hour when the task will run.

- The third column specifies the day of the month when the task will run.

- The fourth column specifies the month when the task will run.

- The fifth column specifies the day of the week when the task will run.

- The sixth and last column specifies the task to run, along with its parameters.

As you noted, you may want to have a range of possible values, or using and all-like wildcard, for example. if you want to run a job every hour, it will be not practical, to explain all the 24 hours of a day. Fortunately, cron has some wildcards, basically the hyphen, and the asterisk. The first is used to specify a range, where the latter means "all."

So, for example, if we want to specify that a task needs to be run every day at 10 PM at night, we can use the following expression:

```
0 22 * * * task_to_run
```

So for example to launch our task at midnight we can schedule it as follows:

```
0 0 * * * /path_to_pdi/kitchen.sh /rep pdirepo /user admin /dir / /job KettleJob
```

The following bullet list, will help a lot in understanding it:

* * * * * command to execute

Where:

- The first position specifies the minute (0-59), or all of them (*).

- The second asterisk specifies the hour (0-23) or all of them (*).

- The third position specifies the day of the month (1-31), or all of them (*).

- The fourth position specifies the month (1-12) or all of them (*).

- The fifth position specifies the day of the week (0-6), starting on Sunday and ending on Saturday, or all of them (*).

- And the last, is the command to execute.

■ **Note**   For those that have some problems with the command line, there is a graphical editor for cron called gnome-schedule. It is bundled in many systems including Ubuntu. Obviously, it is not always possible to have a GUI in the server, but if that is not possible, you can still use the tool in another computer, and then do a crontab -l command to list the content of the cron file, and copy it straight to the server after doing a crontab -e, which will open the current crontab file for the user logged in, in a text editor (usually vi).

Running Database Maintenance Tasks

Running automated tasks opens a new window. We usually do not run only productive tasks, but also some maintenance tasks need to be run from time to time to ensure all systems operate under optimum conditions. You can have some housekeeping tasks, recurrent tasks to improve performance, and other tasks. We are going to talk for a while about maintenance tasks, more precisely, database maintenance tasks.

As we saw, our datawarehouse runs flawlessly. But the content of our tables in the database changes from time to time. It may be useful to run some maintenance tasks against it to ensure it keeps running in an optimal manner. Apart from proactive work from the DBA, or whoever oversees the administration part, we can schedule some recurrent tasks: to gather statistics to ensure that our queries are well as we saw in Chapter 7, perform integrity checks to confirm we have no corrupted data, and the backup task. Let's start with probably the most important one, the backup task.

Backing Up Our MySQL/Maria DB Database

To back up our database (MySQL or MariaDB), there is a bundled tool that helps. The tool is called mysqldump, and basically it is a database dumper, that is, it puts in a file all the content of the database. Depending on the parameters we specify to the program, we can get the structure, the data, or the structure + the data. Usually we will use the latter, as in case of a restore, we need both things.

```
The usage of mysqldump is the following:
bibook@bibook:~$ mysqldump
Usage: mysqldump [OPTIONS] database [tables]
OR     mysqldump [OPTIONS] --databases [OPTIONS] DB1 [DB2 DB3...]
OR     mysqldump [OPTIONS] --all-databases [OPTIONS]
For more options, use mysqldump --help
```

Let's first create an export directory that we will back up to a safe location:

```
bibook@bibook:~$ sudo mkdir /opt/export
bibook@bibook:~$ sudo chmod 777 /opt/export
```

Then we can run mysqldumpl and put the target export in the newly created folder. Again, remember to back up this folder properly; otherwise if you need to restore the database you'll be in trouble. Let's start first with the staging. Although for the staging we only need the structure, as the data is supposed to be volatile in this database, we can back up the data as well.

```
bibook@bibook:/opt/export$ mysqldump -u bibook -p --create-options staging > /opt/export/
db_staging_initial.sql
Enter password:
```

And after entering the password the backup will be ready. We can do an ls -la to list the backup file and confirm the backup has been made successfully:

```
bibook@bibook:/opt/export$ ls -la /opt/export
total 376
drwxrwxrwx 2 root    root      4096 Jan 28 14:46 .
drwxr-xr-x 4 root    root      4096 Jan 28 14:41 ..
-rw-rw-r-- 1 bibook bibook 375954 Jan 28 14:46 db_staging_initial.sql
```

Then, we need to do the same for the data warehouse database, dwh:

```
bibook@bibook:/opt/export$ mysqldump -u bibook -p --create-options dwh > /opt/export/db_dwh_
initial.sql
Enter password:
bibook@bibook:/opt/export$ ls -la
total 644
drwxrwxrwx 2 root    root      4096 Jan 28 14:48 .
drwxr-xr-x 4 root    root      4096 Jan 28 14:41 ..
-rw-rw-r-- 1 bibook bibook 274328 Jan 28 14:48 db_dwh_initial.sql
-rw-rw-r-- 1 bibook bibook 375954 Jan 28 14:46 db_staging_initial.sql
```

As you will notice, it is a bit odd that our staging area backup file is greater than our datawarehouse. For the first days, this can be quite possible, as there is data that we are still not incorporating in the data warehouse, and our data warehouse only contains few data, but over time, this should not happen.

Now it is time to schedule this in our task schedule. We decided for a daily backup, so to ensure we know which backup belongs to each day, we need to timestamp the file name. This may not be necessary if you are moving the file outside the system, but if you want to leave a copy, it is necessary to avoid overwriting. Again, this strategy depends on the size of your datawarehouse and other factors.

If we are in a Unix machine, let's edit our crontab:

```
crontab -e
```

and type the following:

```
0       0       *       *       *       mysqldump -u bibook -p --create-options dwh >
/opt/export/db_dwh_$(date -d "today" +"%Y%m%d%H%M").sql
```

With that, we will schedule a backup every day at 00.00 and the date timestamped will be appended to the file. Then verify that all is in place with a crontab -l command. Tomorrow you should see the first backup there.

Perform Checks and Optimizations in the Database

We have seen how to back up our database in case of a disaster. Let's see what kind of proactive work we can do to avoid issues like loads hung, or processes taking too long due to poor performance.

To check the status of our database tables and analogous to the SQL CHECK TABLE tablename command, we have a command-line tool named mysqlcheck. The syntax is very similar to mysqldump:

```
bibook@bibook:/opt/export$ mysqlcheck -u bibook -p --database dwh
Enter password:
dwh.t_f_sales                          OK
dwh.t_l_category                       OK
dwh.t_l_currency                       OK
dwh.t_l_cust_country                   OK
dwh.t_l_customer                       OK
dwh.t_l_emp_department                 OK
dwh.t_l_emp_level                      OK
dwh.t_l_employee                       OK
dwh.t_l_month                          OK
dwh.t_l_parent_category                OK
dwh.t_l_product                        OK
dwh.t_l_quarter                        OK
dwh.t_l_status                         OK
dwh.t_l_year                           OK
dwh.t_r_time                           OK
```

It may be interesting to schedule a task to run that command from time to time (maybe every day before the backup) or once a week, depending on the activity on your database, and make sure the output is OK.

If we detect any table(s) that need to be repaired, we can use the same command-line tool to do it.

```
mysqlcheck -u bibook -p --auto-repair --check --database dwh
```

▨ **Note**   For more options and different command switches of the mysqlcheck tool, you can read the latest information of the tool here: `https://dev.mysql.com/doc/refman/5.7/en/mysqlcheck.html`

Apart from checks of integrity, we can also perform other types of maintenance tasks. These include optimizing tables and indexes in our database and collecting fresh statistics about the data inside our tables, which will help the database optimizer in choosing the right execution plan, when presented a query.

There are basically two tools that we can use, which in fact are symbolic links to the same mysqlcheck tool we saw previously. These two are mysqlanalyze and mysqloptimize. Let's have a look at both.

The analyze tool analyzes the distribution of the data in a table. As explained previously, knowing a little bit about the data present on a table helps a lot the database engine. Without that information, the optimizer works blind, and makes assumptions that could turn out to be very inaccurate. To help the optimizer, we can analyze the tables in advance. The general rule is only to analyze a table again if more than % of the data has changed, so there is no need at all to run this command on a daily basis.

We can schedule for example a weekly job to analyze all tables in the datawarehouse. Since tables are read locked, we must ensure this operation is done off-peak hours, usually after the ETL execution. The command and the expected output is as follows:

```
bibook@bibook:/opt/export$ mysqlanalyze -u bibook -p --database dwh
Enter password:
dwh.t_f_sales                          OK
dwh.t_l_category                       OK
dwh.t_l_currency                       OK
```

```
dwh.t_l_cust_country                        OK
dwh.t_l_customer                            OK
dwh.t_l_emp_department                      OK
dwh.t_l_emp_level                           OK
dwh.t_l_employee                            OK
dwh.t_l_month                               OK
dwh.t_l_parent_category                     OK
dwh.t_l_product                             OK
dwh.t_l_quarter                             OK
dwh.t_l_status                              OK
dwh.t_l_year                                OK
dwh.t_r_time                                OK
```

This command-line tool is analogous to the ANALYZE TABLE tablename command that can be run from the MySQL or Maria DB client.

■ **Note**   More help for the analyze command can be found here: `https://dev.mysql.com/doc/refman/5.7/en/analyze-table.html`

Last, but not least, we can optimize tables. If our database has a lot of data coming in and out, we need to reorganize our tables from time to time, to make sure the space is used wisely. Having tables not properly organized means not only the tables will take much more space than the needed on disk, but also, as a consequence, when we retrieve that data, the database will take longer to serve because it has more places where to look. So we are not only speaking about space savings but also speeding up the queries.

The command to reorganize a table is mysqloptimize. We can run the command to optimize the entire list of tables inside a database. The call and the expected output is shown here:

```
bibook@bibook:/opt/export$ mysqloptimize -u bibook -p --database dwh
Enter password:
dwh.t_f_sales
note       : Table does not support optimize, doing recreate + analyze instead
status     : OK
dwh.t_l_category
note       : Table does not support optimize, doing recreate + analyze instead
status     : OK
dwh.t_l_currency
note       : Table does not support optimize, doing recreate + analyze instead
status     : OK
dwh.t_l_cust_country
note       : Table does not support optimize, doing recreate + analyze instead
status     : OK
dwh.t_l_customer
note       : Table does not support optimize, doing recreate + analyze instead
status     : OK
dwh.t_l_emp_department
note       : Table does not support optimize, doing recreate + analyze instead
status     : OK
dwh.t_l_emp_level
note       : Table does not support optimize, doing recreate + analyze instead
```

```
status    : OK
dwh.t_l_employee
note      : Table does not support optimize, doing recreate + analyze instead
status    : OK
dwh.t_l_month
note      : Table does not support optimize, doing recreate + analyze instead
status    : OK
dwh.t_l_parent_category
note      : Table does not support optimize, doing recreate + analyze instead
status    : OK
dwh.t_l_product
note      : Table does not support optimize, doing recreate + analyze instead
status    : OK
dwh.t_l_quarter
note      : Table does not support optimize, doing recreate + analyze instead
status    : OK
dwh.t_l_status
note      : Table does not support optimize, doing recreate + analyze instead
status    : OK
dwh.t_l_year
note      : Table does not support optimize, doing recreate + analyze instead
status    : OK
dwh.t_r_time
note      : Table does not support optimize, doing recreate + analyze instead
status    : OK
```

As you see, depending on the engine we use with the database, the operation done can change slightly. In our case, we are working with InnoDB tables. When using the InnoDB engine, the command does a re-create and analysis of the table instead. This operation can take considerable time, especially if the tables are larger. We advise you to run this command during a maintenance window, or at point in time when we can afford to have our database offline for a while.

As in previous tools, this command is analogous to the OPTIMIZE TABLE tablename command that can be launched from the MySQL or Maria DB client.

■ **Note**   More help for the optimize command can be found here: https://dev.mysql.com/doc/refman/5.7/en/optimize-table.html

---

Conclusion

We are almost finishing our journey, and while we have already seen the core and foundations of designing a BI solution, this chapter added extra value by presenting some maintenance tasks that are very important to have the platform up and running.

Adding controls in our ETL jobs is an essential task. People expect to know when things go wrong. And they should be informed and take fast action to solve the issues that may appear from time to time. We also saw, how to improve database performance and do some basic maintenance tasks in the database, which will improve the health of our solution.

We also saw how to program tasks to run in a recurrent manner. During your solution life cycle, you will have to design plenty of these, so it is a good time to start getting some practice with it. Using the task schedulers bundled with the operating systems may be good, but at some point, you will need to create more complex dependencies between tasks, and you will likely need to either develop some scripts or give a go at more complex tools, but at this point both the cron and the Windows scheduler are good tools to start with.

In the following chapters, we will see how to structure more than one environment to work with; and how to ensure the environments interact between them by not losing developments or data, and how to implement some order, which is especially necessary when more than one person is working on the same things at the same time. We will also see how to deploy the solution we built on the cloud, how to do a small capacity planning, and choose a solution that will help us to scale if our company starts performing better and better.

CHAPTER 11

▪ ▪ ▪

Moving to a Production Environment

Following our recommendations and deployment steps until here, you will have an environment that will have all required elements to work, the relational database in place, an ETL system that brings information from the source ERP, a BI platform that allows you to analyze the information in an easy way, and an MOLAP platform that helps you in next year's target definition. But we are quite sure that you won't stop here. You will want to add new analysis, new fields, new calculations, new attributes, or new hierarchies to your system. It is possible that you want to add volume, more granularity, daily detail instead of monthly, or arrive to the same level of information than your transactional ERP system. You can require multiple modifications in your analytic system that can interfere in your database structure and your already created reports that are being used by multiple customers. In order to ensure that you have reliable data to offer to anybody analyzing it in your system, the best scenario is to have different environments for data analysis and for new developments, in order to avoid undesired affectations.

We will also evaluate some considerations to take into account when you move to a productive environment, such as server dimensionality, backup definition, monitoring, security management, high availability, or disaster recovery configuration. As in the rest of the book, some of these subsections could be a matter of writing a whole book dedicated to each, but we are doing just a brief introduction as a matter of explaining you the purpose of each section and some recommendations but without entering into detail about each subject.

Multienvironment Scenario

The most common status in all installations is to have more than one environment for each component of your server layout. In fact this is a general rule for all systems; this is not something pure from BI. Once you start having multiple users accessing a system to any functionality that you can think of, you cannot just test operations, new software versions, new developments, or configuration changes in this system because you can cause affectation to the daily work of the whole company. In case of BI, once the system is working and users are accessing, we will also want to have different environments to avoid affectations on our modifications to the system used by end users. A normal situation is to have from two to five scenarios of different systems, but the most common approach is to have three areas of work. We consider that especially in BI, three environments are enough to comfortably work with all the required functionalities.

- **Production**: This environment will be used for the real operation on a daily basis for all end users of the platform. It will be the main environment in terms of usage, sizing, and availability requirements.

© Albert Nogués and Juan Valladares 2017
A. Nogués and J. Valladares, *Business Intelligence Tools for Small Companies*,
DOI 10.1007/978-1-4842-2568-4_11

- **Simulation / User Acceptance Testing / Integration / Testing:** You can find multiple nomenclatures for this environment but these are the ones we have found more often in the customers we have visited. Here some key end users will be able to validate that the new requirements developed meet with the specifications required by them.

- **Development:** In this environment developers will perform the modifications required for each user requirement (or user story; remember Scrum concepts from Chapter 2).

We will see within the next sections how to build up these scenarios, some considerations to take into account, and recommendations of usage based on what we have been finding during our experience.

Deploying a Productive Environment

To achieve the objective of having multiple environments, you will have different options; if the servers/components you have used for installing all the components are performing correctly, you can consider them as productive ones and then implement new environments to work as simulation and development ones. This can be a valid procedure if you mitigate its main inconveniences:

- **Hardware capacity:** If you have been building this platform in servers without enough capacity because you were just trying, you should reconsider this strategy as far as production capacity must fit with performance requirements of your environment. We will talk in the next section about server capacity plans.

- **Test and error:** It is quite possible that during installation of required tools development, you have been doing multiple test installations, trying different components and solutions, doing test during the data modeling, the BI tool development, the ETL process, and it is highly recommended to use a clean environment as a productive one; so if you haven't been so clean and you have multiple test objects in your platform, maybe it is better to go for fresh servers and install and deploy there just the required components.

- **Security:** It is also usual that security policies required by your company in development environments are not as strict as the ones required in production environments so it is possible that you require reviewing accesses, installation procedures, or service users used to run the different applications that we will be running.

- **Network separation:** Related to the previous topic, it is possible that your company security policies require having separate networks for productive and development environments, so if you haven't asked specifically for then when you created your servers, they could have been created in the development network and they cannot be moved to the production network.

All these topics can affect you in a greater or lesser extent depending on your company security policies, network complexity, or server requirements for your BI solution, so in a general way we prefer to implement from zero the production environment with enough capacity, just with required components for the analysis and following all recommendations and policies that require a productive environment. Anyway, a good possibility is starting by using your existing system with all these constraints in order to know which usage of the platform is being done by users, which performance requirements it has, and gathering statistics from the existing system to create a productive environment correctly dimensioned.

Server Capacity Plan

One of the main topics to take into account when you are going to prepare the production environment is to provide it with the required capacity to cover your performance and storage expectations. Creation of a good capacity plan is something quite complex because it requires the analysis of multiple variables such as ETL periodicity, parallel processes required, type of reports executed, quantity of reports executed, number of expected users, number of parallel users, maximum parallel executions in the peak hour; and most of these variables are quite complex to predict when you are creating a system from zero. You need to take into account also crossed dependencies. Your ETL system will affect the source and target databases, and the target database capacity must take into account ETL performance expectance and report execution. You can have locations in different countries with different time frames and you can have parallel ETL processes during the night of a country affecting to daily report executions during the morning of another country. Of course it is easier to predict all these variables if you have previous data from an existing system, as commented in the previous section, but it is quite possible that you don't have them so let's try to define a list of steps to follow in the capacity plan. Let's see first general considerations and then we will enter in variables that can affect each of the components in the platform. In general terms (in fact these considerations are not only for BI, if you are managing other type of systems) we will need to consider the following:

- Main objective of a capacity plan is to define how many servers are needed (in case of going for a clustering environment) and capacity of each of them in terms of RAM Memory, CPU, and Disk space.

- We need to focus on maximum peak usage to have a correct dimensioned platform.

- It is highly recommended (despite sometimes it is not possible) to agree with users about future projects and incoming requirements.

- If you don't have these requirements when doing the capacity planning, review it each time that a new project comes to your system.

- Data history required to save in the datawarehouse is a critical parameter to know that can affect all the components, DB sizing, and ETL and BI processor capacity.

- The usage of existing systems to gather some data from unitary tests will allow you to extrapolate the results for the overall needs of the production environment.

Database server capacity planning

In order to define the capacity planning for the Database component, you will require gathering information about different KPIs:

- Average row size for bigger tables (usually Fact ones): you can gather this information from the existing system.

- Expected number of rows per month/year: if in the existing system you have real data you will be able to know how much space will be used despite the fact that you may not have the whole history loaded in your database. If this is not the case and you have just some sample data user to build up your application, you will require the analysis of the source environment to try to understand how big the number of received rows based on extractions or direct queries over ERP database can be.

- Expected datawarehouse history: as commented in the previous section datawarehouse history will affect directly in the size used by fact tables.

- Take into account that you will possibly require temporary tables to save different process steps, dimensions, and lookup tables that will have a default growing ratio that shouldn't be that much as fact tables; you need to have space enough available for some copies of tables for backup, working tables, and so on.

- You will require also space available in disk for backup, temporary tablespace for running queries, and system and catalog size for technical schemas.

- In terms of CPU required you will need to analyze peak of parallel processes running on the database for the ETL process and peak of reports from the BI running during the daily hours.

ETL server capacity planning

You ETL system will provide you the information available in the database during through some load processes containing different update/insert statements over the database schema. In order to correctly analyze which requirements this system has, you will need to know:

- Data load period: Requirements are not the same if information is loaded in a daily, weekly, or monthly basis.

- Data load window: We can have multiple dependencies on our daily processes and we need to know business expectancy in terms of data availability in order to match with business requirements. Explained with an example, if you are going to implement a daily reporting system for your sales force team and they start visiting customers at 9:00 in the morning you will require having your daily load for this process before 8:00 if the automatic reporting process takes one hour to run. In a daily sales analysis, you will want to have data about invoiced goods until the day before, so your process will not be able to start earlier than 0:00. But also you can have some dependence on some billing process that ends at 2:00, a backup job of the ERP or your database that ends at 3:00, or whatever other process that could be linked. So you will need to ensure that your ETL process fits in the time window that you will have available.

- Data amount: You will need to estimate the amount of data that will be moved by the system in order to correctly define its capacity. For this analysis you can take as reference current ETL data amount and its throughput (rows per second).

- Parallel processes: You must take into account all parallel process that will be executed by the ETL in order to define the server capacity to define the maximum rows per second you will need to manage.

BI server capacity planning

In the case of BI server, as far as it is the direct interface with the users, you will need to center your efforts on analyzing the user execution profile.

- Number of users: It is relevant for the capacity planning to know how many users are expected to be in our system and how many of them will be executing reports in parallel in the system (and usually in the database).

- Number of reports: You will need to estimate number and size of daily reports required by your users in an overall daily analysis and also focusing on the peak execution time.

- Execution time expectance: This value can affect also to the database, as far as depending on the BI strategy, it is quite possible that the biggest part of the execution time lies in the database server.

■ **Note** A critical point to analyze when defining our network topology that can affect to end-user experience is the network latency. In general it is highly recommended that all components on our network topology (ETL, BI, and Database) are quite closer in terms of network latency. But in case of the latency between BI server and end users can be more complicated to ensure because you will have users connecting from multiple sources and environments. It is especially critical when you are connecting from external connections to ensure that VPN performance is good enough to give the customer a correct end-user experience.

MOLAP server capacity planning

In case of the MOLAP database server we will require you to apply a mix of considerations between the database server and the BI server, but taking into account that the size defined will be multiplied also by the amount of scenarios that we will have. Usually it will require less parallel loads and queries than a datawarehouse database but we will need to analyze also number of users, number of reports, and number of parallel query executions.

■ **Note** Performing a good capacity planning can be very difficult and become a critical process but you can minimize this process criticality if you can choose scalable servers, with the possibility of adding disks, CPUs, and RAM slots or virtual servers that you can easily adapt to incoming needs. In next chapter we will see also another alternative that is the usage of cloud systems to locate your servers.

Licensing and Support Costs

Let's imagine a possible situation. You have gone live with your BI system one year ago. Your system is performing correctly; you have a set of daily loads defined in your ETL that are running perfectly on time, with data accuracy, without any incident for more than six months but suddenly your ETL process is returning an exception. Nothing has changed, data is correctly saved, there are no inconsistencies in data, and anything that you can check or evaluate seems perfectly fine and you don't find any useful entry in your research in Google. Your sales force users have implemented a system based on this process that they use each day in their visits to customers to prepare commercial conditions and agreements that they can use in their negotiation and they require extract information for their visits of the day. They will press you a lot to have available the information that the ETL process in error is delaying. What can you do now? Who to ask about it? One possibility is to ask support from some consultancy company that can help you to solve it but maybe they don't have availability during some days. Another possibility is to hire support of the vendor to help you in this situation to solve or workaround the issue.

Another situation. Let's imagine that your BI system implemented with Microstrategy Desktop or QlikSense has been a real success, you have a lot of users that have it installed in their laptops, so they can use it for free but they cannot share the information among them, execution performance is not good enough, and they require implementing some data security or job prioritization to attend first to managers in your organization. These kinds of actions can be solved if you move to a server installation instead of using just desktop clients to launch your reports, so in this case maybe you need to think about purchasing a license so you can move to a licensed server-based environment.

Licensing cost can be an important bill to consider but it is possible that you require moving to this scenario if your solution grows enough and becomes important enough for the company. In this case you will need to evaluate the ROI (Return of Investment) that this solution will provide so you can justify this movement.

■ **Note** We strongly recommend you evaluating licensing and support costs of the tools that you are going to install for free as an important variable when you are choosing each tool. Maybe you see some tool quite attractive when it is for free, but then you cannot get budget from company management to move to a licensed environment and you need to rebuild the solution using another tool just for this cost.

Adding Environments

Once you have your production environment defined, it is time to create the rest of the environments that you will require for your daily work. And these are mentioned in the introduction at least two more, development where we will create all new objects and modifications on existing ones and simulation or test where we will test those modifications that have been done in development with some significant data to validate that they are performing as expected. Depending on the size of the project and the number of developers and users accessing to the system, it is possible that we require a fourth environment that we can name Hotfix, Patch, Incidental, or Support; at the end the name is not relevant, the relevant thing is the usage. We have seen also installations with a fifth environment, a Sandbox used to test new features and versions or capabilities. And we have also found some installations with an Integration environment where users can do the tests of new developments before moving them to UAT. You can see some characteristics of all of them with deeper detail (but not so much, so don't worry!) in Figure 11-1. In this figure we can see some Accesses granted column where there are some roles. We will see these roles in detail at the end of the chapter.

Environment	Purpose	Access	Data	Capacity
Production	This is the main environment that will contain real and complete information of the company where all end users will connect daily to execute reports	End Users Key Users Product Owners Support Team	It will contain full history of real data	100%
Test	In this environment key users will be in charge of validate that developments are performing correctly. It is used as last validation environment before applying any functional or technical modification to the production system. It can be used for performance tests to don't affect production environment	Key Users Development Team	It will contain some significant data from Production environment. Depending on the privacy requirements data could be fake or encrypted data	Between 50% and 100%
Development	This environment will be used by developers to create code, model structure, ETL processes and BI reporting using just sample data to validate that basic operations expected from data processing are correctly developed	Development Team	It will contain sample unitary data for unitary validation	20%
Sandbox	This environment is used to test new functionalities, do Proof of Concept projects, test new products and new capabilities of existing ones, validate new technical components and any other operation that can require some testing. Development and modifications performed here must be used as basis to get software knowledge, document it and replicate it in the rest of environments	Administrator team Key Users	Typically just sample data, but depending on the test that we are doing (a new database engine) could require having significant amount of data	20%
Hotfix	This environment is expected to be used by key users or support team to perform small modifications to test and correct some minor errors without requiring the interaction of development teams and apply the whole transport process for urgent incidents	Support Team Key Users	It must be similar to the Production environment to be able to show relevant information to ensure that corrections are correctly applied	Between 50% and 100%

Figure 11-1. *Environment characteristics*

The creation of a new environment can be done basically with two different approaches, starting from an empty environment and installing all the components and deploying all different objects or cloning environments at one or multiple levels. Depending on the direction of the copy and some other variables we will prefer one or the other option:

- New clean environment: With this strategy you create an empty system; you install all the required components; and you deploy on the different layers, Database, ETL, BI, and MOLAP all the objects required for the analysis, using creation scripts if you have this possibility. In this way you will have a clean installation without any incorrect entry in the system registry; without dummy files; and also you avoid transport test objects, unused objects, backups of data, or any object that is not explicitly required. It is the most efficient strategy in terms of sizing and cleaning but it is the most costly to do because you need to have an exhaustive control on each requirement object. Also it is quite possible that you require changing some configuration objects in the system: of course, server name and IP address but also ODBC connections, configuration files, database names, etc.

- Fully cloned environment: On the opposite side you can clone all the servers with some backup tool restoring the whole server in an empty one. This is the quickest way to create a new working environment, but it can drag forward some problems, unused objects, incorrect installations, software installed and uninstalled that lets some unneeded file in the system, database objects that are not needed, etc. Of course you will have then the possibility to clean whatever that you don't need, but there will be always some objects that you don't know if you can delete such as some library in a shared path of the system.

- Partially cloned environment: An intermediate solution that we think that is a good approach is to install all the software required, configuring all the required connections, names, and parameters, and then move the application objects from one server to another, usually using client tools that allow copy/paste functionalities.

Isolation of Environments

Depending on your company security policies it is possible that your environments are completely separated in different network segments without visibility among them. Isolation of environments allows you to have more flexibility on development systems facilitating the development by opening direct access to the servers for developing, giving administrator rights, or opening Internet access from servers in development as far as with separated networks a security incident in development environment won't affect the production environment. Once you have all your development finished you can transport modified objects to Testing and Production (or your administrator team will do it following your instructions) and nobody apart from the administrator team must have elevated access to any server, database, or tool.

This isolation will affect a cloning and maintenance strategy because in separated networks you won't be able to directly move objects from one environment to another; you will need to use a strategy that uses an intermediate system in order to locate temporary objects in movement from one environment to another.

Multienvironment Recommendations

In a multienvironment scenario you will need to take into account that your work is not just to develop the requirements coming from your end users but you need also to follow some strategy that facilitates the movement of objects across environments; otherwise you will be forced to repeat some development work or some adaptation in each object transport. In order to make yours and your administrator's life easier, we propose in the next subsections a set of recommendations to follow on your developments.

Parameterizing

All that can be parameterized should be parameterized. Let's develop a little bit more this assertion. You need to evaluate all that could change across environments to try to extract from the object any reference to a fixed value and use a variable or parameter external to the object itself. Let's give an example: imagine that you are developing an ETL process. You can configure the session to connect to the DevERP server (your development environment for ERP) for reading and to connect to DevDWH (your development datawarehouse server) using DevUSER for writing data. If you do that, you cannot just copy and paste the session to Testing, because after the copy you will require changing the connection string to TestERP and TestDWH. But if you use ${HOST} and ${DATABASE} variables you will be able to substitute them externally in a single configuration file or using environment variables. In Figure 11-2 you can see how to configure a session using variables.

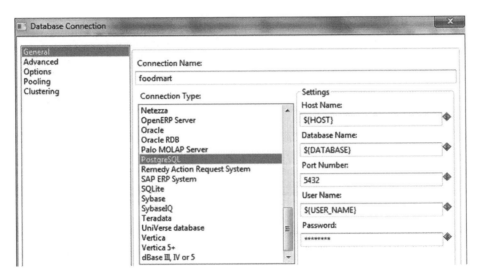

Figure 11-2. *Parameterized connection*

Server Paths

You should try to install and locate files always in the same path inside servers in order to facilitate the reference to files in the different tools, such as source files for ETL, parameter files commented on in the previous section, MOLAP cubes, repositories, or any other file used in any of the tools. Sometimes it is not possible because you can have multiple deployments inside the same server. It can happen because we don't live in a wonderful world with unlimited resources to implement your system and sometimes we can define logical environments sharing the same physical environment. Imagine that you have a single ETL server for Development and Testing environments and you have two deployments of an ETL engine based on different folders. Whenever you want to transport a workflow that is loading a file in a defined path you will need to change it when moving it from development to testing. In this scenario you can try also to use relative paths to refer to that file from the base path of the environment. In other words, you can use relative paths similar to these:

```
./SourceFiles/Area1/Sales_area1.txt
```

Or using variables, which at the end are using the same path name in terms of configuration:

```
$INSTALLPATH/SourceFiles/Area1/Sales_area1.txt
```

Naming Alignment for Connectivity Objects

As we have seen, during the installation and configuration of all the tools you will need to define multiple connectivity objects, such as ODBC connections, JDBC connections, or native connection strings, especially for ETL, BI, and MOLAP tools connecting in some way to the database engine. As in the rest of paths and parameters, it is also convenient here to align names for connectivity objects, so you can use the same object in multiple environments just changing the configuration of the connection in a single object. Let's say that you create an ODBC pointing to your ERP database and you name it ERP_DB and you name your ODBC connection to write in the Sales Datamart database as SALES_DB, no matter if you are defining it in the Development, Testing, or Production environment, pointing to ERP Development, ERP Testing, and ERP Production and Sales Datamart Development, Testing, and Production respectively. You will be able to use the same workflow reading from ERP_DB and writing to SALES_DB in all the environments.

Maintaining Your Environment

Defining a multienvironment strategy has an initial effort to configure and set up all your servers that can be a significant part of your whole BI project, but if you follow the previous recommendations, you will be able to maintain it minimizing the effort required for this task. This effort can be minimal, but it doesn't mean to be low, as you will have multiple tasks to do to keep your system ongoing in a healthy way. Let us introduce you also to some concepts related to the platform maintenance that every single platform requires.

Object Life Cycle

Once you have your multienvironment system initialized you can start with new developments without affecting your real data. The life cycle to add new capabilities in the system is shown in Figure 11-3. Developers will do their work in the Development environment, and once finished required objects will be moved to the Testing environment where Key users can validate that the development is matching with requirements and once validated, required modifications will be transported to production environment. You can see also in this graph two environments more. As commented, Hotfix will be used to solve incidents with some development in a quick way with real or near-to-real data. We could have a Hotfix environment just for BI, or just for ETL systems; in this case they would be linked to the production database. Once we have solved the issue we need to move to the production environment with the required modifications but we need also to apply them into the development system in order to not lose any modification in case of moving some development that affects the same object. It is also quite usual that the Hotfix environment is periodically refreshed with production data and structure.

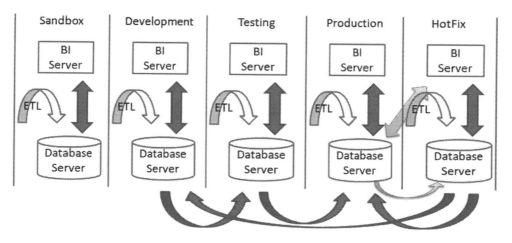

Figure 11-3. *Multienvironment schema*

■ **Note**   Changing objects in different environments rather than the expected one can cause regression errors when you overwrite a modified object based on a different version of the object. Imagine that you modify a column type in a database table in Production and then you re-create it using the development script because you have added two more columns. It will change back the column type.

Transport Procedure

We strongly recommend you to define and follow always a transport procedure for each tool in your BI environment that will facilitate the task of rolling back changes, investigate sources of possible issues, or just track performed changes in your environment. Procedure itself will depend for sure on the tool that you are using but also it can depend on the object type inside each tool and with the transport options required. You should have a list of steps for each type, defining paths and locations of auxiliary objects and responsibilities of each step. In Figure 11-4 you can see the workflow related to the transport of a database table.

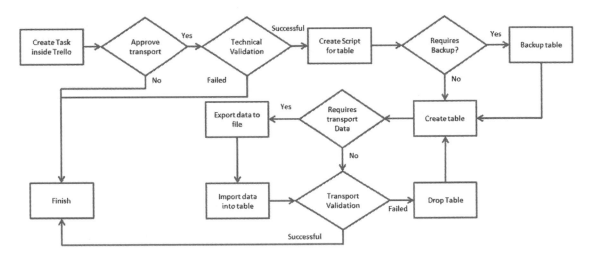

Figure 11-4. *Table transport workflow*

In this workflow we can see different steps:

1. We are considering that we will use Trello for task management, so the first step will be to open the task in Trello by the development team.

2. The project owner will approve the transport or not.

3. Admin team will validate if the table follows standards for nomenclature, that it has correct parameters and column types, etc.

4. If table follows the standards we will generate the script for creating the table.

5. After checking if backup is required, we will proceed with it or not (if table doesn't exist it wouldn't make sense at all; if it exists, development team should specify), defining the backup procedure, that we could fix with different strategies also:

    a. Script of the table to replace: defining a location where to save the script and file nomenclature.

b. Copy of the table with another name (_BCK at the end) or in a backup database.

c. Export table using export database utilities, defining also export file nomenclature and location.

6. Then admin team proceeds with table creation.

7. If table requires transport data we should export/import this data using a temporary file that also requires the definition of nomenclature and location.

8. Once transported we need to validate if transport is correct; otherwise we re-create the table and if all is correct we finish the process.

This is just an example of transport procedure but you should define it for all (at least most usual) object types to transport across environments.

■ **Note**   It is important to remark from the transport procedure the inclusion of a backup of modified objects in order to be able to roll back any change done in case that something fails during the transport or due to an incorrect development.

Transport windows

Inside the transport procedure we recommend you also include a transport window where transports should be performed with two main objectives:

- **Minimize impact on end users**: Imagine that you want to transport a change in your main fact table that requires dropping and re-creating it including new fields and reloading the whole history. It would let the data be unavailable and any reports querying it will show nothing until the transport operation finishes. If you have a transport window defined (usually out of business hours or next to be out of business hours) your users can extract the report out of this time frame.

- **Minimize efforts on transports**: Having a defined window to transport objects, you can minimize the administration efforts of the platform by facilitating the administration work organization.

Transport Automation

Going further in your environment it is possible that you arrive to a moment when you need to deliver an increasing quantity of transports due to an increasing number of developments. We recommend that you investigate command-line possibilities for transport automation in order for you to implement an easy way of moving objects and data across environments so you can automate as much as possible this transport task that can be a repetitive task, required but without adding too much value to the overall BI solution itself whose main objective is just to analyze data. Don't misunderstand us; as most technical tasks are required behind any environment, an incorrect transport policy can cause a lot of headaches so it is quite important to define a correct procedure but the final objective of a BI solution is to analyze data, not to correctly transport objects across environments. There are quite a few open source or freeware tools that can help you in these purposes like Puppet, Chef, and Python fabric library. These tools automatize actions, and can repeat the same over a large number of machines. They are especially interesting when you need to replicate that work in different environments or groups of machines.

Testing Procedure

An important part of the transport procedure is to define what will be the testing methodology to validate that your development is correctly performing the business requirements defined. Also it is important to validate that any object or modification that is derived from the development doesn't affect to existing functionalities, usually known as **regression tests**. This last part is usually ignored when you have a completely new development but don't underestimate the capacity of interrelation of developments to affect you in a productive environment. So although you think that a new development doesn't have anything to do with what is in place, don't miss the execution of regression tests to validate that this is completely true. Some considerations when defining a testing procedure are these:

- Set of tests definition: which reports, queries or scripts must be launched to validate all the tests.

- Ensure test execution: it wouldn't be the first time that we find a complete test procedure but that is not followed by the responsibility.

- Define responsibilities: assign a responsible of executing the tests and also a responsibility to analyze the test results. If they are not the same person you can ensure that tests will be executed.

- Testing should include also validations of the methodologies and standards defined in this book such as naming conventions or any other restriction that you want to apply to your system.

Test automation

Similar to the transport procedure, when we talk about testing we can apply the same considerations. Testing is a technical task required but that doesn't contribute directly to the business analysis. So we will be interested in automating as much as we can the execution of testing procedure and the validation of these execution results. As far as we are developing a BI platform it definitively makes sense to automatically include test results in some graphical analysis using our BI platform, saving the results on our database. Interesting free testing tools are AutoIt and Selenium, among others.

Monitoring Tools

Once you have moved to production and your users are accessing on a daily basis the information that you are providing, it is important to deliver a reliable service ensuring that your users can access the information whenever they need it. But this is a task that you will hardly accomplish if you don't have any automatic way of checking that all the related services are up and running and that there is no any imminent risk of having problems due to missing space on disk, overload of used memory, or permanent CPU usage over a defined threshold.

It is quite possible that you already have in your company some monitoring system installed to check the health of any other of your running systems, but if not we recommend you have a look at some open source tools. There are multiple tools in the market but the one that we consider more complete is Nagios, which helps you to monitor main technical KPIs of your servers providing also a graphical interface to analyze the data and the ability of alert you via email or mobile whenever it detects any problem in the server. You can find it at:

```
https://www.nagios.org/projects/nagios-core/
```

High Availability

When you consider your server as productive, you will apply a set of considerations that should provide some reliability to the platform such as avoiding modifications of source code in the production environment, requiring validation of new code in development, and testing environments, limiting the access to production servers to people with administration profiles or using monitoring tools, as Nagios commented some lines before to ensure service availability. In case your service is becoming more critical you could think of implementing some High Availability configuration by delivering a clustering environment.

This kind of clustering configuration only makes sense when you are talking about server installations, so in the case of the BI tools seen in this book that in their free version provide a client installation, it wouldn't make much sense to talk about clustering. In the case of the BI tool you should move to a commercial version in order to have this capability.

In the scenario that we are promoting across the entire book it makes much more sense to talk about clustering when we are referring to the database component. Both MariaDB and MySQL offer the possibility of configuring their services in cluster. To ensure that you know what a cluster is, let us quickly introduce you to cluster concept. At the end this is just two or more servers (also known as nodes) accessing the same data that allows you to access this data from both nodes. In this way the possibility of having both nodes down is much lower than if you have just one server, so you will have available the information a higher percentage of the time. This is basically the concept of High Availability.

There are mainly two types of clustering configuration:

- **Active-Passive**: One server is giving service and the secondary node is stopped but in alert to check if the primary server goes down, so it must start giving service to the users. This option just offers high availability but doesn't increase the performance of your environment.

- **Active-Active**: Both nodes on the server are giving service so you will have the possibility of accessing one or the other node. This option is also giving you more capacity because you can use both servers at the same time to access your data.

In Figure 11-5 you can see two examples of clustering configuration, the first one based on a cluster just of the Database component in an Active-Passive approach, where the BI server connects to the primary node of the database cluster and the secondary node is down waiting for primary node failures. If a failure occurs, the secondary node becomes Active and the primary node switches off as passive while the BI server automatically connects to the secondary node. Both nodes must be synchronized in order to not lose any data in case of failure. That kind of configuration usually requires some kind of shared disk to store the information of the database. In the second example of Figure 11-5 you can see double clustering, at BI server and at Database server level. In this case Database servers are both up and running and giving service indistinctly to both BI servers. DB servers must be also synchronized in order to keep coherence between them as far as both can be accessed from both BI servers. Users can also connect to one node or to the other. In order to do this connection transparent for the user it is quite usual to have a load balancer component that redirects user requests to one or to another node depending on the server's load.

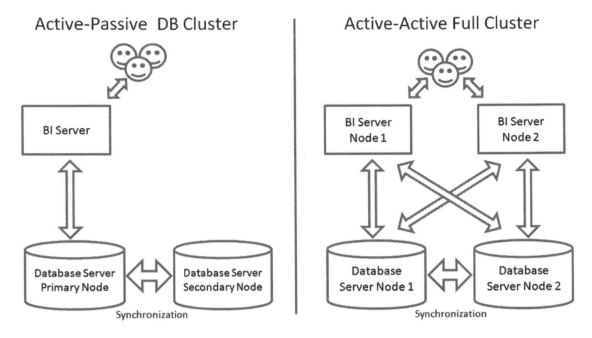

Figure 11-5. *Cluster configuration examples*

Using Testing as Disaster Recovery of Production

If your application has become critical enough, it is possible that you require accessing the information provided by your BI environment in almost each possible condition because it is considered as Mission Critical in your application portfolio. If this is the case, you need to cover not only the possibility of a server going down (secondary node would cover the availability) but also the possibility that the whole data center goes down, in case of natural disaster, accident, power off, terrorism, or just a big network failure. In order to follow a Disaster Recovery policy you need to have some environment available in a different data center than the one that is locating the production environment. It means that you will require having some hardware available just in case of needing it, so it is possible that you have unused hardware just waiting if there is some big incident in the production data center.

In order to take credit for this unused hardware is a common strategy to use it to locate testing environment, so you can use it for user validation or stress tests. Of course this implies that in case of disaster you will lose your Testing environment; but of course it is something that everybody could accept in that case. Anyway, there are some considerations that your Disaster Recovery environment must match in order to be a real alternative to the production environment:

- **Complete environment**: We need to have all components available in our DR location in order to ensure complete service to your users. It seems to be obvious but it wouldn't be the first time that we see a DR with the capacity to locate the database and the BI component but without any server available for the ETL service.

- **Similar capacity**: Some of the dimensionality parameters that we have been analyzing during the capacity planning must be similar to that in your productive environment. You could survive in case of disaster with less CPU capacity, but of course you will need at minimum the same disk space than your production environment to be able to restore the whole infrastructure.

- **Updated information**: You will need to replicate the production information to the Testing environment. Maybe not directly to the testing environment but to some device in the same location than testing. In other words your production backup system must be in the same data center as the testing environment or it must have some replication there.

- **Minimum distance**: There are no exact specifications for that but it is recommended that both datacenters are separated with a minimum distance in order to ensure that the disaster doesn't affect both locations. This is especially advisable to avoid natural disasters.

- **RTO and RPO parameters definition**: Recovery Time Objective and Recovery Point Objective define the maximum time that your company can accept having the service down and the maximum data loss that be considered as acceptable. In a BI system it is quite usual to find definitions of RTO/RPO of 24/72, what means that you will require the BI service back in 24 hours allowing having a data loss of 72 hours, in other words, when your service is back you will need to launch some recovery ETL process that fills up information from the previous 3 days.

Backing Up Your Platform

We have commented during previous sections that in order to have disaster recovery available you will need to have some backup data available in the secondary data center, but until now we have hardly spoken about backup. Let's proceed to quickly explain some recommendations about backup configuration in a BI environment. First we need to define the backup type required for each component.

BI and ETL servers usually have a software installation that doesn't change until a new version or patch is installed and a repository where it saves the objects used inside the tool. This repository usually can be file based or database based and it is quite possible that in this second case the repository is located in the same DB server. In this case we will be interested just if keeping a snapshot of the server after each installation, and a daily backup for the repository objects, which usually occupy a small amount of space compared with the whole server or the whole datawarehouse database.

In the case of the database, the backup policy will vary depending on the nature of the data. Usually database servers will require daily backups in order to be able to recover the previous day's image in case of database corruption or disaster. But it is also possible that some of your input data has a transactional character, in other words, that it is not loaded in a daily process during the night but it can also be updated during all day at any time, so your backup policy must provide you the possibility of recovering a database status at the latest moment possible. On the other hand, you can have some data that is refreshed monthly so that doesn't make any sense at all to have daily backup copies because they will contain exactly the same information until next month.

When you are defining the backup policy there are also more parameters to fix, not only the backup periodicity. You cannot have infinite copies of the database waiting just in case a recovery is needed in the future. The parameter that fixes the life of each backup is usually known as data retention. It is usual to have different data retention mixed with different periodicity. On the other hand, most backup tools allow you to have full and incremental backups, so a whole copy of objects that you want to back up or just the difference since the previous backup. Also it is important to define which will be the backup device and the location. In Figure 11-6 you can see an example of backup policy per environment with different data retention.

Environment	Periodicity	Type	Data Retention	Device	Datacenter
Production	Daily	Incremental	2 weeks	Tape + Disk	Same
Production	Weekly	Full	2 months	Tape + Disk	Disaster Recovery
Production	Monthly	Full	2 years	Tape	Same
Testing	Weekly	Full	2 months	Tape + Disk	Same
Testing	Monthly	Full	2 years	Tape	Same
Development	Weekly	Code Only	2 months	Disk	Same

Figure 11-6. *Backup policy per environment*

Historification Process

Over time your database will grow. This is something that you need to take into account. Despite the fact that you haven't created new developments with new data requirements, something that we consider near to impossible, all (at least most of) your fact tables will be time based, so each day, each month, each year, you will have new data to add to your system increasing tables and database size. Also in lookup tables, you will have new products, new customers, new years; otherwise, your company will die soon, and it is quite likely that there will also be vegetative growing for your lookup and dimension tables. This growing will affect directly to the space used, something that can be just covered by purchasing more disk and adding it to the servers, but it can also affect the performance degradation because not all queries that were working perfectly fine when you had 2 years of history in your datawarehouse will continue to work well when faced with 5 years of data. On the other hand, you don't need to have all the possible history available in daily used tables so you can implement some cleaning process that moves, compresses, or directly removes old data. This process is the one that we know as the historification process.

■ **Note**   In the historification process it is quite useful to follow a strategy of partitioned tables inside the database; it will facilitate the task of removing data, compressing or moving it.

You can have multiple strategies for this historification process depending on which objective you are trying to achieve; basically you will have two: cleaning spaces and/or improve performance. On the other hand there are some other criteria that you need to meet. It is possible that due to your business users you need to have available a minimum amount of data in spite of the fact that you are not interested on querying in a daily basis, because there are some legal restrictions to keep data online available during a minimum period of time or a set of years, or they might ask you to keep some aggregated data in some table for 10 years while in the highest detailed table you only require to keep it 2 years. We show you in this list just some of the strategies, but at the end you need to adapt your historification process to your server capacity and to your business requirements. Some of them can be combined:

- **Data deletion**: This is the simplest way of saving space and keeping the amount of information controlled. You agree with your users what is the amount of data required and each year you remove data older than this limit after the previous backup.

- **Movement to other tables**: you can move your data to another set of tables and then develop some customization to access it when you want to access it from the BI interface, such as defining some views that join old information that are only accessed when users want to access that old information.

- **Movement to other database**: If you have a way of joining information from multiple databases, either with your database engine or your BI tool, this allows you to cross and query information as if it was a single data source, you can think on moving this information out of the main database.

- **Cheaper storage**: In combination with previous options, you can save old data in cheaper storage with lower performance empowering recent data by saving it in the most powerful storage.

- **Data compression**: If your database engine allows you to do that you could configure old partitions to be compressed to save space but being available for querying them.

- **Read-only**: It is possible that you want to limit not only the information available for querying but also to limit the amount of data updated by your ETL process. So you could define partitions or tables containing old information to be read-only.

- **Table based**: You can think of a strategy that removes old data from big detailed tables and keeps wider history in aggregated tables. Also in some cases such as snapshot analysis, the ones that cannot be aggregated by time, as stock or product distribution, you can have different granularity depending on the snapshot age. You can have a daily stock snapshot for the last two months, and just a monthly snapshot on the last day of the month for last two years.

■ **Note**   From the scope of tools that we have seen in this book, the database component is not the only one that can be affected by this historification process. You can apply it also to MOLAP databases by creating multiple cubes or deleting the scenario or time members of the outline.

Security

Moving to a productive environment usually implies giving access to more users. And giving access to more users usually implies the need for having different security profiles derived from different user profiles. When defining the security in a platform for a user you usually need to answer three questions:

1. What does your user need to do?

2. What objects does your user need to see?

3. What information does your user need to analyze?

From these three questions you can see that we can derive three security types. In order to answer the first question you will define a set of **functionalities** that will cover user needs and that must be adapted to user knowledge. With the second question you will be able to set up the **access by object**, which tables and views he needs to access in the database, which reports, documents, metrics, filters, or attributes that he must see in the BI tool, etc. With the third question you will be able to apply **filters** to restrict at database level or in the BI tool (depending on the options that it offers you) which data he has access to see.

But let's analyze it using some examples. Regarding functionalities we can have restrictions of user type due to the license that we have, which requires restricting the access to some functionalities, or we could think of a scenario where we have some kind of restriction; imagine that due to security policies we are not allowed to save data in shared folders and our tool by default allows us to do it so we need to restrict this functionality to users in order to follow security standards. You can imagine also a situation based on user profiles; think of a business user without technical knowledge that doesn't know at all what a database is or the meaning of the acronym SQL but your tool allows you to develop reports using free SQL or to modify the generated SQL to adapt it to your needs. It wouldn't make sense to allow this profile of users to modify that, because in the best scenario he won't use it and in the worse scenario we will end up with dramatic consequences.

It is easier to understand the access by object. If you have your financial and your sales departments and you want to apply departmental restrictions you will allow your sales users to see just sales reports and your financial users to see finance reports. Usually access by objects has three levels of access: read-only, which allows the user to see the report or to see data in the database; modification permissions that allow the user to change data in the table or modify the definition of the report; and full control that allows the user to delete, create, modify, or perform any action over the object. This nomenclature is not common for all tools but usually the security concept is similar.

Finally if we are talking about filtering the data that each users has access to, we can think of multiple utilities for that but one of the most clear ones for us is the use case of developing a report for the sales force of a company. The basic level of the sales force team should see only sales of their assigned customers, but supervisors of this team can see information related to them and their dependent team, and the sales force manager can see the overall sales of the company. In order to apply data security you can implement it mainly in three ways:

- Filtering data in the database: Following with the previous example, you will have your fact tables and then a relationship table that has the relation between user and the customers that he has assigned. Then you will have a view related to each table that will join the fact table and the relationship table and a condition that filters the field ID_USER (let's use this field name) from the relationship table must be equal to the connection user. On the other hand, you will require denying direct access to the tables, only allowing access to the views. In this way the users that connect to the database will see only data for their customers.

- Filtering data in BI reports: You can define the same filter by users at the report level so it will be applied to the SQL when the report is executed. In order to have an environment really secure you cannot allow your users to edit the report so they can remove the filter or create reports from scratch.

- Applying security filter to the users: You will apply the filter by sales force or any other attribute to the user so each report that it executes will include this filter in the SQL automatically. In order to use this option, which is easier to manage, you will require your tool to allow you the usage of security filters.

Security Management, Using Roles and Groups

In order to facilitate maintenance of the security you can use roles and/or groups; nomenclature and exact meaning will depend on the tool, in order to grant permissions to similar users at one time. Let us correct this sentence. We strongly recommend you the usage of roles and groups to grant any permission to your users. Direct grants to users should be something rare that only has some sense in rare user cases. So our recommendation in this case is quite easy and clear... you should use always user groups and roles to grant permissions.

But you can be thinking, why do I need to use a user group if there is only one user inside this group? You can be thinking of a security group named General Management and you have only one general manager who needs access to all data in the company. Our answer is easy. The first reason is that you never know if this one-to-one condition will be always true; maybe you decide to use this group to include all your steering committee. On the other hand, your general manager can change so you would be required to reassign all the security granted to your previous general manager to the new one. So it leads us to the second reason; security maintenance is much easier using groups in spite that the initial situation could be easier to be managed directly with users when you have a few users to assign.

User Role Definition

We will have different functions in our project, some of them related to the roles we saw in Chapter 2 related to Agile roles, some of them out of the project development, as key users and end users. We will be able to define different roles related to the functions that each user will have, and then apply security accordingly granting functionalities, accesses, and data security depending on the role. In this security definition, especially for the functionality and object access, we will need to take into account the environment; access will be different in the Development environment than in Testing or Production. Let's see the most usual roles in a BI environment so we can then set the security based on their functions.

- **Key User**: He will be the responsible one to organize all requests related to the BI project. This user can be the same as the product owner seen in Scrum. We have usually different key users by datamart, as far as the nature of the information is different for each one. You will have a key user for Sales, another key user for Finance, another key user for Operations, etc. It is possible that he is also responsible for creating reports and publishing them for the rest of users, organizing folders and information inside the tool, defining the security that must be applied for the rest of users, defining roles and permissions, etc. He will have full control of the environment.

- **End User**: This is the widest group of users and also you can have multiple user types inside this group, from analysts that use ad hoc reporting, ad hoc metric creation, and navigation capabilities to investigate across data, to just final recipients of information that receive static reports in pdf format.

- **Developer**: This user will be in charge of implementing the required structure for the analysis so he will require access in the development environment to all levels of the BI platform, ETL, Database, and BI tools in order to be able to design requested objects in all the layers.

- **Administrator**: This user will be in charge of moving objects across environments, monitor environments, implementing user security, installing and configuring the platform, and in general all technical management tasks required for the correct performance of the platform.

If your environment grows up then it can appear new roles such as service managers, application support, platform owners, etc. There are some interesting methodologies such as ITIL standards that define a set of rules to follow up for IT infrastructure maintenance. But this topic is a matter of multiple documentation, courses, and certifications that are out of the scope of this book.

Security Matrix

Once you know which roles will be involved in your project is interesting to define what permissions they will have by environment so we recommend that you implement a matrix like the one shown in Figure 11-7 with the granularity and detail that your project could require.

Type	Permission	Key User			End User			Developer			Administrator		
		Dev	Test	Prod	Dev	Test	Prod	Dev	Test	Prod	Dev	Test	Prod
Database	Access		Read	Read				Modif	Read		Full	Full	Full
Database	Configuration										Yes	Yes	Yes
Database	Tables		Select	Select				Create	Update		Create	Create	Create
Database	Views		Select	Select				Create	Update		Create	Create	Create
Database	Procedures		Execute	Execute				Create	Execute		Create	Create	Create
ETL	Access		Read	Read				Modif	Read		Full	Full	Full
ETL	Editor							Create	Read		Create	Create	Create
ETL	Monitor		Read	Read				Execute	Execute		Execute	Execute	Execute
BI	Access		Read	Modif				Modif	Read		Full	Full	Full
BI	Configuration										Yes	Yes	Yes
BI	Design tools		Read	Read				Create	Modif		Create	Create	Create
BI	Report creation path		Public	Public			Personal	Public	Public		Public	Public	Public
BI	Drilling		Yes	Yes			Yes	Yes	Yes		Yes	Yes	Yes
BI	Report Distribution		Yes	Yes			Yes	Yes	Yes		Yes	Yes	Yes

Figure 11-7. *Security matrix*

Auditing

We would like to comment on a final topic related to a productive environment: there is the need for ensuring that all the rules that we have defined in previous sections are met. In order to ensure it, we recommend that you define and execute a set of audit processes that ensures correctness in all the related platform guidelines in terms of naming conventions, locations, performance checks, security definitions, and defined methodologies for development, transport, or any other running process. We think that there are two main groups of audits, the ones that pretend to ensure that we are applying the correct security to our system and the ones that pretend to ensure that we are following correctly the guidelines or best practices defined in our system. We won't do an extended analysis of auditing processes, but we would like to remark that it is important to follow some processes that ensure that all the rules that we have defined (or that somebody asks us to follow) are followed. We recommend you also to automate as much as you can this audit process and the actions derived from it.

Security Audits

As you can imagine this group of audits are the most relevant ones to implement in order to avoid data leaks from your company. Without being exhaustive let's analyze some security advices that you should try to verify in your environment:

- First of all you should have some auditing process running in your database saving at least all the connections in the system in order for you to check who was connected in each moment to the system. Validating this log should be the first audit that you should implement in order to be able to track whatever information that you require.

- The usage of generic users can be needed to implement processes such as ETL, connections between the BI tool, and the datawarehouse to connect the MOLAP system to the database, but we strongly recommend you limit generic users for technical processes; each manual connection to the system should be done with a personal user and monitored in a log system.

- You should validate that when a user leaves the company or he doesn't require access to the BI system his user is removed, or at least disabled, from the system. The best approach to do that is to use some authentication system already implemented in the company or at least synchronized with it, such as Active Directory, LDAP, Novell, or any other authentication method that your company is using. In this way when the user is disabled to access generic systems, he will be also disabled to access your system. But anyway, you should validate that this is true.

- Development of database models, ETL processes, reports, and dashboards doesn't require doing it over real data. A general rule is that developers cannot access production data, as far as they are not business users, so you should avoid granting anybody access that doesn't need it. We know from our experience that sometimes you cannot analyze an incident or a mismatch in data without access to it, so it is usual to define a role as project manager or some application support team that has this kind of access; but for initial development you don't need usually real data. On the other hand, in combination with the previous advice, if your development team is externally provided by a consultancy company, it is possible that your control process over external people is different than the one over internal teams, so you need to validate from time to time that external users still require access.

- Sometimes shared folders are required to share information across different groups of users, to locate flat files to load them into the database or to extract information there. You should validate also who is accessing these folders avoiding grants to everyone/public group.

- If you need to define a data security model based on database views or the BI tool, ensure that nobody can jump over security restrictions accessing to the whole project, ensure the robustness of your security development, and check from time to time that nothing has been changed accidentally.

Auditing Best Practices

Usually they are not so important in terms of security but defining some rules and ensuring that they have been followed can save you much time and headaches. Validate that your rules are met and can facilitate your managing of stuff. Let's comment also on some examples of best practices that should be audited.

- From time to time you can be required to define a cross-environment connection, development of the BI tool accessing the production database due to unavailability of development database environment due to a platform upgrade, your development ETL environment connected to the production database to validate the performance of a load with real volume of data or some other requirement that can be managed with this kind of workaround. Crossed access between environments can be useful at one point to do some check or temporary connection but you should ensure that your connections are back once this temporary period has finished. In the previous section we have commented about developers accessing real data, having crossed environments could also cause a security issue.

- In Chapter 5, we talked about naming conventions to define DB objects. Following naming conventions seems to be something arbitrary that we could skip but if the rule is there to follow it. Imagine that we define that based on space used in the database we split the cost of the platform across different departments. Naming convention for tables was that the third part of the table name defined the area, SAL for sales, FIN for finance, or OP for operations. If the environment is not defined in a table we cannot know who to charge this cost and our IT department will require assuming it. Or we define a maximum database name of 10 characters and we have a backup script that has a limitation based on this length to define the backup file name. We create a database of 20 characters and then the backup name is not correct and backups are overwritten keeping only one available instead of the history required.

- Imagine that we have a limited load window of 4 hours during the night to launch all ETL processes to deliver information on time. We will be interested in auditing all process load times in order to detect which ones must be improved.

- Defining some rules to locate files can facilitate the tasks of doing backups, transporting objects across environments, keep clean environments, or upgrading software versions. You should check that files are correctly located in order to not miss any object to copy across environments or to not overload a filesystem or a disk because you are saving big log files in an unwanted location.

These have been just some examples about auditing process, but what we would like to remark about is the need for auditing that things are correctly done. Sometimes it is easy to follow some rules during the implementation but then over time, with modifications and changes on teams, it is more difficult to keep the attention on having a clean environment.

Conclusion

Within this chapter we have seen how to apply some rules that allow us to define a set of environments to work with for the different roles that we will have in our company. We will have an environment reliable for end users because they will have an environment with modifications under control that delivers the information required for their daily work. We will have also an environment available for developers to perform their work without affecting end-user information. We will have an environment to test and validate modifications done. We will ensure that the environment is secure enough by granting the required security in each environment. We will add a control layer in transports to validate that they are done without affectation and we will have a maintenance policy to control all our environments by auditing that our rules are followed. In the next chapter, the last one, we will take a quick glance at a related topic, to understand how to implement a cloud environment, so without requiring an up-front investment on hardware and just renting cloud capacity. It is related because it will show also an easy way to define those multiple environments that we have been evaluating during this chapter.

■ ■ ■

## CHAPTER 12

Moving BI Processes to the Cloud

Nowadays there is a trend to reduce the investment in machines, especially when it comes to small companies. Due to this, several top companies saw a business opportunity. Among them, Amazon, Google, and Microsoft. These companies realized that most companies can't/don't want to afford expensive purchases in hardware that may become obsolete just after a few years. They turned themselves into cloud services providers, which means, basically, that you can rent a part of a machine or an entire machine and run whatever you want there, without having to purchase expensive hardware.

Currently, things have gone further and not only you can rent infrastructure from them, but also some other services. All three companies, for example, sell some of their databases as a service, so instead of having a DBA to tune the database, the administer can do proactive work to ensure it runs smoothly.

In this chapter, we will see how to upload our BI processes, basically the database and the ETL to the cloud. Looks easy, right? Well, it is. However, we can't start directly creating our resources. There are some configurations that we must be aware of before we start creating our virtual servers and this is the first thing we need to see. But before that, let's elaborate about our cloud solution a little bit, and present the possible options.

Deciding our Cloud Provider

As we have seen, we can choose between a wide range of cloud providers. You will find many on the Internet. Any of them will work; simply choose the one that suits you better or that you find easier to use. Of course, the price is something important when making the decisions, but we have not seen very big differences between all the providers analyzed. The billing is another aspect to consider: some of them invoice per usage, usually by hour; and some of them invoice a fixed monthly rate, so this is something to think about. For this book, we have evaluated two solutions, which are the most used: Amazon AWS and Microsoft Azure.

Initial Considerations

Before starting to create and launch our cloud instances we should think about several aspects, the most important ones being the following:

- Sizing of the platform. Number of servers we need, dimensioning them appropriately to support the load with a solution that provides us performance and scalability.

- Decide which services will go to the cloud and which ones will remain local.

- Security, access control to the servers and integration to the own company intranet, especially if we keep some servers on premises.

© Albert Nogués and Juan Valladares 2017

A. Nogués and J. Valladares, *Business Intelligence Tools for Small Companies*,
DOI 10.1007/978-1-4842-2568-4_12

- Data Location.

- Tracing and Logging of all services running in the cloud using health checks.

- SLA and the possibility (or not) to hire some support plan offered by the cloud vendor.

- Disaster Recovery (We would skip this one as this is a small development, but in a serious one, this should be planned carefully.)

- Calculate the running cost of the platform, and calculate the savings (if there are any).

We are going to see a little bit of these aspects before entering in the actual implementation details

Sizing of the Platform: Performance and Scalability

Moving your developments and infrastructure to the cloud isn't easy. The first thing that may come to your mind is what the technical requirements of the platform are. You already started to consider in your own mind about terms of capacity and usage and started to think about the required resources. Well, there are more things to factor, but clearly this is one. One of the advantages of a cloud solution is that you don't need to go big, nor make any massive investment at the beginning.

Most cloud platforms offer a pay per use solution. This is probably the clear choice when you first start your investment in a BI solution. If the project ends successfully and adds value to your company, you can always expand your analysis and go further. But if for whatever reason, the project does not have the expected results, you do not lose all this investment in hardware, but only your development effort. This is an important aspect to consider, as the investment entry can be an important barrier in such small companies.

Sizing the platform can be complex, but at this point, you would be better choosing a pay per use solution, with no fixed up-front payment to even reduce the risk further. Then a standard/small instance will serve to host the datawarehouse and the ETL server. You can decide to separate both machines, which is a good solution especially if you will run intensive processing tasks. The option of moving the transactional tool to the cloud may also be viable. If that is the case, we recommend hosting it on a separate machine.

The Physical Design

Deciding which servers will go in the cloud and which ones will remain local is also part of this stage. All the physical design comprises mostly sysadmins work: deciding the IP ranges, the firewalls, the security access, the operative systems, the networking of the platform, and many others. This is clearly tied to the sizing of the platform, but it also entitles other questions. Will we have all our environments in the cloud (Development, Simulation, and Production) or we will keep some of these locally? In our opinion, if the company is not too big, and you can afford it, we would choose a full cloud solution.

Having a mixed architecture looks to be one of the newest trends in mid-sized companies that can afford having an on-premises installation. Some of them have their testing environments in the cloud whereas the production environment is hosted onsite. While this solution may suit you as well, it involves having more work in maintaining the environments up. Think about backup policies, security, the network configuration also goes more complex, as all the internal networks need to be blended, so cloud resources can access or "see" on-premises resources and the other way round. To avoid these complications, we will deploy three environments in the cloud: development, simulation, and production; and ideally all three should sit on different networks to avoid communication between them, in order to avoid human errors, that is, writing data from test to production or things like that.

Security

Security is always paramount. Not only because you can have sensitive data that you don't want unauthorized people to have access to but also for legal reasons. Every year, jurisdiction in terms of privacy gets tougher. You can be in major trouble if you do not secure your data. Most cloud providers offer some sort of security bundled with their deployments. Usually this entitles a firewall that can be configured to block the ports you want, certificates to access your machines when remotely connecting, and roles of permissions to assign to the different actors interacting with the platform, so you can limit their access and restrict the permissions to the work they should do.

Integrating this cloud network with your already running infrastructure can represent a challenge too. You need to add new entries in the DNS servers, add rules to your existing firewalls, and modify your routing tables so the new subnets get routed from inside. This can also affect security, not only in the cloud but in your organization so you must review any change carefully.

Data Location

Choosing where the data resides is important too. Ideally you will prefer to store the data as close as possible to your customers (and in that aspect, you consider the customer yourself or your BI department). Most cloud solutions offer different geographical zones to choose from, where to store your data. Whether they may not be in your country, it is likely that you will find a region closer to home. This is important, especially in real-time applications, but can make an impact too in a BI deployment. Inside the same geographical zone, you usually have different locations. These are used for redundancy and as a disaster recovery. Bear in mind that some of these features come to a cost, though.

Health Checks

All cloud solutions, provide as well, some dashboards to check performance. Make sure you review the metrics shown there and take appropriate action if there are unexpected results. You usually will have the possibility of adding alerts and configure thresholds in these dashboards to make sure you are informed of any unexpected behavior.

SLA and Support Plans

Any cloud provider should guarantee a minimum service operation level. In computing, this is usually regulated by a Service Level Agreement (or SLA). This is a document where the most common evaluation metrics are described and the provider agrees to guarantee them to a specific value. It is your responsibility to ensure these agreements are accomplished and if not, ask for compensation.

As in the case of the SLA, some cloud providers give the user the option to purchase extra support. Sometimes this is in means of better SLA agreements, more responsivity from the technical teams in case of trouble, or advanced support from qualified engineers.

While going on these extra packages may be expensive, you need to consider them once you have a system put in production. If users or the company depends on having the system available 100% of the time, and the unavailability of it causes loses to the company, you must take appropriate action to ensure these problems do not appear, or when they appear are solved as quickly as possible. But probably at the beginning, you won't be interested in them.

Calculate the Running Cost of the Platform

Calculating the cost of the platform is complicated. Fortunately, most cloud providers help you in publishing web calculators, where at least, you can try to calculate, how much it will cost you your cloud deployment. Obviously, development efforts are not considered with these calculators but at least you can anticipate the money needed to invest in infrastructure. We will see a sample later in this chapter when analyzing one of the cloud providers.

A First Look at the Physical Design

After some talks with our sysadmin we have decided that for our first environment (test environment) we need the following:

- A machine where to run the transactional system (in our case Odoo, and this can be already deployed on premises, so it is a matter of deciding for or against moving it to the cloud).

- A machine where to run our datawarehouse database.

- A machine where to run the ETL engine. If using Pentaho, we need to install the engine somewhere. Also, this machine will. Host the crontab for the jobs, so this will be our job executor.

- A machine where to run our reporting engine, if needed. If working, for example, with Power BI, the only option is to use Microsoft cloud. If working with Qlikview Desktop, we may use Qlik cloud or install a server in our deployment. Since we are talking about a test environment, we may decide against any server deployment and limit the option to desktop clients only, connecting directly against the datawarehouse.

Once this is clear, we need to decide a few aspects. These are:

- Do we want to run all our machines inside a virtual private cloud (VPC)?

- Which ports do we need to open?

- Which type of instance will we will choose for each machine?

The answer to these questions is subjective and depends entirely on your platform choice. For illustration purposes we decided to go for the following:

- **Create a VPC**. Though this may not be the easiest choice in most cases, as it will make our environment a bit more complex, it is required by the instance type we plan to use. More about it later. Not creating a VPC means that each machine will need to have a public IP associated so we can connect from the outside. If we go for the VPC option, all machines can be part of the same private subnet, and then you may want to link this subnet with your internal network or put a machine to do the NAT between your network and the isolated private network in the cloud.

- Create **a Machine with approx 2GB of RAM** to run our datawarehouse, and only one CPU. We don't need too much power at this point as this is a test database.

- Create **another 2 GB machine** where to run Pentaho data integrator and the Odoo.

Now it is time to see how to implement this efficiently while keeping the budget low. Let's have a look of the three cloud providers, their prices, and current offers and see how we can deploy the platform.

Choosing the Right Cloud Provider

There are many providers to choose from but we will restrict the list to three: AWS (Amazon Web Services), Microsoft Azure, and Google cloud. All three offer some sort of free service, usually limited to a number of hours or a period of time. Let's have a look at their proposals, starting with the biggest one, Amazon Web Services.

Amazon Web Services (AWS)

Amazon web services, or AWS, is the most common cloud solution you will find. They have more than 500 products and/or services (and counting) offered to customers grouped by several categories like Databases, Analytics, Networking, Storage, Computer, Mobile, Development, Internet of things...

■ **Note** For up-to-date information on new services and possibilities, check out the following AWS blog: https://aws.amazon.com/en/new/?nc2=h_l2_cc

---

We will focus on the Compute instances and the Database instances, and we will have a look at the Analytics offer at the end of the chapter. In order to start launching instances, we first need to register for an account here: https://aws.amazon.com/?nc1=h_ls

The good news is that AWS provides a free layer to test the service. It is possible that we can adapt our needs to this free layer, but this will largely depend on the requirements of our platform. Currently AWS free layer is explained here: https://aws.amazon.com/free but to make things easier, as they offer quite a lot of services that we don't need at this point, let's sum up the offers that may be interesting to us:

- 750 Hours per month of Amazon Compute Instance (EC2). This will be used for running one machine, the ETL machine, and the job launcher.

- 5 GB of S3 Storage. We can use them for backups or static files.

- 750 Hours per month of Amazon RDS for running a database. This will be used for the Datawarehouse database.

- 1 GB of SPICE for AWS Quicksight. This is brand new and it is an analytic tool developed to create easily reports and dashboards from data in your AWS cloud.

Some of these offers run forever, like the Quicksight one, but the most interesting ones, the EC2 instance, the RDS instance and the S3 storage, will run for free for a year only. So watch out as at some point; you will be billed. It is time to sign up at AWS and start building our test platform. Let's go!

Implementing the dev Environment in AWS

To implement our solution in AWS, we will start by creating the EC2 instance to hold the ETL. For this, and assuming that you already registered with AWS, you need to go to your AWS management console. This is the central place where all of the services or applications are launched. At first it may be a bit overwhelming as we already said there are hundreds of services. But right now, we only need to focus ourselves in the EC2 instances.

Sign into the AWS console by going to the following address: https://aws.amazon.com/console/ once logged, we will see our main AWS console. There are a few things that we need to get familiar with, and we need to explain first. Figure 12-1 shows these.

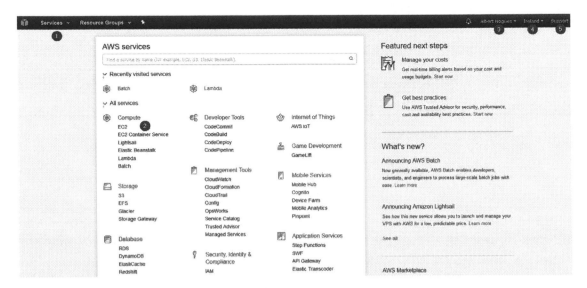

Figure 12-1. Main AWS console window with important points highlighted

The point 1 marked in the Figure 12-1, shows the menu where to access all your services. Accessing them by this menu or by clicking on the name of the service in the main window, like point 2 shown on the picture is the same. But we find easily to access them by the menu as all is ordered and organized.

The name of the account owner, as shown in point 3, gives us access to several important sections of AWS. Inside them, apart from the account details and some other configurations we have access to the billing dashboard and the security credentials. These two aspects show us how much money we have spent in the current billing cycle and past ones, and where this money has been spent. This is very important to track to keep your costs in line with what you expect and do not find any surprises at the end of the billing cycle. It also lets you configure alarms based on thresholds to make sure you get alerted in case the expenses start to rise over a previously defined threshold.

The second topic important to note, inside the same point 3 in the picture, is the security credentials menu. From here we can decide whether to use IAM permissions, which are Identity and Access management, an application to assign roles to accounts, so you can control which permissions are given to specific users and what is the solution recommended if more than one person needs to manage the environment, which should be the case in most of deployments. It is possible, however, to stick to a single management account, which will hold permission for everything. In this place, you can also create the API keys, in case you want to interact with the AWS API, or run services from external tools (Scripts, API calls, external providers…).

Another section important to note, labeled as the 4th point is the Region Selection we are on. In the picture you see, I am using the Ireland region. You can navigate through all possible regions, but make sure that all your services sit on the same zone for better performance. Also, take note that each zone may have different prices on the services so make sure you check them. For this example, we will deploy everything in the Ireland region.

Finally, the fifth important point is the Support menu. From here you can contact AWS support, and browse the forums and the documentation webpages for more information or help.

Launching an EC2 instance

To launch an EC2 instance we need to go to the EC2 Service, inside the Compute menu. Then we will see an assistant asking us a few questions to deploy the instance. Here we can configure and choose which type of instance we want, including the hardware specs, the operative systems, assign any IP address, and the security details (ports, keys to access the machines ...). The main screen can be seen in Figure 12-2.

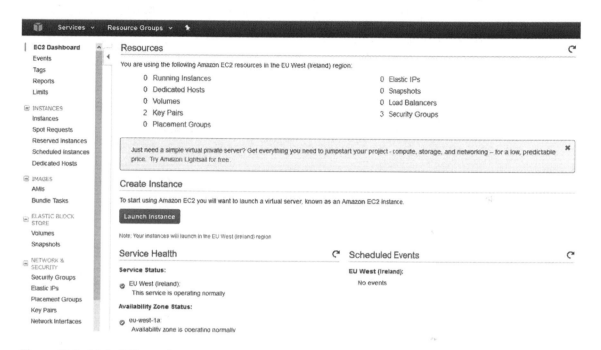

Figure 12-2. *Main EC2 panel*

It is now time to start defining our first instance; for this just click on the blue button called Launch Instance. If we use the assistant, this will be comprised of seven screens. These are:

- The first one, which is the operative system selector. If you intend to use the free tier offer, we can use one of the operative systems labeled with free tier in the picture. In our case, we have decided to use the Amazon Linux AMI, but there is no problem in using the Ubuntu AMI or any of the others.

- The second screen shows the instance type. Here the first problem comes, because the free tier instance only has 1 GB of ram. Here, we can decide to stick to 1GB of RAM, which likely will make it very difficult for Odoo to coexist with the PDI or use a bigger one not covered by the free tier. In our case we will stick to the t2.micro instance as we plan to install Odoo, but once installed and created, the test database shuts it down and starts PDI only. Even with that PDI will not run very fluently with only 1 GB of ram but the only option to solve this problem is to move to a paid instance.

- The third screen is to configure our instance. At this point we need to create a VPC as this instance requires it. We leave 1 as the number of instances to launch, and make sure we do not click the spot instance check box. In the networking section, we hit on the create new VPC button, which will launch a new page. In this new page click on the Create VPC blue button and specify the following parameters: BiBook as the Name Tag, 172.16.1.0/24 as the CIDR IPv4 Block, and leave the remaining options by default. This will create a subnet in the 172.16.1 range, which is a private range of IP addresses. Once created close the new window, go back to the assistant, and hit the refresh button. We should see our VPC there. We select it.

- In the subnet section we need to do something similar. Create a new subnet listing all the IP addresses in the range we choose for our VPC. Once done, go back to the assistant and refresh it. The new subnet range should appear on the list to be selected. Select it.

- In the auto assign public IP section we leave the default option, which is Disable. If we need a public IP to connect later there, which at this point is not the case, we can always assign one (and pay for it).

- If using IAM roles, we can specify the one we want there. Since I am not using them, leave None selected.

- Leave the remaining options as by default, including the shutdown behavior. Make sure **NOT to select Terminate,** as if chosen, when our machine is stopped all will be destroyed, and there will be no possibility of recovering the content.

- In the Network interfaces section, make sure a new interface is added, called eth0. Here you can specify a private IP inside the range we defined previously or leave AWS to pick one for you. I am leaving the default and AWS will pick one for me, but for ease of use it may be interesting to choose one that you will easily remember.

- The fourth screen is the storage section. By default, 8 GB are assigned for the operative system partition. While this may be good, the free tier includes up to 30 GB of free SSD storage. So, we will put 30 instead of 8. You can configure multiple disks but be warned that you will need to pay for them. However, instead of having them assigned to only one partition we may choose to leave the 8 GB for the operative system, and 22 for a data partition. This is up to you. In my case for simplification, I am assigning all the 30 GB to one disk only.

- The fifth screen lets us define a few tags to identify our machine. I defined a key named Name and a value of BiBookI1.

- The sixth screen is a very important one. Here we assign a Security group to our instance. The security group contains the inbound and outbound rules for our firewall. We can create one or reuse one created previously. By default, it asks us for a set of IP addresses to connect to the ssh port (22). If we leave the default 0.0.0.0, this will make the instance available only on port 22 (SSH) to the entire world. Ideally we only want to open them to some IP addresses, like the ones from our company, so we can amend the 0.0.0.0 for the set of values we want. I also gave BiBookI1 name to this security group and put BiBook Security Group I1 as a description.

- The last step is a review step. We make sure we filled all the options properly and we can start our instance! The result should be similar to Figure 12-3.

Step 7: Review Instance Launch

Your instances may be accessible from any IP address. We recommend that you update your security group rules to allow access from known IP addresses only.
You can also open additional ports in your security group to facilitate access to the application or service you're running, e.g., HTTP (80) for web servers. Edit security groups

▼ AMI Details

Amazon Linux AMI 2016.09.1 (HVM), SSD Volume Type - ami-70edb016

Free tier eligible
The Amazon Linux AMI is an EBS-backed, AWS-supported image. The default image includes AWS command line tools, Python, Ruby, Perl, and Java. The repositories include Docker, PHP, MySQL, PostgreSQL, and other packages.
Root Device Type: ebs   Virtualization type: hvm

▼ Instance Type

Instance Type	ECUs	vCPUs	Memory (GiB)	Instance Storage (GB)	EBS-Optimized Available	Network Performance
t2.micro	Variable	1	1	EBS only	-	Low to Moderate

▼ Security Groups

Security group name   BiBook11
Description   Bibook Security Group I1

Type ⓘ	Protocol ⓘ	Port Range ⓘ	Source ⓘ
SSH	TCP	22	0.0.0.0/0

▶ Instance Details

▼ Storage

Volume Type ⓘ	Device ⓘ	Snapshot ⓘ	Size (GiB) ⓘ	Volume Type ⓘ	IOPS ⓘ	Throughput (MB/s) ⓘ	Delete on Termination ⓘ	Encrypted ⓘ
Root	/dev/xvda	snap-9d6b5a4e866a... 30		gp2	100 / 3000	N/A	Yes	Not Encrypted

Figure 12-3. *Review the EC2 instance options*

After we finish the assistant, we will be asked to select a key pair or create a new one. These are the pair of private and public keys. The public key will be set to the machine, whereas the private will be downloaded to your computer. Make sure you don't lose it, as you need it to connect to your machine. Since we do not have any so far, let's select create a new key pair and name it with BiBookKey. Then click on Download Key pair and your private key, a .pem file, a text file containing your private key, will be downloaded to your machine. After this, you're ready to click on the Launch Instance blue button. This process will take a few seconds. After this, click on the View instances blue button and you will be redirected to your EC2 AWS instances page. As you can see, even before launching we already have a private IP address. In my case, the machine name is: ip-172-16-146-1.eu-west-1.compute.internal and the associated private IP address: 172.16.146.1.

To be able to connect to it, we wait a few seconds until the instance is initialized and we will be ready to connect. Obviously, at this point we only have a private IP address so there is no way we can connect to it. We can start deploying our private network inside but this is not the main aim of this book. To enable connection to the machine, at this point we will create an Elastic IP (Public IP) and attach it to the network. This includes some extra cost, but not so much. Right-click on the instance, select Networking, and then Manage IP Addresses. On the new dialog, click on Allocate an Elastic IP. This will open a new window, select VPC in the scope, and click on Allocate. The system will return us a message like the following: New address request succeeded. Elastic IP: 34.250.160.101. This is our public IP address that we will use to connect. Hit on close.

A new window will open. The elastic IP window. Now we need to associate our public IP to our instance. Fill the options with the only possible option and click on Associate. At this point we already have all we need to connect to our instance.

Connecting to our EC2 instance

Now that we have a public IP we can connect to our instance. If we are on a Linux/Unix system we can use ssh and specify our pem key. Something like this: ssh -i "BibookKey.pem" ec2-user@ec2-34-250-160-101. eu-west-1.compute.amazonaws.com. Feel free to change the DNS name by the public IP.

If we are on Windows, there is an extra step we need to do first. This involves converting the pem format to convert our PEM file into a ppk file usable by putty; otherwise when trying to connect using putty we will receive an error like the following: Unable to use key file "C:\Users\Albert\Downloads\BibookKey.pem" (OpenSSH SSH-2 private key).

To do this conversion we need PuttyGen, available in the same webpage as Putty. Open PuttyGen, click on Load an existing private Key file, and something similar to Figure 10-4.

Figure 12-4. *PuttyGen options to convert the private key*

Now hit the Save private key button, as pointed by the arrow in Figure 12-4. This will create a ppk file, which works with Putty. Give it a meaningful name, and do not lose it. It's now time to open putty and connect to our instance. Write the public IP address in the Host name box in the Session Section and then move to the Auth section inside Connection > SSH. In the private key file for authentication, click on browse and choose the recently created ppk file. After this we can connect.

If you haven't specified a default login user, we need to provide the user. This key is for the bundled user ec2-user. So specify the user and hit return. Once done, we will be presented a screen like the following:

```
login as: ec2-user
Authenticating with public key "imported-openssh-key"

       __|  __|_  )
       _|  (     /   Amazon Linux AMI
      ___|\___|___|

https://aws.amazon.com/amazon-linux-ami/2016.09-release-notes/
[ec2-user@ip-172-16-1-67 ~]$
```

Good news. We are done creating our EC2 instance. Now it is time to install the programs like those shown in previous chapters of this book. After this, let's move to create our RDS instance to hold our MySQL datawarehouse.

Launching a RDS Database

Having our instance up and running is only part of the tasks needed to set up our environment. We still need the core part, which is our datawarehouse. For accomplishing this, we have few options. We can launch another instance and install a database by ourselves. Fortunately, there is a better solution in a cloud environment, which is asking our cloud provider to provide a managed database instance. This avoids the hassle of installing and configuring one. We have a central dashboard where we could make or schedule backups and nothing else. We do not have to worry about setting complex parameters in the configuration file, installing it, or maintaining it. All we need to do is to launch an RDS instance and we are ready to start using it.

Go to the Databases section and browse for RDS. Click on Get Started Now. This will open the assistant to create an RDS instance. We have several technologies to choose from, and this is your choice. If we pretend to have Odoo running in the cloud as well, we can create a PostgreSQL one for the Odoo metadata and a MySQL/MariaDB one for the datawarehouse. For illustration purposes we will choose MariaDB for the datawarehouse. The assistant moves to the second step of the process.

In this second step we get asked what type of MariaDB instance we want. There is an option called Dev/Test that is included in the free usage tier. Select this one and click on Next Step.

In the third step we can configure our database. If we want to stick with the free tier, make sure you select db.t2.micro instance class, which offers a database with 1 vCPU and 1GB of RAM. We allocate 5GB of storage for the time being, choose the version we desire or leave the one by default, and assign BiBook name as the db instance identifier and we set our desired username and password.

The fourth step includes some advanced configuration that's needed to be filled correctly. Let's start with the network and security section.

In the network and security section, it is very important that we fill these options appropriately, or we will end up with a useless instance that we can't connect to. Select the VPC we created previously, named BiBook. In the subnet group, select Create new DB subnet group. In the **publicly accessible** drop-down list we can decide if we want to assign our instance a public IP or not. If we assign a public IP, we will be able to connect from outside, whereas if we say no, we will only be able to connect from the inside of our VPC. Since we do not plan to connect to the VPC from the outside, **we select NO**. Make sure, if you want to interact directly with the database from external computers, like your company ones, select Yes, and give a public IP address to the instance. For performance purposes, in the availability zone, we will select the same where our instance was created, in our case, eu-west-1c. As a last step in this section, make sure you select BiBookI1 as the VPC security group.

On the Database options, write DWH as the database name, and leave 3306 as the port for the database. Leave the other options as specified by default. Change the backup policies according to your company needs and once done, review the maintenance section and change the values provided by default by ones that suit you, to configure for example, the upgrade maintenance windows for the weekends. Once done, review all is ok and click on the Launch DB instance button.

At this point we will receive a very weird error. DB Subnet Group doesn't meet availability zone coverage requirement. Please add subnets to cover at least two availability zones. Current coverage: 1

We need to go again to our VPC configuration, subnet section, and create a new subnet in another availability zone, for example, eu-west-1b and put the CIDR to 172.16.2.0/24. End up naming it BiBookRDS. Refresh and hit on the LaunchDB instance. At this point all should be ok and a pop-up similar to this one will appear: A new DB Subnet Group (default-vpc-3fca455a) was successfully created for you in vpc-3fca455a.

Hit on View our DB instances button and have a look at the instance already created. The status column will show creating, so we need to wait a while. In the meantime, we will go back to our EC2 instance in Putty and install the MySQL client.

■ **Note**   If you get an error, similar to the following one, VPC must have a minimum of 2 subnets in order to create a DB Subnet Group. Go to the VPC Management Console to add subnets, this means that in our VPC we have only defined a subnet and you consumed all of the IP addresses on this VPC. There is no easy fix for this, so the solution involves terminating your EC2 instance and deleting the subnets and then creating them back appropriately.

For installing the client, we can use the following sentence:

```
[ec2-user@ip-172-16-1-67 ~]$ sudo yum install mysql
Loaded plugins: priorities, update-motd, upgrade-helper
... (Output truncated)
Resolving Dependencies
--> Running transaction check
---> Package mysql.noarch 0:5.5-1.6.amzn1 will be installed
--> Processing Dependency: mysql55 >= 5.5 for package: mysql-5.5-1.6.amzn1.noarch
--> Running transaction check
---> Package mysql55.x86_64 0:5.5.54-1.16.amzn1 will be installed
--> Processing Dependency: real-mysql55-libs(x86-64) = 5.5.54-1.16.amzn1 for package:
    mysql55-5.5.54-1.16.amzn1.x86_64
--> Processing Dependency: mysql-config for package: mysql55-5.5.54-1.16.amzn1.x86_64
--> Running transaction check
---> Package mysql-config.x86_64 0:5.5.54-1.16.amzn1 will be installed
---> Package mysql55-libs.x86_64 0:5.5.54-1.16.amzn1 will be installed
--> Finished Dependency Resolution

Dependencies Resolved

...  (Output truncated)

Installed:
  mysql.noarch 0:5.5-1.6.amzn1

Dependency Installed:
  mysql-config.x86_64 0:5.5.54-1.16.amzn1    mysql55.x86_64 0:5.5.54-1.16.amzn1
  mysql55-libs.x86_64 0:5.5.54-1.16.amzn1

Complete!
```

After this it is time to go back to the RDS menu and look for the endpoint of your newly created database. It should be something like the following: Endpoint: bibook.c3nefasein6d.eu-west-1.rds. amazonaws.com:3306

Before attempting to connect we need to modify our Security Group. If you remember, only port 22 was allowed to connect. If you hover the mouse next to the endpoint name, over the No inbound Permissions, you will see a button to edit the security group. Go there and add a new inbound rule for port 3306, and open it to the range of IP addresses you want to trust. The new inbound rules should be as show in Figure 12-5.

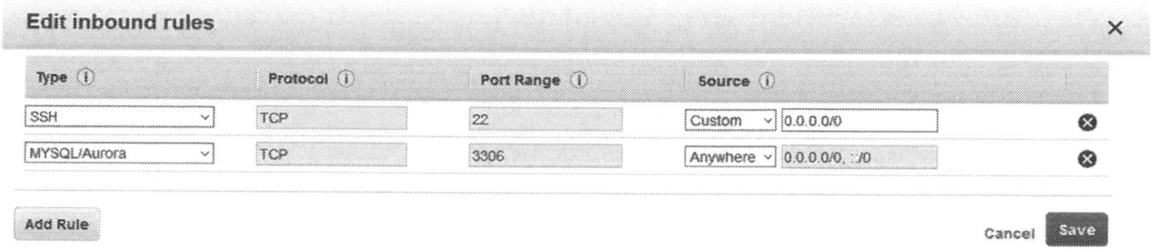

Edit inbound rules   ✕

Type ⓘ	Protocol ⓘ	Port Range ⓘ	Source ⓘ	
SSH ⌄	TCP	22	Custom ⌄ 0.0.0.0/0	✖
MYSQL/Aurora ⌄	TCP	3306	Anywhere ⌄ 0.0.0.0/0, ::/0	✖

Add Rule   Cancel   Save

Figure 12-5. Adding a new inbound rule to allow connection to the datawarehouse database

After adding the new rule, click on the Save button and go back to your RDS instances page. Now instead of No inbound permissions you should see something like (authorized). We can now connect to our instance, so go back to the EC2 and type the following (replacing the hostname by your RDS instance, the root username by the username you choose on the assistant and password keyword by your chosen password).

```
mysql -h bibook.c3nefasein6d.eu-west-1.rds.amazonaws.com -u root -ppassword
```

All going well, the client should connect and the MariaDB banner should appear:

```
Welcome to the MySQL monitor.  Commands end with ; or \g.
Your MySQL connection id is 28
Server version: 5.5.5-10.0.24-MariaDB MariaDB Server

Copyright (c) 2000, 2016, Oracle and/or its affiliates. All rights reserved.

Oracle is a registered trademark of Oracle Corporation and/or its
affiliates. Other names may be trademarks of their respective
owners.

Type 'help;' or '\h' for help. Type '\c' to clear the current input statement.

mysql>
```

Issue a show databases command and make sure your DWH database appears on the list:

```
mysql> show databases;
+--------------------+
| Database           |
+--------------------+
| dwh                |
| information_schema |
| innodb             |
| mysql              |
| performance_schema |
+--------------------+
5 rows in set (0.01 sec)
```

As we expected, all clear. We have our platform now set up in the cloud! Now it is time to create the required tables and start populating them.

Calculating Prices with the AWS Calculator

Having the platform in the cloud costs you money. In order to calculate or forecast how much it can cost to you, AWS offers a free online calculator. The calculator can be found in the following link: `https://calculator.s3.amazonaws.com/index.html`

Don't get overwhelmed with that you see! You only need to fill a few spots. First of all, we need to choose the geographic region. In our case, it is Europe/Ireland, but in other cases it can be different. Choose the one you plan to use, which should be the one closest to you.

After filling the region, simply click on the plus sign in the Compute: Amazon EC2 Instances. Put any familiar name, adjust the instance type you want to use, and the billing option. In our case, we haven't chosen to do any up-front payment (though there are juicy discounts), so we will be selection On Demand (No Contract). On elastic IP select 1, and leave the remaining boxes as 0 as despite the fact we will be assigning it, there won't be too many remaps, and the first few are free.

The Data transfer section is a very important one. Amazon bills from data coming in and coming out from their servers. However, inbound data transfer is free whereas outbound data transfer is billed after 1 GB per region has been consumed. So, if you pretend to download a lot of data from the server to your computer, this may not be a good solution for you, and it is better to look at other services that offer bandwidth at a cheaper price. Since we do not plan to let people connect directly to that server, only developers will have access to it for testing, so we plan a very small consumption, let's say 10 GB. The inbound data will be the data coming from the petitions from the clients that should be not very big, and in any case, is for free. If you look now in the second tab, the Estimate of your monthly bill you will see some money being allocated to different concepts. If you choose a free micro instance, you will see a discount as well. The total bandwidth will only cost us less than a dollar, so we are ok. We still need to add the database RDS instance, so look at the left pane, fifth tab, and move to RDS.

Click on the plus sign again, add a fancy description, fill in the details of your instance, and adjust the bandwidth consumption. In the database case pay attention as most usage will be uploading data so this is free. Then, if we have been wise enough and placed the RDS in the same availability zone as our ETL and Reporting application engine (if needed), we won't have to pay Intra Region Data Transfer. So, we only need to account from the data transferred to the server until our reporting client tools. Most reporting tools push calculations to the database engines and this saves a lot of bandwidth, and in our case money, as only top-level calculations that need to be displayed or browsed in the dashboard go through the network. In any case this is the place where most network consumption will be done. This clearly depends on the amount of information we retrieve from the server, the users accessing it, and how your dashboards are designed, and at what level of detail the information is present. To go very safe, we assume 50 GB/month, but this is clearly a subjective guess.

You may want to add other services that you plan to use, but at the end the bill is less than $40 a month. You can also play around and see the impact of renting better and powerful servers, and check the impact that will have on your bill. If you have the free tier, you will see that sticking to micro instances, you will only have to pay for the bandwidth and the IP address leasing during the first year. So, it is an excellent way to get you started. Full charges that we calculated are shown in Figure 12-6.

⊖	Amazon EC2 Service (Europe)		$ 18.30
	Compute:	$ 14.64	
	Elastic IPs:	$ 3.66	
⊖	Amazon RDS Service (Europe)		$ 14.45
	DB instances:	$ 13.18	
	Storage:	$ 1.27	
⊕	AWS Data Transfer In		$ 0.00
⊖	AWS Data Transfer Out		$ 5.31
	Europe (Ireland) Region:	$ 5.31	
⊕	AWS Support (Basic)		$ 0.00
Free Tier Discount:			$ -30.35
Total Monthly Payment:			$ 7.71

Figure 12-6. *Calculating AWS costs with the AWS Monthly calculator*

Having a Go with AWS Quicksight

In the past few months, Amazon has released AWS Quicksight. Quicksight is a fast cloud reporting and dashboarding solution that is extremely easy to use. Quicksight can use multiple data sources like Excel and CSV files, Databases hosted in AWS including RDS, Aurora, Redshift and even the S3 Storage.

But not only cloud. You can point Quicksight to on-premises relational databases too or even connecting to some other cloud applications like Salesforce.

The SPICE engine is an in-memory, columnar engine, which is very powerful and fast. It can scale to loads of users and replicate data over different places. Quicksight also offers a very beautiful visualization interface and is very easy to use. As most bi tools, integration with mobile devices is possible with iPhone support recently added, and Android versions on its way. You can share and work in a collaborative environment as in many others of the bi tools and create stories like the ones you can create with QlikSense.

It is also planned in the near future to integrate with most BI vendors like Tableau, Qlik, and TIBCO, so with these tools you will be able to access data in the SPICE engine, and work with it directly in your preferred BI tool, enhancing the processing engines built in with these tools.

Quicksight is free for personal use of up to 1 GB of data in the SPICE engine and cheap plans are available for enterprise usage. As always with all AWS services, you pay more as you demand more. So apart from a small fee per user, the space used in the SPICE engine is also used as a billing metric, starting with a quarter of dollar per gigabyte per month. A small screen of how Quicksight report looks is featured in Figure 12-7.

Figure 12-7. *Quicksight dashboard screen*

As you can see in the picture, there is a similar menu on the left featuring all the Dimensions and Facts that can be used from your data source, and a large canvas in the right. With the plus sign on the main top menu bar, you can add visualizations, which are graphics or tabular information to the canvas. Then you drag and drop the desired dimensions and facts from the left to the new added visualization and a graphic is generated. You can change at any time the shape and type of visualization by selecting one of the available ones in the Visual Types list, in the bottom part of the screen. With a few mouse clicks you have a clean and tidy interface featuring some meaningful graphs to analyze.

After testing Quicksight a little bit, we find it is very similar to QlikSense in terms of visualization, so for users already experienced with QlikSense they will find it useful. Obviously, it hides most of the complexity behind the standard BI tools as there is no scripting editor, and all the information available is built in this screen. In that aspect, it is also similar to the associative engine of Qlik. Clearly a tool to test by those less experienced with reporting and dashboarding tools who want to escape from the complexity of defining tricky calculations in scripts and want to have an easy way to look out on their data with only a few clicks.

Using Software from the AWS Marketplace

Another approach to deploy a solution in the AWS cloud is by using a software appliance. A software appliance is a solution built by some vendor that includes all the setup you need for that tool. Then the vendor, depending on what type or size of platform you choose, selects for you the best infrastructure from AWS that fits the needs of the tool. The final cost of running the appliance will be the cost of the AWS resources required to run the platform, alongside with the costs of the tool in terms of licensing or any other fee you may incur.

You can browse AWS appliances in its marketplace here: `https://aws.amazon.com/marketplace/` but there is a category featuring all the business intelligence tools, with more than 350 appliances in the following link: `https://aws.amazon.com/marketplace/b/2649336011?ref_=header_nav_ category_2649336011`.

In the market, you will see applications like Tibco Spotfire, Tableau, and Qlik, as well as ETL tools like Informatica and databases like Exasol, Teradata, Oracle, MySQL, MariaDB, and others. You simply select the one you want, the environment will create the required resources in the AWS console and as soon as they are up, you will get charged. When you do not need the platform anymore, you can shut it down and delete it, and you will not be charged anymore, as you would do with any other AWS resource.

Microsoft Azure

Implementing a similar solution in Azure is perfectly possible. The way Azure works it is like AWS. Microsoft solution is the second biggest cloud provider, so it is an excellent choice as well. The name of the services change a little bit, but they are easy to find. A deployment of a server involves several things, but the first one is to set up a subscription, which usually will be a pay-for-use subscription, very much like AWS. We can start by browsing the following URL: `https://portal.azure.com`.

Creating a Machine in Azure

Once the subscription is set up (and the credit card linked) then you will be able to create a resource group, attached to specifically one of the subscriptions you created in the previous step. All resources you create, by default, will be linked to the resource group you choose and in turn, these linked to the specific subscription you choose. That resource group, as in AWS, is linked to one of the geographic regions available. Figure 12-8 shows how to create a resource group.

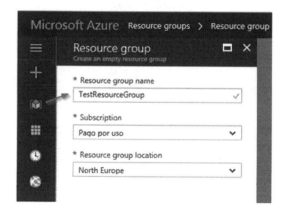

Figure 12-8. *Creating a resource group*

As you see in the previous figure, the resource group will be created in the North Europe geographic zone. Once we have the resource group set, we can create a new machine. To accomplish this, in the left pane menu, the sixth button is called Virtual Machines (Classic). If we press this button the pane will unfold in the main screen, and will ask us which type of operative system we want for our machine. For our example, we will select Ubuntu, then a few flavors of Ubuntu appear onscreen and we will select an LTS version, because these tend to be more stable and have extended support. At this point the last LTS available in Azure for Ubuntu is 16.04, so we will select this one. In the deployment model drop-down box, we will select Classic. Once this is selected, then we start configuring our machines. This is split in four steps.

The first step is configuring the basic settings: name of the machine, the username to log in, the authentication to access to the machine that in Azure can be apart from a SSH public key, a password in contrast to the AWS defaults that only lets you to connect with a public key. Then you need to assign this machine to a specific subscription, and if you already created beforehand the resource group you can assign it as well to a resource group. If you didn't create the resource group in advance, you can do it here automatically, and this machine will be added to the new resource group. The location can be specified as well in this step. Some locations can be chosen depending on the subscription you have defined, so if that is the case, choose another location.

The second step shows the sizing and the pricing of the machine. By default, the assistant tries to choose a few possible options, but you can switch to all options by clicking View all in the top-right corner. The recommended instances will suit us in most cases as these are general purpose instances. The assistant chooses three for us by default, and they can be seen in Figure 12-9.

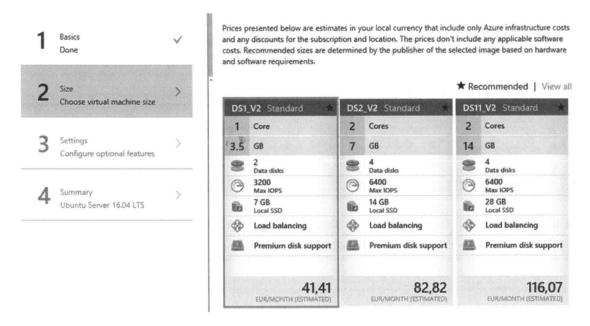

Figure 12-9. Instances suggested by the assistant

Depending on the power you need, the first instance may be enough. Despite that only a 7gb is shown onscreen, the instance has 2 disks, which if magnetic they are up to 50 Gb disks for the same price as the 7GB SSD. Premium storage is based on SSD, so in the next screen you can choose which one you want to stick with. In any case, you can always add more storage disks later at a decent cost.

If you're looking for a cheaper option, you can select view All and move down, and you will find the basic instances, namely A0 and A1 that cost between $12 and $17 per month and are good choices if the use of the machine is light. These two machines consist of 0.75 GB and 1.5 GB of memory, and obviously are only suited to very light workloads.

In the third step, you can configure a few more aspects of the new machine. Basically, the storage and the networking are configured in this step. The cloud service this machine will belong to is chosen in this step. Also, you need to specify the endpoints, or which ports will be open in this machine. By default, only port 22 (ssh) is allowed to connect. Feel free to open any endpoints that you need. As a last step, you can define a high-availability set, but we have decided against it at this point.

The last step is a review of all the previous steps. If all is ok, proceed to create your machine.

Creating a Database in Azure

Creating an SQL Database in Azure is not different from what we have seen so far. In the left menu, either click on the plus green sign, then on Databases and lastly on SQL Database; or click directly on the SQL button, which is the fifth button under the plus sign. Any of these steps will create you a SQL Server database. However, it is possible to choose other database engines, but for this you must go to the plus sign, then databases, and before clicking SQL Database, you click on See All in the top-right corner, and then use the search databases field to find the desired database, as they are not easy to find. See Figure 12-10.

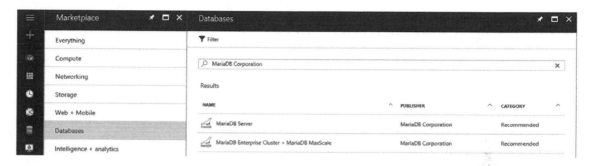

Figure 12-10. *Browsing for Maria DB databases, as they do not feature in the main Databases page*

Once found, click on MariaDB Server, and follow a similar menu to the create virtual machine we described in the previous point. The same needs to be done if you want a MySQL instance.

■ **Note**   As with AWS, the pricing system gets complicated. Fortunately, as its counterpart, Microsoft has a web calculator to aid with all the pricing. The calculator is in the following URL: `https://azure.microsoft.com/en-us/pricing/calculator/`

Google Cloud

Google cloud is the third largest provider. Usually Google cloud is regarded as having some cheaper prices than its rivals. Also, they have a very good sign-in offer, especially if you are still testing or doing a proof of concept. At this point, they offer $300 of free credit for 12 months with all Google cloud resources included. This is an excellent offer to start with.

We do not have enough space in the book to cover a third cloud provider but we want to encourage you to try them as well. With correct machine planning the credit can last enough to test a BI solution on the cloud. The free trial, as well as the platform can be accessed from here: `https://console.cloud.google.com/freetrial?pli=1&page=0`.

For replicating the previous infrastructure we set up with AWS or Azure, you need to focus on the Compute Engine service for creating the virtual Machines and the Cloud SQL service to create either a MYSQL or PostgreSQL database (unfortunately no support yet for MariaDB, but this should not be a big problem).

As with the other two competitors Google also has a pricing calculator. The calculator is available here: `https://cloud.google.com/products/calculator/` and contains all resources in the Google cloud.

Vendor-Based Cloud Solutions

Apart from the three cloud providers we have seen in this chapter, there are some vendor-specific cloud services. These services usually offer their software as a Service (SaaS). From the ones we have seen in the book, the most interesting ones are the following:

- Oracle cloud: `https://cloud.oracle.com/home`

- Qlik sense cloud: `https://www.qlikcloud.com/`

- Microstrategy cloud: `https://www.microstrategy.com/us/platform/cloud`

- Tableau cloud: `https://www.tableau.com/products/cloud-bi`

- Salesforce.com is a CRM tool, but nowadays it features many more applications: marketing, BI tools, customer service, community and chat applications, and many more. You can have your CRM in the cloud, and elaborate dashboards and reports using the data generated by the application from within it. If you want to know more and discover Salesforce cloud tool you can visit their website in the following link: `https://www.salesforce.com`

Using a cloud solution from one of these vendors instead of the desktop tool will save us some maintenance and administration tasks. For example, version upgrades are done automatically in the cloud for all these solutions. With this the administrators move the burden of some of the heavier administration tasks to the cloud administrators. This simplifies the management of these applications. Also in most of the cases, the monitoring can be configured directly from the tool, so you get alerts when the tool is not available, and the tool administrators perform automatically the actions to have the tool back again. This leaves only the development part to you, as well as the user creation and a few more administration or tweaking aspects to you, while the majority of the administrator tasks are freed from you.

Conclusion

In this chapter, we have seen how to implement a BI environment in the cloud. We have seen in depth the main provider: AWS and to some extent two of its competitors, namely, Azure from Microsoft and Google Cloud from Google. We have seen how to approximate the costs of running an entire platform in the cloud; which advantages and inconveniences it has; which are the important things we need to pay attention to; and the considerations we need to be aware with when having infrastructure in the cloud, like security, monitoring, and control of the costs. At the end of the chapter, we saw how some BI vendors have created a specific cloud environment for their products, and the advantages a solution like this over standard deployments can have in terms of administration and maintenance.

CHAPTER 13

■ ■ ■

Conclusions and Next Steps

Well we are almost finished – just a few pages more of suffering. But if you have arrived then here maybe you are not suffering so much; so we are really pleased that you are reading these pages. We haven't done it so badly... But now that we have arrived at this point, what more can we do? This should be the question after you have completed successfully the implementation of your BI system with basic analysis of data located in your database, accessed from your BI system in a multiple environment that can be located fully or partially in the cloud. We expect you to have followed the book following also the examples, downloading and installing the proposed software (and maybe other options that you have heard about, especially for the BI front end there are multiple options with free versions of commercial tools), learning about the different tools while you play following our instructions so you can finalize this book having at least the initial Proof of Concept done. Of course we are aware that if you have read this book from end to end while testing things, you won't have a productive multienvironment or all your servers in cloud; we understand that the first option for this kind of test is just a laptop to install all the required components to evaluate them.

In other words, it is quite possible that you can re-take the latest chapters after a while once you have shown to your stakeholders some examples about what BI can do for you and your company, so once everybody agrees in going ahead with the BI project you can think of moving this sandbox environment to a real one with multiple servers, maybe purchasing some licenses to get better performance and functionalities on the BI tool side; or thinking in a cloud solution to locate all the stuff required for a BI platform, as far as implementing a working solution doesn't consist just in the required development to see data but also in the related management to keep it ongoing.

Your question in this moment could be, is there anything more that you should consider taking into account for the next steps and the future? The answer is yes, we have always something more to implement, improve, and evaluate. We are in an amazing scenario for data analysis, with new features and concepts that appear every day and you need to keep yourself in alert mode if you want to profit from all the features that Business Intelligence technologies and related software can offer. We don't want to extend this chapter too much but we would like to mention some recommendations regarding documentation that we haven't commented about at all during the book. On the other hand, we would like to mention two trending topics that we think that will expand across organizations during the next year, Business Process Management Software (BPMS) and Big Data.

Documentation

Our recommendations for documentation goes in line with some sentences already commented on during Chapter 2 when we were talking about the Agile Manifesto. In our opinion, documentation is important to have but not the main objective of a project. We would like to make just some recommendations:

- Save some time to document all important things while you are performing the actions: It can seem something that takes you time during installation, configuration, or development but if you see it with an overall perspective you will see that you

© Albert Nogués and Juan Valladares 2017
A. Nogués and J. Valladares, *Business Intelligence Tools for Small Companies*,
DOI 10.1007/978-1-4842-2568-4_13

are saving time as far as you are not opening again installation menus to take print screens, you don't miss parameters in the documentation that are relevant, you can remark that things that have been more complicated to resolve, etc.

- Avoid exhaustive documentation: Our experience is that exhaustive documentation will never be read. You need to document only those things that have been relevant during your operations, avoiding the ones that are already documented in the installation guide or that you can check more easily by accessing the system than by accessing the documentation. You don't require a documentation saying that MONTH_ID column of T_L_MONTH table is a NUMBER(6). You can easily check it in the database.

- Documental repository: Use a single shared place to centralize all the documentation in a single platform where anybody that needs access can get all required documents for his daily work. Avoid the usage of local resources to save your documents. You have multiple platforms available for free: Google Drive, Dropbox, One Drive, etc.

- Versioning: If possible, use a tool that allows you to save previous versions of a document without requiring version control inside the documents. It is the easier way to manage the documentation.

- Better multiple small documents than a single big one: This is more related to our preferences but we see that it is easier to manage small documents explaining a single topic than big documents containing hundreds of topics inside. In this way you will have few versions of each document as far as each topic will require fewer modifications than a big one with hundreds of pages.

BPM Software

Business Process Management software products offer you help in improving your business results focused on improving your business processes based on continuous improvement theories. But let's first explain what a business process is so you can understand how a BPM system works. A business process is a set of activities or tasks related to obtain a product or a service from which your company to obtain a benefit. The benefit can be direct, if your company produces shoes: the manufacturing of the shoe is a process, the design of the shoe follows another process, the procurement department follows their own purchasing processes; or it can be indirect, your financial department follows a closing process, the IT department follows its own processes to create a user, create an email address, or develop a BI improvement. All companies and departments and areas inside each company have their own processes. With a BPM tool you can analyze these processes, model them, document them, organize them, and in this way try to improve them by detecting bottlenecks, inefficiencies, or undesired dependencies. During this book, in concrete examples in Chapter 11 talking about transports, we have already modeled a process, the transport of a database table; you can see this in Figure 11-4. This is just an example from the IT department of your company inside a BI platform maintenance system.

A process can be also a concatenation of smaller processes. Following with this example, the transport procedure is just a process inside the chain of a full BI development, where you need to get users requirements, analyze them, model a solution, create the database tables, create the ETL process, create the BI interface report, transport the whole thing across environments, and validate it. So the idea in a BPM analysis is to start with a high-level process as shown in this second example and then split and enter in the detail of that process steps that are taking more time or are unclear.

> ■ **Note** We consider BPM software highly related with BI software as far as they have multiple similarities. Both are focused on analysis, both are focused in Business Intelligence in the widest term of the concept, both are used to improve performance of the company; and the basic difference is that BI is helping to understand data and BPM is helping to understand processes. A company that uses and obtains benefits from a BPM system is considered more mature technologically than a company using just BI; BPM is a step beyond the technology maturity range. BI also supports BPM as far as with a BPM system; we propose changes on the way that the company is doing the things so after applying the change you can use BI to analyze the results of BPM actions.

There are multiple BPM open source software in the market and also free versions of commercial ones. We have used Bonita soft, but you can find also Zoho, Processmaker, and multiple other software products that can help you to implement a BPM system. But if you are reading this book you will be possibly implementing a BI solution, so going for a BPM implementation can be still far away. This could be a matter for a whole book so let's stop here. If in the future we decide to write another book, this would be a good candidate for the subject!

Big Data

BigData implementations have already started to come and they came to stay for us, at least in the foreseeable future. If you still don't know what is behind this fancy name, we are going to give you a first introduction to BigData.

What we all understand about BigData is a set of tools that act altogether as a framework. These tools let us to process different types of workloads: Batch workload, Real Time, Semi Real time... and different types of data: Structured, semi-structured data, unstructured data, images, videos...

Organizations have started to accumulate data, and processing and analyzing this data is usually out of the scope of traditional BI tools, as much more power is required. When this situation is met, costs start to become prohibitive because scaling vertically is very expensive. Imagine having a computer with 1 TB of RAM memory. That will cost an awful lot of memory. But in turn, imagine 8 computers with 128 GB of RAM memory each. Probably much cheaper, right? Well maybe you are reading this book 10 years after we are writing it and you have available petabyte RAM laptops, but nowadays 1TB of RAM is a big server. In fact, it is possible that in 10 years everything that we are commenting in this section is already old but it is nowadays trending in BI. So, that's what BigData is all about: processing important quantities of information, from different sources and different shapes, with a set of broad tools.

In essence, it follows the 4 v's: Volume, as you have a large volume of data to process; Velocity, as you need to speed the processing of this data using multiple machines at once (cluster); Variety in the data sources, as we already explained in the previous chapter; and Veracity as the data processed needs to be worked to provide not just meaningful value for the company, but also it needs to be trustful, so people that make decisions at your company can rely on it.

> ■ **Note** There is a good set of infographics about some facts of the 4v's on the IBM website here: http://www.ibmbigdatahub.com/infographic/four-vs-big-data and here: http://www.ibmbigdatahub.com/infographic/extracting-business-value-4-vs-big-data

What is more interesting from the BI point of view is the different technologies in Hadoop that can act as SQL databases (or at least to some extent). The Hive project was the first one to come, but in its original form it is slow. Don't get us wrong, it is a very good tool if you process large batch jobs, but when you need to resolve ad hoc queries, probably from a reporting tool, the response time isn't acceptable. To solve this, some other engines like Spark, which is one of the newest and most important additions to the Hadoop stack, offers faster response times.

Impala, another SQL engine on top of Hadoop, makes extensive use of the RAM memory in the cluster machines, which makes it a very fast tool and one of the best suited to resolve analytic queries. So it is a good candidate tool to use in conjunction with reporting tools, and most of them already support it.

On the other side, Apache Kudu, a project which pretends to create a distributed relational database on top of Hadoop, allowing transactions, data modification, and mimics many of the traditional relational database capabilities. It is a new tool, just in beta stages but some organizations have already started to implement it as at this point, and it can work very well as a cold storage area, for example, to store historical data that is accessed with less frequency, whereas the newest one, will still be inside our relational database. With a strategy like this one, we can free space in our day-to-day database, while keeping our old data, to enable us to do comparative analysis, but in a different machine or engine.

But that's not all. If you want to do data mining, machine learning, streaming information, connect to social networks and process data in real time, you have tools that support these scenarios in Hadoop. This makes the whole Hadoop thing a very interesting platform, broader in its use and very capable for all needs, and yet, there is much more to come in the following months or years.

In addition, storage pricing trends, as well as the fact that Hadoop can run on commodity hardware, and the price drop of the machines, either on premises or in the cloud, promise to bring further improvements to the area. Every day, more and more companies are storing much more information, which probably at this point they have not yet started to analyze, but are starting to store it for doing future analysis and this is a trend that shows no sign of stopping.

Well, finally after all these different topics that we have seen we are done – that's all, folks! We wish you the best possible luck in you BI implementation and we hope that you have enjoyed this reading. And more than enjoy, we hope that you have get some knowledge from it. And more than getting knowledge, we hope that this knowledge helps you in getting a successful BI system in place.

Index

▓ E

▓ Q

Get the eBook for only $5!

Why limit yourself?

With most of our titles available in both PDF and ePUB format, you can access your content wherever and however you wish—on your PC, phone, tablet, or reader.

Since you've purchased this print book, we are happy to offer you the eBook for just $5.

To learn more, go to http://www.apress.com/companion or contact support@apress.com.

Apress®